Control System
Design Guide

Third Edition

Control System Design Guide

A Practical Guide

Third Edition

George Ellis
Danaher Corporation

ELSEVIER
ACADEMIC
PRESS

Amsterdam   Boston   Heidelberg   London   New York   Oxford
Paris   San Diego   San Francisco   Singapore   Sydney   Tokyo

Elsevier Academic Press
525 B Street, Suite 1900, San Diego, California 92101-4495, USA
84 Theobald's Road, London WC1X 8RR, UK

This book is printed on acid-free paper. ∞

Library of Congress Cataloging-in-Publication Data
Ellis, George (George H.)
   Control system design guide: a practical guide/George Ellis.—3rd ed.
      p.   cm.
   ISBN 0-12-237461-4 (hardcover : alk. paper)
   1. Automatic control.   2. System design.   I. Title.
   TJ213.E5625 2003
   629.8'3—dc22   2003023742

British Library Cataloguing in Publication Data
A catalogue record for this book is available from the British Library

ISBN: 0-12-237461-4

For all information on all Academic Press publications
visit our website at www.academicpress.com

Printed in the United States of America
04   05   06   07   08   09   9   8   7   6   5   4   3   2   1

To my loving wife, LeeAnn, and to Gretchen and Brandon, who both make us proud.

Contents

Preface

The basics of control systems were developed in the first half of the 20th century. Our predecessors aimed a cannon or warmed a bath using many of the same concepts we use. Of course, time and technology have generated many refinements. Digital processors have changed the way we implement a control law, but in many cases they haven't changed the law itself. Proportional integral differential (PID) control works about the same today as it did four or five decades ago.

Control systems are broadly used and are thus well integrated into our educational system. Courses are offered at most engineering universities, and a few disciplines even require students to undergo modest training in the subject. Given the longevity of the principles and the number of trained engineers engaged in their use, one might expect most of the trade's practitioners to be comfortable with the basics. Unfortunately, that does not seem to be the case.

Over the past several years, I've had the opportunity to teach a total of about 1500 engineers through a daylong seminar entitled "How to Improve Servo Systems." These are motivated people, willing to spend time listening to someone who might provide insight into the problems they face. Most are degreed engineers who work in industry; roughly half have taken at least one controls course. A few minutes into the seminar, I usually ask, "How many of you regularly apply principles of controls you learned at school?" Normally, fewer than one in ten raises a hand. It's clear there is a gap between what is taught and what is used.

So why the gap? It might be because the subject of controls is so often taught with an undue emphasis on mathematics. Intuition is abandoned as students learn how to calculate and plot one effect after another, often only vaguely understanding the significance of the exercise. I was one of those students years ago. I enjoyed controls and did well in all my controls classes, but I graduated unable to design or even tune a simple PI control system.

It doesn't have to be that way. You can develop a feel for controls! This book endeavors to help you do just that. Principles are presented along with practical methods of analysis. Dozens of models are used to help you practice the material, for practice is the most reliable way to gain fluency. A goal of every chapter is to foster intuition.

What's New in This Edition?

This third edition of *Control System Design Guide* includes several improvements over the previous edition. First, *ModelQ*, the modeling environment from the second edition, has been rewritten to create *Visual ModelQ*; the preprogrammed models have been replaced with a fully graphical modeling environment. You should find it easier to follow what is being modeled. Second, two chapters have been added, both concerning observers: Chapter 10 is a general presentation of observers; Chapter 18 focuses on observers in motion-control systems. I hope these presentations will convey the power of these remarkable software mechanisms as well as the ease with which they can be implemented. Also, a question set has been added to the end of almost every chapter, with answers provided in Appendix G.

Organization of the Book

The book is organized into three sections. Section I, Applied Principles of Controls, consists of ten chapters. Chapter 1, Introduction to Controls, discusses the role of controls and controls engineers in industry. Chapter 2, The Frequency Domain, reviews the *s*-domain, the basis of control systems. Chapter 3, Tuning a Control System, gives you an opportunity to practice tuning; for many, this is the most difficult part of commissioning control systems.

Chapter 4, Delay in Digital Controllers, culls out the fundamental difference in the application of digital and analog controllers, the contribution of instability from sample delay. Chapter 5, The *z*-Domain, discusses *z*-transforms, the technique that extends the *s*-domain to digital control. Chapter 6, Six Types of Controllers, covers practical issues in the selection and use of six variations of PID control. Chapter 7, Disturbance Response, provides a detailed discussion of how control systems react to inputs other than the command. Chapter 8, Feed-Forward, presents techniques that can substantially improve command response. Chapter 9, Filters in Control Systems, discusses the use of filters in both analog and digital controllers. Chapter 10, Introduction to Observers in Control Systems, is a general presentation of observers.

Section II, Modeling, has three chapters. Chapter 11, Introduction to Modeling, provides overviews of time- and frequency-domain modeling methods. Chapter 12, Nonlinear Behavior and Time Variation, addresses how to deal with nonlinear operation when using linear control techniques. Unfortunately, this subject is missing from most texts on controls, although significant nonlinear effects are common in industrial applications. Chapter 13, Seven Steps to Developing a Model, gives a step-by-step procedure for developing models.

Section III, Motion Control, concentrates entirely on motion control using electric servomotors. Chapter 14, Encoders and Resolvers, discusses the most common feedback sensors used with electric servomotors. Chapter 15, Basics of the Electric Servomotor and Drive, reviews the operation of these motors. Chapter 16, Compliance and Resonance, is dedicated to one of the most widely felt problems in motion control, instability resulting from mechanical resonance. Chapter 17, Position-Control Loops, addresses the control of position, since the great majority of applications control position rather than velocity or torque. Chapter 18, Using the Luenberger Observer in Motion Control, focuses on observers in motion-control systems.

Reader Feedback

I have endeavored to right the errors of the second edition; for those errata that slip through into this edition, corrections will be posted at qxdesign.com. Please feel free to contact me at george.ellis@DanaherMotion.com or gellis@qxdesign.com.

Acknowledgments

Writing a book is a large task and requires support from numerous people, and those people deserve thanks. First, I thank LeeAnn, my devoted wife of more than 25 years. She has been an unflagging fan, a counselor, and a demanding editor. She taught me much of what I have managed to learn about how to express a thought in writing. Thanks also to my mother, who, when facts should have dissuaded her, was sure I would grow into someone of whom she would be proud. And thanks to my father, for his unending insistence that I obtain a college education, a privilege that was denied to him, an intelligent man born into a family of modest means.

I am grateful for the education provided by Virginia Tech. *Go Hokies.* The basics of electrical engineering imparted to me over my years at school allowed me to grasp the concepts I apply regularly today. I am grateful to Mr. Emory Pace, a tough professor who led me through numerous calculus courses and who, in doing so, gave me the confidence on which I relied throughout my college career and beyond. I am especially grateful to Dr. Charles Nunnally; having arrived at university from a successful career in industry, he provided my earliest exposure to the practical application of the material I strove to learn. I found him then, as now, an admirable combination of analytical skill and practical application.

I also thank Dr. Robert Lorenz of the University of Wisconsin at Madison, the man most influential in my education on controls since I left college. His instruction has been well founded, enlightening, and thoroughly practical. Several of his university courses are available in video format and are recommended for those who would like to extend their knowledge of controls. I took the video version of ME 746 and found it quite useful; much of the material of Chapter 7, Disturbance Response, is derived from that class.

Thanks to the people of Danaher (manufacturer of Kollmorgen products), my long-time employer, for their continuing support in the writing of this book. My gratitude to each of you is sincere.

Section I
Applied Principles of Controls

Important Safety Guidelines for Readers

This book discusses the normal operation, commissioning, and troubleshooting of control systems. Operation of industrial controllers can produce hazards, such as:

- Large amounts of heat
- High-voltage potentials
- Movement of objects or mechanisms that can cause harm
- Flow of harmful chemicals
- Flames
- Explosions or implosions

Unsafe operation makes it more likely for accidents to occur. Accidents can cause personal injury to you, your coworkers, and other people. Accidents can also damage or destroy equipment. By operating control systems safely, you make it less likely that an accident will occur. *Always operate control systems safely!*

You can enhance the safety of control system operation by taking the following steps:

1. Allow only people trained in safety-related work practices and lock-out/tag-out procedures to install, commission, or perform maintenance on control systems.
2. Always follow manufacturer-recommended procedures.
3. Always follow national, state, local, and professional safety code regulations.
4. Always follow the safety guidelines instituted at the plant where the equipment will be operated.
5. Always use appropriate safety equipment, for example, protective eyewear, hearing protection, safety shoes, and other protective clothing.
6. Never attempt to override safety devices such as limit switches, emergency stop switches, light curtains, or physical barriers.
7. Always keep clear from machines or processes in operation.
8. Provide reliable protection, such as mechanical stops and emergency off switches, so that unanticipated behavior from the controller cannot harm you or anyone else and cannot damage equipment.

Remember that any change of system parameters (for example, tuning gains), components, wiring, or any other function of the control system may cause unexpected results, such as system instability or uncontrolled system excitation.

Remember that controllers and other control system components are subject to failure. For example, a microprocessor in a controller may experience catastrophic failure at any time. Leads to or within feedback devices may open or short together at any time. Failure of a controller may cause unexpected results, such as system instability or uncontrolled system excitation.

This book presents observers, the use of which within control systems poses certain risks, including that the observer may become unstable or otherwise fail to observe signals to an accuracy necessary for proper system operation. Ensure that observers behave properly in all operating conditions.

If you have any questions concerning the safe operation of equipment, contact the equipment manufacturer, plant safety personnel, or local governmental officials, such as the Occupational Health and Safety Administration.

Always operate control systems safely!

Chapter 1

Introduction to Controls

Control theory is used for analysis and design of feedback systems, such as those that regulate temperature, fluid flow, motion, force, voltage, pressure, tension, and current. Skillfully used, control theory can guide engineers in every phase of the product and process design cycle. It can help engineers predict performance, anticipate problems, and provide solutions.

Colleges teach controls with little emphasis on day-to-day problems. The academic community focuses on mathematical derivations and on the development of advanced control schemes; it often neglects the methods that are commonly applied in industry. Students can complete engineering programs that include courses on controls and still remain untutored on how to design, model, build, tune, and troubleshoot a basic control system. The unfortunate result is that many working engineers lay aside analysis when they practice their profession, relying instead on company history and trial-and-error methods.

This book avoids the material and organization of most control theory textbooks. For example, design guidelines are presented throughout; these guidelines are a combination of industry-accepted practices and warnings against common pitfalls. Nontraditional subjects, such as filters and modeling, are presented here because they are essential to understanding and implementing control systems in the workplace. The focus of each chapter is to teach how to use controls to improve a working machine or process.

The wide availability of personal computers and workstations is an important advance for control system designers. Many of the classical control methods, such as the root locus method, are graphical rather than analytical. Their creators sought to avoid what was then the overwhelming number of computations required for analytical methods. Fortunately, these calculations no longer present a barrier. Virtually every personal computer can execute the calculations required by analytical methods. With this in mind, the principles and methods presented herein are essentially analytical, and the arithmetic is meant to be carried out by a computer.

1.1 *Visual ModelQ* Simulation Environment

Most engineers understand the foundations of control theory. Concepts such as transfer functions, block diagrams, the *s*-domain, and Bode plots are familiar to most of us. But how should working engineers apply these concepts? As in most disciplines, they must develop intuition, and this requires fluency in the basics. In order to be fluent, you must practice.

When studying control system techniques, finding equipment to practice on is often difficult. As a result, designers often rely on computer simulations. To this end, the author developed, as a companion to this book, *Visual ModelQ*, a stand-alone, graphical, PC-based simulation environment. The environment provides time-domain and frequency-domain analysis of analog and digital control systems. *Visual ModelQ* is an enhancement of the original *ModelQ*, in that *Visual ModelQ* allows readers to view and build models graphically. Dozens of *Visual ModelQ* models were developed for this book. These models are used extensively in the chapters that follow. Readers can run these experiments to verify results and then modify parameters and other conditions to experiment with the concepts of control systems.

Visual ModelQ is written to teach control theory. It makes convenient those activities that are necessary for studying controls. Control law gains are easy to change. Plots of frequency-domain response (Bode plots) are run with the press of a button. The models in *Visual ModelQ* run continuously, just as real-time controllers do. The measurement equipment runs independently, so you can change parameters and see the effects immediately.

1.1.1 Installation of *Visual ModelQ*

Visual ModelQ is available at www.qxdesign.com. The unregistered version is available free of charge. Although the unregistered version lacks several features, it can execute all the models used in this book. Readers may elect to register their copies of *Visual ModelQ* at any time; see www.qxdesign.com for details.

Visual ModelQ runs on PCs using Windows 95, Windows 98, Windows 2000, Windows NT, and Windows XP. Download and run the executable file *setup.exe* for *Visual ModelQ V6.0* or later. *Visual ModelQ* installs with both a User's Manual and a Reference Manual. After installation, read the User's Manual. Note that you can access the Reference Manual via Internet Explorer by pressing the F1 key. Finally, check the Web site from time to time for updated software.

1.1.2 Errata

Check www.qxdesign.com for errata. It is the author's intention to regularly update the Web page as corrections become known.

1.2 The Control System

The general control system, as shown in Figure 1-1, can be divided into the controller and the machine. The controller can be divided into the control laws and the power converter. The machine may be a temperature bath, a motor, or, as in the case of a power supply, an inductor/capacitor circuit. The machine can also be divided into two parts: the plant and the feedback device(s). The plant receives two types of signals: a controller output from the power converter and one or more disturbances. Simply put, the goal of the control system is to drive the plant in response to the command while overcoming disturbances.

1.2.1 The Controller

The controller incorporates both control laws and power conversion. Control laws, such as proportional-integral-differential (PID) control, are familiar to control engineers. The process of tuning — setting gains to attain desired performance — amounts to adjusting the parameters of the control laws. Most controllers let designers adjust gains; the most flexible controllers allow the designer to modify the control laws themselves. When tuning, most control engineers focus on attaining a quick, stable command response. However, in some applications, rejecting disturbances is more important than responding to commands. All control systems should demonstrate robust performance because even nearly identical machines and processes vary somewhat from one to the other, and they change over time. Robust operation means control laws must be designed with enough margin to accommodate reasonable changes in the plant and power converter.

Virtually all controllers have power converters. The control laws produce information, but power must be applied to control the plant. The power converter can be driven by any available power source, including electric, pneumatic, hydraulic, or chemical power.

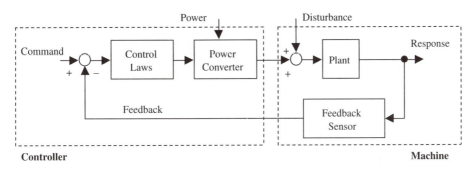

Figure 1-1. The general control system.

1.2.2 The Machine

The machine is made of two parts: the plant and the feedback. The plant is the element or elements that produce the system response. Plants are generally passive, and they usually dissipate power. Examples of plants include a heating element and a motor coupled to its load.

Control systems need feedback because the plant is rarely predictable enough to be controlled *open loop* — that is, without feedback. This is because most plants integrate the power converter output to produce the system response. Voltage is applied to inductors to produce current; torque is applied to inertia to produce velocity; pressure is applied to produce fluid flow. In all these cases, the control system cannot control the output variable directly but must provide power to the machine as physics allows and then monitor the feedback to ensure that the plant is on track.

1.3 The Controls Engineer

The focal task of many controls engineers is system integration and commissioning. The most familiar part of this process is tuning the control loops. This process can be intimidating. Often dozens of parameters must be fine-tuned to ensure that the system lives up to the specification. Sometimes that specification is entirely formal, but more often it is a combination of formal requirements and know-how gained with years of experience. Usually only the most senior engineers in a company are capable of judging when a system is performing well enough.

For some control systems, each installation may require days or weeks to be correctly commissioned. In a complex machine such as a rolling mill, that process can take months. Each piece of the machine must be carefully tuned at the site. So even after the design of the machine is complete, the expertise of a controls engineer is required each time a unit is installed.

Although most controls engineers focus on installation, their job should begin when the machine is designed. Many companies fail to take advantage of their controls expertise early in a project; this is shortsighted. A controls engineer may suggest an improved feedback device or enhancements to a machine that will help overcome a stubborn problem. Ideally, the project manager will solicit this input early, because changes of this nature are often difficult to make later.

The controls engineer should also contribute to the selection of the controller. There are many controls-oriented factors that should be taken into account. Does the controller implement familiar control laws? For digital controllers, is the processor fast enough for the needs of the application? Is the unit appropriate for the support team and for the customer base? The selection and specification of a controller often involve input from many people, but some questions can be answered best by a skilled controls engineer.

What is the role for control theory in the daily tasks of controls engineers? At its root, control theory provides understanding and, with that, intuition. Should the company purchase the controller that runs four times faster even though it costs 25% more? Should they invest in machine changes, and what will be the expected improvement from those efforts? How much will the better feedback device help a noise problem? Understanding controls doesn't guarantee that the engineer will have the correct answer. But a firm grasp on the practical side of controls will provide correct answers more often and thus position the controls engineer to provide leadership in process and product development and support.

Chapter 2

The Frequency Domain

The frequency domain provides intuition on practical subjects traversing the field of control theory. How can responsiveness be quantified? How stable is a system and how stable should it be? How do tuning gains work? How well does a system reject disturbances? The frequency domain is the beginning of control theory.

The competitor to the frequency domain is the time domain. The time domain is, above all, convenient. It is easy to comprehend and easy to measure. The measurements of an oscilloscope are immediately understood. The time domain is often the best way to communicate with customers and colleagues. Thus, the controls engineer should be fluent in both time and frequency. They are two corroborating witnesses, furnishing together a clearer understanding than either can alone.

This chapter will present the frequency domain, beginning with its foundation, the Laplace transform. Transfer functions are presented with examples of common control elements and plants. Bode plots, the favored graphic display in this text, are presented next. The chapter will discuss two important measures of control system performance, stability and response, and then conclude with a *Visual ModelQ* experiment.

2.1 The Laplace Transform

The Laplace transform underpins classic control theory [32,33,85]. It is almost universally used. An engineer who describes a "two-pole filter" relies on the Laplace transform; the two "poles" are functions of s, the Laplace operator. The Laplace transform is defined in Equation 2.1.

$$F(s) = \int_0^\infty f(t)e^{-st}dt \qquad (2.1)$$

The function $f(t)$ is a function of time, s is the Laplace operator, and $F(s)$ is the transformed function. The terms $F(s)$ and $f(t)$, commonly known as a *transform pair*, represent the same function in the two domains. For example, if $f(t) = \sin(\omega t)$, then $F(s) = \omega/(\omega^2 + s^2)$. You can use the Laplace transform to move between the time and frequency domains.

The Laplace transform can be intimidating. The execution of Equation 2.1 is complex in all but the simplest cases. Fortunately, the controls engineer need invest little time in such exercises. The most important benefit of the Laplace transform is that it provides s, the Laplace operator, and through that the frequency-domain transfer function.

2.2 Transfer Functions

Frequency-domain transfer functions describe the relationship between two signals as a function of s. For example, consider an integrator as a function of time (Figure 2-1). The integrator has an s-domain transfer function of $1/s$ (see Table 2-1). So it can be said that

$$V_o(s) = \frac{V_i(s)}{s} \tag{2.2}$$

Similarly, a derivative has the transfer function s; differentiating a time-domain signal is the same as multiplying a frequency-domain signal by s. Herein lies the usefulness of the Laplace transform. Complex time-domain operations such as differentiation and integration can be handled with algebra. Dealing with transfer functions in the time domain (that is, without benefit of the Laplace transform) requires *convolution*, a mathematical process that is so complicated that it can be used only on the simplest systems.

2.2.1 What Is s?

The Laplace operator is a complex (as opposed to real or imaginary) variable. It is defined as

$$s \equiv \sigma + j\omega \tag{2.3}$$

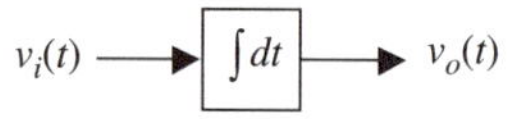

$$v_i(t) \longrightarrow \boxed{\int dt} \longrightarrow v_o(t)$$

Figure 2-1. Integrator.

TABLE 2-1 TRANSFER FUNCTIONS OF CONTROLLER ELEMENTS

Operation	Transfer Function
Integration	$1/s$
Differentiation	s
Delay	e^{-sT}
Simple filters	
Single-pole low-pass filter	$K/(s + K)$
Double-pole low-pass filter	$\omega_N^2/(s^2 + 2\zeta\omega_N s + \omega_N^2)$
Notch filter	$(s^2 + \omega_N^2)/(s^2 + 2\zeta\omega_N s + \omega_N^2)$
Bilinear-quadratic (bi-quad) filter	$(\omega_D^2/\omega_N^2)(s^2 + 2\zeta_N\omega_N s + \omega_N^2)/(s^2 + 2\zeta_D\omega_D s + \omega_D^2)$
Compensators	
Lag	$K(\tau_L s + 1)/(\tau_P s + 1), \ \tau_P > \tau_L$
PI	$(K_I/s + 1)K_P$
PID	$(K_I/s + 1 + K_D s)K_P$ or $K_I/s + K_p + K_D s$
Lead	$1 + K_D s/(\tau_D s + 1)$ or $[(\tau_D + K_D)s + 1]/(\tau_D s + 1)$

The constant j is $\sqrt{-1}$. The ω term translates to a sinusoid in the time domain; σ translates to an exponential ($e^{\sigma t}$) term. Our primary concern will be with steady-state sinusoidal signals, in which case $\sigma = 0$. The frequency in hertz, f, is defined as $f \equiv \omega/2\pi$. So for most of this book, Equation 2.3 will simplify to Equation 2.4:

$$s_{\text{STEADY–STATE}} = j2\pi f \tag{2.4}$$

The practical side of Equation 2.4 is that the response of an s-domain transfer function to a steady-state sine wave can be evaluated by setting $s = j2\pi f$.

2.2.1.1 DC Gain

Often it is important to evaluate the DC response of a transfer function, in other words, to determine the output of the transfer function subjected to a DC input. To find the DC response, set s to zero. For example, we discussed before that the transfer function for differentiation is $V_o(s) = V_i(s) \times s$. What happens when a DC signal is differentiated? Intuitively, it produces zero, and that is confirmed by setting s to zero in this simple equation.

2.2.2 Linearity, Time Invariance, and Transfer Functions

A frequency-domain transfer function is limited to describing elements that are linear and time invariant. These are severe restrictions and, in fact, virtually no real-world system fully meets them. The criteria that follow define these attributes, the first two being for linearity and the third for time invariance.

1. *Homogeneity.* Assume that an input to a system $r(t)$ generates an output $c(t)$. For an element to be homogeneous, an input $k \times r(t)$ would have to generate an output $k \times c(t)$, for any value of k. An example of nonhomogeneous behavior is saturation, where twice as much input delivers less than twice as much output.
2. *Superposition.* Assume that an element subjected to an input $r_1(t)$ will generate the output $c_1(t)$. Further, assume that the same element subjected to input $r_2(t)$ will generate an output $c_2(t)$. Superposition requires that if the element is subjected to the input $r_1(t) + r_2(t)$, it will produce the output $c_1(t) + c_2(t)$ [32,80].
3. *Time invariance.* Assume that an element has an input $r(t)$ that generates an output $c(t)$. Time invariance requires that $r(t - \tau)$ will generate $c(t - \tau)$ for all $\tau > 0$.

So we face a dilemma: Transfer functions, the basis of classic control theory, require linear, time-invariant (LTI) systems, but no real-world system is completely LTI. This is a complex problem that is dealt with in many ways, some of which are detailed in Chapter 12. However, for most control systems, the solution is simple: design components close enough to being LTI that the non-LTI behavior can be ignored or avoided.

In practice, most control systems are designed to minimize non-LTI behavior. This is one reason why components used in control systems are often more expensive than their noncontrol counterparts. For most of this text, the assumption will be that the system is LTI or close enough to it to use transfer functions. Readers who are troubled by this approximation should consider that this technique is commonly applied by engineers in all disciplines. For example, Ohm's law, $v = iR$, is an approximation that ignores many effects, including electrical radiation, capacitive coupling, and lead inductance. Of course, all those effects are important from time to time, but few would argue the utility of Ohm's law.

2.3 Examples of Transfer Functions

In this section, we will discuss the transfer functions of common elements in control systems. The discussion is divided along the lines of Figure 2-2.

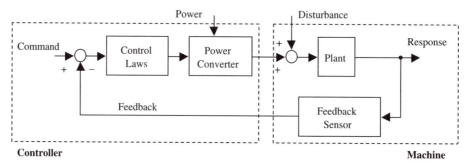

Figure 2-2. Elements in the control system.

2.3.1 Transfer Functions of Controller Elements

The controller elements divide into the control laws and power conversion. Examples of operations used in control laws are shown in Table 2-1. Note that Appendix A shows the implementation of many of these laws using operational amplifier (op-amp) circuits.

2.3.1.1 Integration and Differentiation

Integration and differentiation are the simplest operations. The s-domain operation of integration is $1/s$ and of differentiation is s.

2.3.1.2 Filters

Filters are commonly used by control systems designers, such as when low-pass filters are added to reduce noise. The use of filters in control systems is detailed in Chapter 9. Table 2-1 lists the s-domain representation for a few common examples.

2.3.1.3 Compensators

Compensators are specialized filters. A compensator is a filter that is designed to provide a specific gain and phase shift, usually at one frequency. The effects on gain and phase either above or below that frequency are secondary. Table 2-1 shows a lag compensator, a proportional-integral (PI) compensator, a proportional-integral-differential (PID) compensator, and a lead compensator. The principles of compensators are described in Chapters 3, 4, and 6.

2.3.1.4 Delays

Delays add time lag without changing amplitude. For example, a conveyor belt with a scale can cause a delay. Material is loaded at one point and weighed a short distance later; that distance causes a delay that is inversely proportional to the belt speed. A system that controls the amount of material to be loaded onto the conveyor must take the time delay into account. Since microprocessors have inherent delays for calculation time, the delay function is especially important to understanding digital controls, as will be discussed in Chapters 4 and 5.

A delay of T seconds is defined in the time domain as

$$c(t) = r(t - T) \qquad (2.5)$$

The corresponding function in the frequency domain is

$$T_{\text{delay}}(s) = e^{-sT} \tag{2.6}$$

2.3.2 Transfer Functions of Power Conversion

Power conversion is the interface between the control laws and the plant. Power can come from electric, pneumatic, hydraulic, and chemical sources. The transfer function of the conversion depends on the device delivering the power. In electronic systems, the power is often delivered through pulse modulation, a family of near-linear methods that switch transistors at high frequency. Seen from the frequency domain, modulation inserts a short delay in the control loop. Modulation also injects nonlinear effects, such as ripple, which are usually ignored in the frequency domain.

2.3.3 Transfer Functions of Physical Elements

Physical elements are made up of the *plant*, the mechanism or device being controlled, and the feedback sensor(s). Examples of plants include electrical elements, such as inductors, and mechanical elements, such as springs and inertias. Table 2-2 provides a list of ideal elements in five categories [32]. Consider the first category, *electrical*. A voltage *force* is applied to an impedance to produce a current *flow*. The impedance is generated by flow in three ways: inductance, capacitance, and resistance. Resistance is proportional to the current flow, inductive impedance is in proportion to the derivative of current flow, and capacitive impedance is in proportion to the integral of flow.

The pattern of force, impedance, and flow is repeated for many physical elements. In Table 2-2, the close parallels between the categories of linear and rotational force, fluid mechanics, and heat flow are evident. In each case, a forcing function (voltage, force, torque, pressure or temperature difference) applied to an impedance produces a flow (current, velocity, or fluid/thermal flow). The impedance takes three forms: resistance to the integral of flow (capacitance or mass), resistance to the derivative of flow (spring or inductance), and resistance to the flow rate (resistance or damping).

Table 2-2 reveals a central concept of controls. Controllers for these elements apply a *force* to control a *flow*. For example, a heating system applies heat to control a room's temperature. When the flow must be controlled with accuracy, a feedback sensor can be added to measure the flow; control laws are required to combine the feedback and command signals to generate the force. This results in the structure shown in Figure 2-2; it is this structure that sets control systems apart from other disciplines of engineering.

The plant usually contains the most nonlinear elements in the control system. The plant is often too large and expensive to allow engineers the luxury of designing it for LTI operation. For example, an iron-core inductor in a power supply usually saturates; that is, its inductance declines as the current approaches its peak.

TABLE 2-2 TRANSFER FUNCTIONS OF PLANT ELEMENTS

Electrical

Voltage (E) and current (I)

Inductance (L)	$E(s) = Ls \times I(s)$	$e(t) = L \times di(t)/dt$
Capacitance (C)	$E(s) = 1/C \times I(s)/s$	$e(t) = e_0 + 1/C \int i(t)dt$
Resistance (R)	$E(s) = R \times I(s)$	$e(t) = R \times i(t)$

Translational mechanics

Velocity (V) and force (F)

Spring (K)	$V(s) = s/K \times F(s)$	$v(t) = 1/K \times df(t)/dt$
Mass (M)	$V(s) = 1/M \times F(s)/s$	$v(t) = v_0 + 1/M \int f(t)dt$
Damper (c)	$V(s) = F(s)/c$	$v(t) = f(t)/c$

Rotational mechanics

Rotary velocity (ω) and torque (T)

Spring (K)	$\omega(s) = s/K \times T(s)$	$\omega(t) = 1/K \times dT(t)/dt$
Inertia (J)	$\omega(s) = 1/J \times T(s)/s$	$\omega(t) = \omega_0 + 1/J \int T(t)dt$
Damper (b)	$\omega(s) = T(s)/b$	$\omega(t) = T(t)/b$

Fluid mechanics

Pressure (P) and fluid flow (Q)

Inertia (I)	$P(s) = sI \times Q(s)$	$p(t) = I \times dq(t)/dt$
Capacitance (C)	$P(s) = 1/C \times Q(s)/s$	$p(t) = p_0 + 1/C \int q(t)dt$
Resistance (R)	$P(s) = R \times Q(s)$	$p(t) = R \times q(t)$

Heat flow

Temperature difference (J) and heat flow (Q)

Capacitance (C)	$J(s) = 1/C \times Q(s)/s$	$j(t) = j_0 + 1/C \int q(t)dt$
Resistance (R)	$J(s) = R \times Q(s)$	$j(t) = R \times q(t)$

The inductor could be designed so that saturation during normal operation is eliminated, but this requires more iron, which makes the power supply larger and more expensive.

Most of the impedances in Table 2-2 can be expected to vary over normal operating conditions. The question naturally arises as to how much they can vary. The answer depends on the application, but most control systems are designed with enough margin that variation of 20–30% in the plant should have little impact on system performance. Of course, parameters sometimes vary by much more than 30%. For example, consider the inertia in rotational motion. An application that winds plastic onto a roll may see the inertia vary by a factor of 50 between an empty and a full roll. We will deal with variations in gain of this magnitude in Chapter 12.

2.3.4 Transfer Functions of Feedback

The ideal transfer function for a feedback device is unity. The ideal current sensor measures current with perfect accuracy and without delay. Of course, feedback elements are not perfect. First, consider delay. Feedback devices inject some delay between the true signal and the measurement. As we will discuss in succeeding chapters, that delay can be the limiting item in system performance. This is because the dynamics of the control loop cannot be more than about a fifth as fast as the dynamics of the feedback sensor. Sluggish feedback supports sluggish control. Often the dynamics of the feedback sensor can be modeled as a low-pass filter. For example, a typical force transducer acts like a low-pass filter with a bandwidth of around 20 Hz.

$$T(s) = \frac{2\pi 20}{s + 2\pi 20} \tag{2.7}$$

Sensors also contribute offset and ripple. These factors do not directly affect traditional measures of performance such as stability and responsiveness.

2.4   Block Diagrams

Block diagrams are graphical representations developed to make control systems easier to understand. Blocks are marked to indicate their transfer function. In North America, the transfer functions are usually indicated with their s-domain representation. The convention in Europe is to use schematic representation of a step response; Appendix B provides a listing of many North American and European block diagram symbols.

Figure 2-3 shows a block diagram of a simple control loop. It connects a proportional controller with a gain of K to a plant that is an integrator with a gain of 1000. Here, power conversion is ignored and ideal feedback is assumed.

2.4.1 Combining Blocks

You can simplify a block diagram by combining blocks. If two blocks are in parallel, they can be combined as the sum of the individual blocks. If two blocks are cascaded (i.e., in series), they can be represented as the product of the individual blocks. For example, the three blocks in Figure 2-4a can be combined to form the single block of Figure 2-4b.

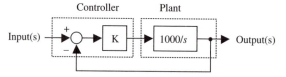

Figure 2-3. Simple control loop.

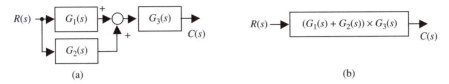

Figure 2-4. (a) A multiblock diagram and (b) its single-block counterpart.

2.4.1.1 Simplifying a Feedback Loop

When blocks are arranged to form a loop, they can be reduced using the $G/(1 + GH)$ rule. The forward path is $G(s)$ and the feedback path is $H(s)$. The transfer function of the loop is $G(s)/(1 + G(s)H(s))$ as shown in Figure 2-5.

The $G/(1 + GH)$ rule can be derived by observing in Figure 2-5a that the error signal ($E(s)$) is formed from the left as

$$E(s) = R(s) - C(s)H(s)$$

$E(s)$ can also be formed from the right side (from $C(s)$ back through $G(s)$) as

$$E(s) = C(s)/G(s)$$

So

$$R(s) - C(s)H(s) = C(s)/G(s)$$

One or two steps of algebra produces $C(s)/R(s) = G(s)/(1 + G(s)H(s))$.

We can apply the $G/(1 + GH)$ rule to Figure 2-3 to find its transfer function. Here, $G = K \times 1000/s$ and $H = 1$:

$$T(s) = \frac{G(s)}{1 + G(s)H(s)} = \frac{1000 \times K/s}{1 + 1000 \times K/s} = \frac{1000 \times K}{s + 1000 \times K} \tag{2.8}$$

If a block diagram has more than one loop, the $G/(1 + GH)$ rule can be applied multiple times, once for each loop. While this produces correct results, the algebra can be tedious. An alternative, Mason's signal flow graphs, can simplify the process.

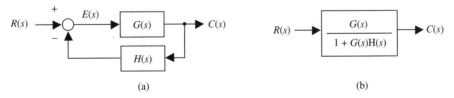

Figure 2-5. (a and b) Simple feedback loop in equivalent forms.

2.4.2 Mason's Signal Flow Graphs

An alternative to the $G/(1 + GH)$ rule developed by Mason [20,64,65] provides graphical means for reducing block diagrams with multiple loops. The formal process begins by redrawing the block diagram as a signal flow graph.[1] The control system is redrawn as a collection of nodes and lines. Nodes define where three lines meet; lines represent the s-domain transfer function of blocks. Lines must be unidirectional; when drawn, they should have one and only one arrowhead. A typical block diagram is shown in Figure 2-6; its corresponding signal flow graph is shown in Figure 2-7.

2.4.2.1 Step-by-Step Procedure

This section will present a step-by-step procedure to produce the transfer function from the signal flow graph based on Mason's signal flow graphs. The signal flow graph

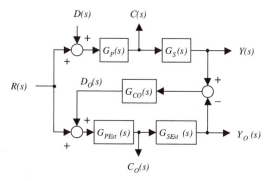

Figure 2-6. An example control loop block diagram.

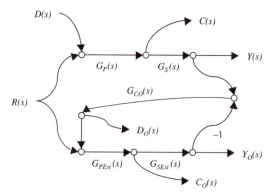

Figure 2-7. Signal flow graph for Figure 2-6.

---

[1]For convenience, this step will be omitted in most cases and the block diagram used directly.

of Figure 2-7 will be used for an example. This graph has two independent inputs, $R(s)$ and $D(s)$. The example will find the transfer function from these two inputs to $D_o(s)$.

Step 1. Find the loops.
Locate and list all loop paths. For the example of Figure 2-7, there is one loop:

$$L_1 = -G_{PEst}(s) \times G_{SEst(s)} \times G_{CO}(s)$$

Step 2. Find the determinant of the control loop.
Find the determinant, Δ, of the control loop, which is defined by the loops:

$\Delta =1 -$ (sum of all loops)
  $+$ (sum of products of all combinations of two non-touching loops)
  $-$ (sum of products of all combinations of three non-touching loops)
  $+ \ldots$

Two loops are said to be touching if they share at least one node. For this example there is only one loop:

$$\Delta = 1 + G_{PEst}(s) \times G_{SEst}(s) \times G_{CO}(s)$$

Step 3. Find all the forward paths.
The forward paths are all the different paths that flow from the inputs to the output. For the example of Figure 2-7, there is one forward path from $D(s)$ to $D_o(s)$ and two from $R(s)$:

$$P_1 = D(s) \times G_P(s) \times G_S(s) \times G_{CO}(s)$$
$$P_2 = R(s) \times G_P(s) \times G_S(s) \times G_{CO}(s)$$
$$P_3 = R(s) \times G_{PEst}(s) \times G_{SEst}(s) \times -1 \times G_{CO}(s)$$

Step 4. Find the cofactors for each of the forward paths.
The cofactor (Δ_K) for a particular path (P_K) is equal to the determinant (Δ) less loops that touch that path. For the example of Figure 2-7, all cofactors are 1 because every forward path includes $G_{CO}(s)$, which is in L_1, the only loop.

$$\Delta_1 = \Delta_2 = \Delta_3 = 1$$

Step 5. Build the transfer function.
The transfer function is formed as the sum of all the paths multiplied by their cofactors, divided by the determinant:

$$T(s) = \frac{\sum_K (P_K \Delta_K)}{\Delta} \tag{2.9}$$

For the example of Figure 2-7, the signal $D_o(s)$ is

$$D_O(s) = R(s)\frac{(G_P(s)G_S(s) - G_{PEst}(s)G_{SEst}(s))G_{CO}(s)}{1 + G_{PEst}(s)G_{SEst}(s)G_{CO}(s)} + D(s)\frac{G_P(s)G_S(s)G_{CO}(s)}{1 + G_{PEst}(s)G_{SEst}(s)G_{CO}(s)}$$

Using a similar process, $C_O(s)$ can be formed as a function of $C(s)$ and $R(s)$:

$$C_O(s) = R(s)\frac{G_{PEst}(s)(1 + G_P(s)G_S(s)G_{CO}(s))}{1 + G_{PEst}(s)G_{SEst}(s)G_{CO}(s)} + D(s)\frac{G_{PEst}(s)G_P(s)G_S(s)G_{CO}(s)}{1 + G_{PEst}(s)G_{SEst}(s)G_{CO}(s)}$$

As will be discussed in later chapters, a great deal of insight can be gained from transfer functions of this sort. Using Mason's signal flow graphs, transfer functions of relatively complex block diagrams can be written by inspection.

2.5 Phase and Gain

A transfer function can be used to calculate the output for just about any input waveform, although it can be a tedious process for nonsinusoidal signals. Fortunately, for most frequency-domain analyses, we will use a sine wave input. The sine wave is unique among repeating waveforms; it's the only waveform that does not change shape in a linear control system. A sine wave input always generates a sine wave output at the same frequency; the only difference possible between input and output is the gain and the phase. That is, the response of an LTI system to any one frequency can be characterized completely knowing only phase and gain.

Gain measures the difference in amplitude of input and output. Gain is often expressed in decibels, or dB, and is defined as

$$\text{Gain} \equiv 20 \times \text{Log}_{10}(\text{OUT/IN}) \qquad (2.10)$$

where OUT and IN are the magnitudes of the output and input sine waves, respectively. This is shown in Figure 2-8. Phase describes the time shift between input and output. This lag can be expressed in units of time (t_{DELTA}) but more often is expressed in degrees, where 360° is equivalent to one period of the input sine wave. Usually, the output is considered as lagging (to the right of) the input. So phase is defined as

$$\text{Phase} \equiv -360 \times F \times t_{\text{DELTA}}° \qquad (2.11)$$

For example, if the control system was excited with a 1-V, 10-Hz sine wave and the output was 0.7 V and lagged the input by 12.5 msec:

$$\text{Gain} = 20 \text{ Log}_{10}(0.7/1.0) \cong -3 \text{ dB}$$
$$\text{Phase} = -360 \times 10 \times 0.0125° = -45°$$

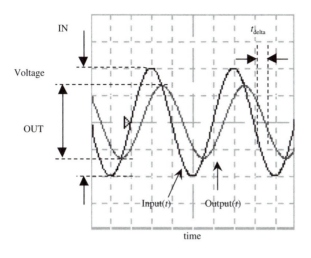

$$\text{Gain} = 20 \, \text{Log}_{10}(\text{OUT/IN}) \qquad\qquad \text{Phase} = -360 \times \text{F} \times t_{\text{delta}}$$

Figure 2-8. Gain and phase.

Frequently, the gain and phase of a transfer function are shown as a *phasor*. For example, the preceding phase and gain are written in phasor form as $-3\,\text{dB}\; \angle{-45°}$.

2.5.1 Phase and Gain from Transfer Functions

Phase and gain can be calculated from the transfer function by setting $s = j \times 2\pi f$, where f is the excitation frequency and $j = \sqrt{-1}$. For example, consider the transfer function of Equation 2.12, which is a low-pass filter:

$$T(s) = 62.8/(s + 62.8) \qquad\qquad (2.12)$$

Now, evaluate this transfer function at 10 Hz (62.8 rad/sec):

$$
\begin{aligned}
T(s)|_{s=10\,\text{Hz}} &= 62.8/(2\pi 10 j + 62.8) \\
&= 62.8/(62.8j + 62.8) \\
&= 0.5 - 0.5j \qquad\qquad (2.13)
\end{aligned}
$$

The response of a transfer function can be converted from a complex number to a phasor by using the magnitude of the number in dB and taking the arctangent of the imaginary portion over the real and adding 180° if the real part is negative:

$$
\begin{aligned}
0.5 - 0.5j &= \sqrt{0.5^2 + 0.5^2} \angle \text{Tan}^{-1}(-0.5/0.5) \\
&= 0.707\angle{-45°} \\
&= -3\,\text{dB}\angle{-45°}
\end{aligned}
$$

2.5.2 Bode Plots: Phase and Gain versus Frequency

Knowing the phase and gain of a transfer function across a wide range of frequencies provides a large amount of information about that function. Bode plots show this information graphically; the gain in decibels and phase in degrees are plotted against the frequency in hertz. The frequency scale is logarithmic and the vertical scales are linear. Figure 2-9 shows a Bode plot of the transfer function of Equation 2.12. The frequency spans from 2 Hz to 500 Hz (see legend just below the plot). The gain is shown in the top graph, scaled at 20 dB per division, with 0 dB at the solid center line. Phase is shown in the bottom graph scaled at 90° per division, again 0° at the center line. Notice that 10 Hz is marked on the Bode plot with the gain (−3 dB) and phase (−45°) shown, agreeing with Equation 2.13.

Determining the appropriate range of frequencies for a Bode plot takes some care. At very high frequencies, control systems essentially stop responding. For example, above some frequency, the energy required to accelerate inertia increases beyond what the system can produce. Beyond that frequency, there is little interest in the system output. Conversely, below some low frequency, the system will act nearly the same as it does at DC. Plotting the system at still lower frequencies would add little information. Correctly generated Bode plots show phase and gain between these frequency extremes.

In the past, Bode plots were constructed using a series of approximations that allowed the response to be graphed with minimal computations. For example, the gain of a low-pass filter could be drawn as two straight lines, intersecting at the break frequency, or −3-dB point. These methods are still used occasionally, although the

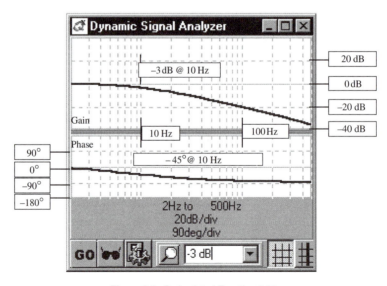

Figure 2-9. Bode plot of Equation 2.12.

availability of computers allows virtually any designer to calculate information for Bode plots rapidly and accurately.

2.6 Measuring Performance

Objective performance measures provide a path for problem identification and correction. This section will discuss the two most common performance measures: command response and stability.

2.6.1 Command Response

Command response measures how quickly a system follows the command. In the time domain, the most common measure of command response is settling time to a step. Although few systems are subjected to step commands in normal operation, the step response is useful because responsiveness can be measured on a scope more easily from a step than from most other waveforms.

Settling time is measured from the front edge of the step to the point where the feedback is within 5% (sometimes 2%) of the commanded value. For example, Figure 2-10a shows the step response to ±10-RPM command. The excursion is 20 RPM, so the system will be settled when the command is within 5% (1 RPM) of the final command (10 RPM), that is, when the system reaches 9 RPM. That time, shown in Figure 2-10a, is about 0.01 sec. Figure 2-10b shows a sluggish system; it

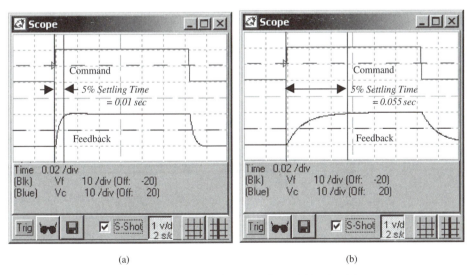

(a)   (b)

Figure 2-10. Step response of (a) a responsive and (b) a sluggish system.

requires 0.055 sec to traverse the same speed change. So by this measure, the system in Figure 2-10a is about five times more responsive. Incidentally, the control systems in Figure 2-10a and b are the same, except in Figure 2-10b the controller gains were turned down by a factor of 5. As expected, higher gains provide faster response.

Response can also be measured in the frequency domain by inspecting the gain plot. Most control systems demonstrate good command response at low frequencies but become unresponsive at high frequencies. At low frequencies, the controller is fast enough to govern the system. As the frequency increases, the controller cannot keep up. Thinking of the transfer function, this means the gain will be approximately unity (1) at low frequencies but will be much less than unity at high frequencies. Consider a power supply that is advertised to produce sine wave voltages up to 100 Hz. We would expect that power supply to be nearly perfect at low frequencies, such as 1 Hz, to begin to struggle as the frequency increased to its rated 100 Hz, and to produce almost no voltage at high frequencies, say, above 10 kHz. Translating this to a Bode plot, we expect the gain to be unity (0 dB) at low frequencies, to start falling at midrange frequencies, and to continue falling to very low values at high frequencies.

Figure 2-11a shows the Bode plot for the system evaluated in Figure 2-10a. The frequency range spans from 1 Hz to 500 Hz. At low frequency, the gain is unity (0 dB). As the frequency increases, the gain begins to fall. At the highest frequency shown, the gain has fallen one division, or −20 dB, which is equivalent to a gain of 10%. As the frequency increases, the gain will continue to fall. Closed-loop responsiveness is commonly measured in the frequency domain as the *bandwidth*, the frequency at which the gain has fallen −3 dB, or to a gain of about 70%. In Figure 2-11a, that frequency is about 52 Hz.

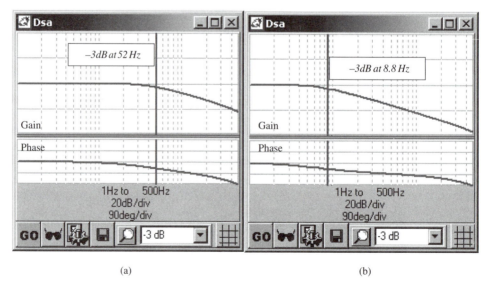

Figure 2-11. Frequency response (Bode plot) of (a) a responsive and (b) a sluggish system.

Figure 2-11b shows the Bode plot for the system of Figure 2-10b. Recall that here, the controller gains were turned down by a factor of 5. Now the Bode plot shows the bandwidth has fallen to 8.8 Hz. The bandwidth, like the settling time, shows the original (high-gain) system to be much more responsive.

Settling time is related to bandwidth. Bandwidth represents approximately τ, the *time constant* of the system according to Equation 2.14:

$$\tau \approx 1/(f_{BW} \times 2\pi) \tag{2.14}$$

where τ is in seconds and f_{BW} is the bandwidth in hertz. With a simple approximation, the system is modeled as $e^{-t/\tau}$. For a step command, the passing of one time constant implies that the error is reduced to 27% (e^{-1}) of its value at the start of the time constant. The time to settle to 5% requires approximately three time constants ($e^{-3} \approx 0.05$). The settling time to 2% requires about four time constants ($e^{-4} \approx 0.02$). So,

$$t_{5\%} \approx 3/(f_{BW} \times 2\pi) \tag{2.15}$$

$$t_{2\%} \approx 4/(f_{BW} \times 2\pi) \tag{2.16}$$

Thus, given a bandwidth of 52 Hz, you would expect a 5% settling time of 0.009 sec, and for a bandwidth of 8.8 Hz, you would expect a settling time of 0.054 sec. This correlates well with the examples in Figure 2-10a and b.

When measuring response, the magnitude of the command (step or sine wave) must be small for the results to be meaningful. This is so the system will remain in the linear range of operation during the measurement of response.

2.6.2 Stability

Stability describes how predictably a system follows a command. In academics, considerable time is spent differentiating unstable systems (those that self-oscillate) from stable systems. However, in industry, this difference is of little consequence; a system near the edge of stability is not useful even if it does not self-oscillate. The focus for working engineers is measuring how stable a system is or, more precisely, how much margin of stability the system possesses.

In the time domain, stability is most commonly measured from the step response. The key characteristic is overshoot: the ratio of the peak of the response to the commanded change. The amount of overshoot that is acceptable in applications varies from 0% to perhaps 30%. Figure 2-12 shows the step response of two controllers. On the left is the system from Figure 2-10a; the overshoot is negligible. On the right, the overshoot is almost one division of a two-division excursion, or about 50%. Both systems are, strictly speaking, stable. But the margin of stability for the system on the right is too small for most applications.

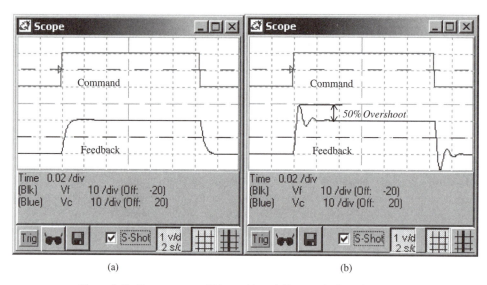

Figure 2-12. Step response of (a) a stable and (b) a marginally stable system.

Stability can also be measured from a system Bode plot; again the information is in the gain. As discussed earlier, the gain will be at 0 dB at low frequencies for most control systems and will fall off as the frequency increases. If the gain rises before it starts falling, it indicates marginal stability. This phenomenon is called *peaking*. The amount of peaking is a measure of stability. For practical systems, allowable peaking ranges from 0 dB to perhaps 4 dB. The systems that were measured in the time domain in Figure 2-12 are measured again in Figure 2-13 using the frequency domain. Note that Figure 2-13b, with 6 dB of peaking, corresponds to the scope trace in Figure 2-12b, with 50% overshoot.

2.6.3 Time Domain versus Frequency Domain

For the systems shown in Figures 2-10 through 2-13, the correlation between time and frequency domains is clear. Settling time correlates to bandwidth; overshoot correlates to peaking. For these simple systems, measures in either domain work well. The natural question is why both measures are needed. The answer is that in realistic systems the time domain is more difficult to interpret. Many phenomena in control systems occur in combinations of frequencies; for example, there may simultaneously be a mild resonance at 400 Hz and peaking at 60 Hz. Also, feedback resolution may limit your ability to interpret the response to small-amplitude steps. In real-world control systems, gleaning precise quantitative data from a step response is often impractical.

Finally, observe that time-domain measures often rely on the step command. The reason is that the step command has a large frequency content. It has high frequencies

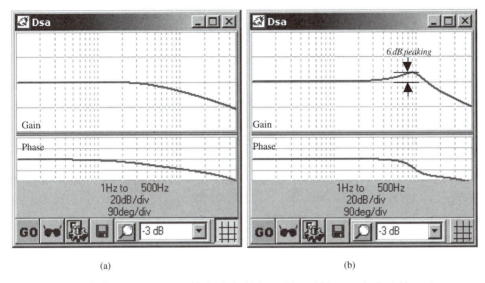

Figure 2-13. Frequency response (Bode plot) of (a) a stable and (b) a marginally stable system.

from the steep edges and low frequencies because of the fixed value between edges. Like a Bode plot, step commands excite the system at a broad range of frequencies. In a sense, the step response is an inexpensive Bode plot.

2.7 Questions

1. Evaluate the following transfer functions at 100 Hz.

   a) $\dfrac{C(s)}{R(s)} = \dfrac{1000}{s + 1000}$

   b) $\dfrac{C(s)}{R(s)} = \dfrac{s + 200}{s + 1000}$

   c) $\dfrac{C(s)}{R(s)} = \dfrac{s + 1000}{s + 200}$

   d) $\dfrac{C(s)}{R(s)} = \dfrac{s^2 + 200s + 1000}{s^2 + 500s + 1500}$

2. Evaluate the transfer functions of Question 1 at DC.
3. What is the gain and phase of a single-pole low-pass filter (a) at its bandwidth and (b) at 1/10 its bandwidth? (c) Repeat for a two-pole low-pass filter with $\zeta = 0.7$ (Hint: for $\zeta = 0.7$, bandwidth = the natural frequency, ω_N).

4. What is the approximate relationship between settling time and bandwidth when settling to 0.5% of the final value?
5. a. Estimate the frequency of ringing in Figure 2-12b.
   b. Estimate the frequency of maximum peaking in Figure 2-13b, the corresponding Bode plot.
   c. What conclusion could you draw?
6. Find the transfer function of the following block diagram using Mason's signal flow graphs.

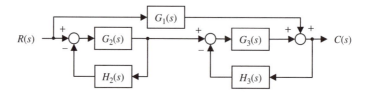

7. Which of the following functions are not linear, time-invariant (LTI)? For those that are not, state at least one of the LTI rules they violate:
   a. Integration
   b. Differentiation
   c. Addition
   d. Subtraction
   e. Multiplication of a variable by a constant
   f. Multiplication of a variable by itself (squaring)
   g. Division of variable by a constant
   h. Division of constant by a variable
   i. Exponentiation

Chapter 3

Tuning a Control System

Tuning is the process of setting controller gains to achieve desired performance. In demanding applications, tuning can be difficult. Control systems have inherent limitations in response and stability; tuning is a challenge when the machine or process requires all the performance the controller has to give.

Tuning is unfamiliar more than it is complex. There are few tasks in everyday life that bear resemblance to the process. Human motor control is so sophisticated that it makes it difficult for the average person to comprehend the comparatively simple industrial controller. One example that gives a hint is the motion that occurs when a person lifts an object after miscalculating its weight. Most people can recall picking up an empty can thought to be full and the jerky motion that follows. Of course, the mind quickly recovers and smoothes the motion. That recovery, often called "adaptive behavior," is universal among people but almost unseen in industrial controllers.

The goal of tuning is to determine the gains that optimize system response. High gains increase responsiveness but also move the system closer to instability. The phenomenon of instability comes from an accumulation of delays around a feedback loop. Reducing delays within the control loop is a sure way to make room for higher gains. This is why fast sampling and high-speed sensors are required for the most responsive applications.

This chapter will investigate the causes of instability, including a discussion of two of its quantitative measures: phase margin and gain margin. A detailed example of a proportional-integral (PI) control system will demonstrate the main principles. Following this, the development of a tuning procedure based on frequency *zones* is presented. The chapter ends with discussions of plant variation, multiple control loops, and saturation, the most prevalent nonlinear behavior.

3.1 Closing Loops

Consider for a moment why closed-loop control is required. Most plants are inherently unpredictable because the control variable is the integral of the applied signal. Refer to

Table 2-2 and notice that in each case the applied signal is integrated to produce the control variable. Controlling such a variable without a feedback loop would imply that a signal could be manipulated by setting its derivative. This is generally impractical because even small deviations in plant gain or disturbances will accumulate in the integral.

In a closed-loop system, the controller will first attempt to command the control variable through its derivative (there is no alternative) and then measure the result and adjust out the error. The loop allows tight control of a variable, although only the derivative can be manipulated. Unfortunately, that loop is also the source of instability. As will be shown, when signals traverse the loop, inherent phase lags in the control system can add enough delay to cause instability.

3.1.1 The Source of Instability

Control loops must have negative feedback. In Figure 3-1, if the sign of the summing junction is reversed from "−" to "+," instability results. Many people have had the experience of reversing the leads of a tachometer, thermistor, or voltage sensor; the system runs away because positive feedback causes the correction signal generated from the loop to move in the wrong direction. Command positive response and the response indeed is positive. But the inverted feedback signal causes the summing junction to act as if the plant response was negative. The controller then generates more positive command; the system yields still more positive response. The cycle continues and the system runs away.

Incorrect wiring is not the only way to produce a sign change in the loop. It can also be produced by accumulating a sufficient amount of phase lag around the loop. Unlike that caused by reversing feedback wiring, instability that results from the accumulation of phase lag usually occurs at just one frequency. This is why unstable control systems oscillate; they do not run away as when feedback leads are reversed. It is counterintuitive to think that instability at a frequency, especially a high frequency, presents a serious problem. The natural response is to wonder, "Why not avoid exciting the control system at the oscillatory frequency?" The problem is that all

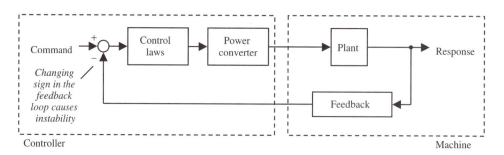

Figure 3-1. Instability comes from positive feedback.

systems have noise, and noise contains virtually all frequencies. If a system is unstable at any frequency, nature will find that frequency, usually in a few milliseconds.

The example of Figure 3-1 is modified in Figure 3-2 to demonstrate how phase lag accumulates. This control system is excited with a 240-Hz sine wave. Each block generates a certain amount of phase lag. The controller is a PI controller that generates a small lag of 4°. The power converter, which always has some lag, here generates 25° lag. The plant introduces 90° lag, which, as will be discussed, is characteristic of integrators. Finally, the inherent limitation of the feedback element contributes another 61° lag. The sum of all these phase lags is 180°, equivalent to a sign reversal. So a signal at 240 Hz will have its sign changed as it traverses the loop; this is also positive feedback.

For most control systems, there will exist at least one frequency where the phase lag accumulates to 180°, but this alone does not cause instability. To have instability, the loop gain must also be equal to unity. Similar to phase lag, each block in the loop contributes gain. The total gain through the loop is the accumulation of the block gains; when measured in decibels, which are logarithmic units, the loop gain is the sum of the gains of the individual blocks. Consider if, at the frequency of 180° phase shift, the gain around the loop were low, say, 10% (−20 dB); the sign reversal would not produce sustained oscillations because the loop thoroughly attenuates the signal as it traverses the loop. Instability requires two conditions: The sign must be reversed, and the gain through the loop must be unity. In this way, the loop will produce self-sustained oscillations; the signal moves through the loop, is reversed without attenuation, and is added back onto itself; the cycle repeats indefinitely. Figure 3-3 demonstrates the conditions for this phenomenon.

The $G(1 + GH)$ rule from Figure 2-5 provides further insight into this problem. First, note that GH is a mathematical expression for *loop gain*. The forward path, G, includes the control laws, the power converter, and the plant. The feedback path, H, here includes only the feedback. Notice that the loop gain of 0 dB ∠−180° is equivalent to $GH = -1$. Now notice that if $GH = -1$, the transfer function of the loop $(G/(1 + GH))$ will have a zero in the denominator, producing an unbounded large number.

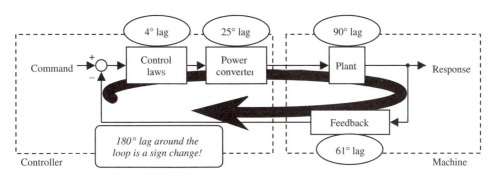

Figure 3-2. Each block in a control system contributes phase lag.

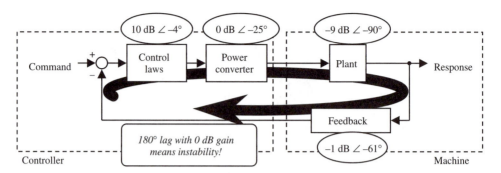

Figure 3-3. Instability comes when the loop gain is 0 dB $\angle -180°$.

3.2 **A Detailed Review of the Model**

The following section will present a Bode plot for each element of the system in Figures 3-1 through 3-3. This is done in preparation for the derivation of the open-loop method, which follows immediately afterward. Figure 3-4 shows the system of Figure 3-1 with a PI control law.

3.2.1 Integrator

The plant at the top right of Figure 3-4 is shown as an integrator, G/s, where G is set to 500. This gain might be that of a 2-mH inductor ($T(s) = 1/Ls$) or a motor with a torque constant (K_T) of 1 Nm/amp and an inertia (J) of 0.002 kg-m^2. In both cases, the gain is $500/s$. From Chapter 2, the phase and gain of any function of s can be evaluated by setting $s = j2\pi f$. So

$$G = 500/(j2\pi f) = -j \times 500/(2\pi f) \tag{3.1}$$

where $j = \sqrt{-1}$.

The gain of j is unity (0 dB) and its phase is 90°, so the phase of G/s, which has a $-j$ term, is fixed at $-90°$. The gain of G/s falls in proportion to the frequency, as can be

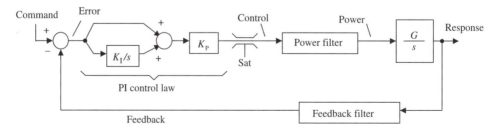

Figure 3-4. PI control system.

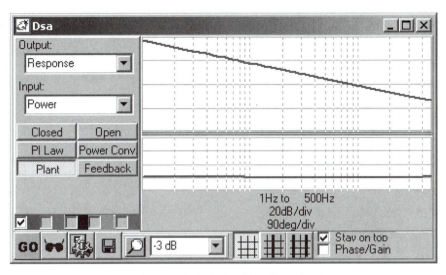

Figure 3-5. Bode plot of ideal integral.

seen in Equation 3.1; the Bode plot of Equation 3.1 is shown in Figure 3-5. Notice that the gain falls 20 dB (a factor of 10) per decade of frequency (also a factor of 10), indicating a proportional relationship between gain and frequency.

You may gain some insight into the integrator by considering how force and its integral, velocity, are related. Figure 3-6 shows this relationship in the time domain;

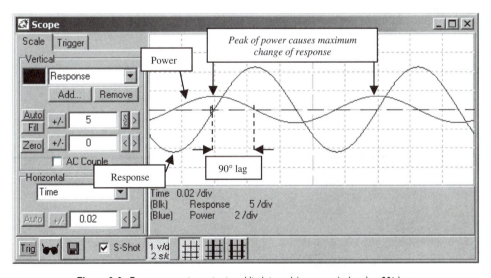

Figure 3-6. Power converter output and its integral (response) showing 90° lag.

the smaller-amplitude signal is force and the larger is velocity. When the force is at its peak, it causes the maximum acceleration. When the force is at its most negative, it causes the maximum negative acceleration. When force crosses through zero, velocity is briefly constant. Now observe that the large signal (velocity) lags the small signal (force) by a quarter cycle, or 90°.

3.2.2 Power Converter

The power converter is modeled as a low-pass filter. Of course, the power converter is much more than a filter. For example, in electronic power converters, complex modulation schemes are employed to convert control signals into power with high efficiency and minimum harmonic content. However, the complex relationships within power converters usually occur at high frequencies; the effects at the lower frequencies where the loop operates are much simpler. For our purposes, the power converter can be modeled as a two-pole low-pass filter, here assumed to have an 800-Hz bandwidth. The Bode plot for such a filter is shown in Figure 3-7.

Notice that although the bandwidth is 800 Hz, the phase of this low-pass filter begins to roll off at just 80 Hz. This is typical of low-pass filters and indicates one of the problems with using them to attenuate noise. The phase rolls off long before benefits of attenuation are realized. Low-pass filters are usually used to achieve attenuation; the price for the control system is phase lag and, through that, reduced stability.

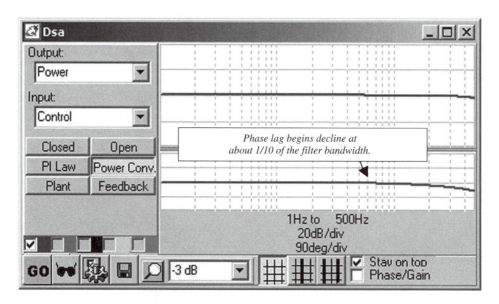

Figure 3-7. Bode plot of power converter modeled as low-pass filter.

3.2.3 PI Control Law

The control law is proportional-integral (PI). This controller is one of the most popular in industry. The proportional gain provides stability and high-frequency response. The integral term ensures that the average error is driven to zero. Advantages of PI include that only two gains must be tuned, that there is no long-term error, and that the method normally provides highly responsive systems. The predominant weakness is that PI controllers often produce excessive overshoot to a step command.

The PI controller output is the sum of two signals, one proportional to the error and the other proportional to the integral of the error. The integral cancels long-term error, but it also injects phase lag. As discussed earlier, the phase lag of an integral is 90°. The phase lag of a proportional gain is 0° (input and output are in phase). The phase lag of a PI controller is always between these two boundaries.

The PI controller here has a break frequency, which is defined as K_I (the integral gain) in rad/sec. Referring to Figure 3-4, the output of the controller is $(K_I/s + 1) \times K_P$. Well below the break frequency, the integral dominates. This is intuitive because the K_I/s term will overwhelm the "1" when "s" is small (i.e., at low frequency). Well above the break frequency, the K_I/s term will diminish and be overwhelmed by the 1. Review the graph in Figure 3-8. Notice that well below the break, the plot looks like an integral: The phase is $-90°$ and the gain falls with frequency. Above the break, the plot looks like a proportional gain: The phase is 0° and the gain is flat with frequency. In the transition, the phase climbs; this behavior is typical of PI control laws.

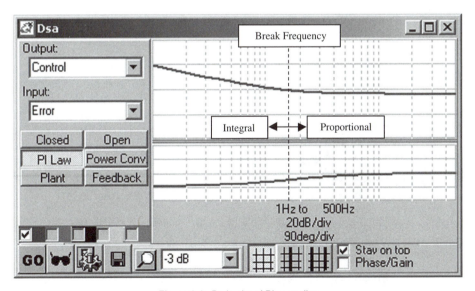

Figure 3-8. Bode plot of PI controller.

The break frequency in this **PI** controller is equal to K_I. Note that K_I was set to 100 when Figure 3-8 was generated. Understanding that K_I is measured in rad/sec, the break should be K_I, or about 15 Hz, which is consistent with Figure 3-8. Although this model provides K_I in rad/sec, you should know that it is common for vendors to provide K_I in internal, product-specific units that, for the user, are often incomprehensible. Sometimes a call to the vendor will allow you to translate from these product-specific units to general units. It is not necessary to know K_I in standard units, but it can be helpful.

3.2.4 Feedback Filter

Like the power converter, the feedback is modeled as a low-pass filter. Filtering effects in feedback devices are common. The feedback mechanism may include inherent filtering, such as thermal mass in a temperature sensor. Explicit filters are commonly added to reduce noise on analog sensors. Sometimes signal processing forms a closed-loop system itself; for example, in motion control, a resolver-to-digital converter includes an internal control system to convert resolver signals to position and velocity signals. These processes degrade the control loop by adding phase lag, and they too can be modeled as filters. Selecting the filter that best represents the feedback device requires careful analysis and, often, coordination with the device manufacturer. For this model, a two-pole filter with a bandwidth set to 300 Hz has been selected to represent the feedback. The Bode plot for this filter is shown in Figure 3-9. Note again that although this filter starts adding phase lag at 1/10 of the bandwidth (about 9° at 30 Hz), significant attenuation occurs only at much higher frequencies.

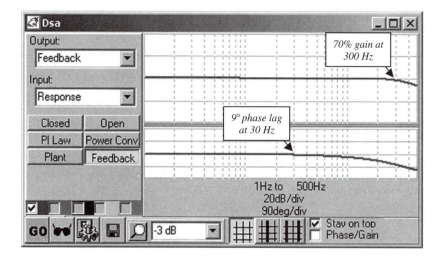

Figure 3-9. Bode plot of feedback filter.

3.3 The Open-Loop Method

Now that the transfer functions of each block have been reviewed, the open-loop,[1] or frequency-response, design [20] method can be derived. The first step is to write the open-loop transfer function, equivalent to GH in Figure 2-5. This is the process of summing the gains and phases of each block cascaded in the loop. For this example, the closed loop ($G/(1 + GH)$) and open loop (GH) are shown side by side in Figure 3-10a and b. The closed-loop Bode plot (Figure 3-10a) demonstrates classic characteristics: a gain of 0 dB and phase of 0° at low frequencies, both rolling off as frequency increases. Compare this with the open-loop gain in Figure 3-10b. The gain here is very high at low frequencies, a characteristic imposed by the two integrators (the plant and the PI controller). As the frequency increases, the gain declines because of the integrators and the filters. Notice that the gain starts high at low frequencies and falls; at some frequency, the gain will necessarily cross through 0 dB. This frequency, called the *gain crossover*, occurs in this example at about 50 Hz.

The open-loop phase is −180° at low frequencies because the two integrators each contribute 90° lag. Notice that the phase rises as the frequency increases: this comes from the PI controller, which has a phase lag that climbs from −90° to 0° as frequency increases (see Figure 3-8). However, phase lag from the two filters begins to dominate at about 50 Hz and the phase turns down. At some frequency, the phase will cross through −180°; this frequency is called *phase crossover*.

You may be surprised to notice that even though this system is quite stable, at low frequencies the open-loop (Figure 3-10b) phase is 180°; this is because the gain is much

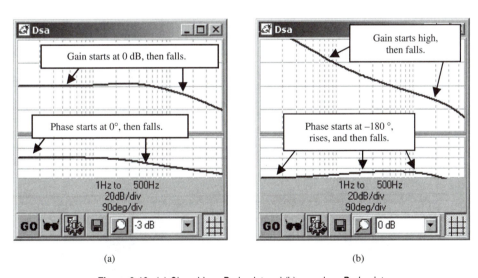

(a)   (b)

Figure 3-10. (a) Closed-loop Bode plot and (b) open-loop Bode plot.

---

[1]The name is a misnomer because there is no need to open the loop to comprehend the method. Originally, the loop was opened, but modern integrating controllers make this impractical.

greater than 0 dB. A system will be stable as long as the open-loop gain is either much greater than or much less than 0 dB when the phase is 180°.

3.4 Margins of Stability

Understanding that instability results when the loop gain is 0 dB $\angle -180°$, the margin of stability can be quantified in two ways. The first measure is *phase margin*, or PM [6,80]. PM is defined by the phase when the frequency is at the *gain crossover frequency* (the frequency where gain = 0 dB); specifically, PM is the difference between the actual phase and $-180°$. Were the PM 0°, the gain would be 0 dB and the phase $-180°$, so the system would be unstable. Figure 3-11a shows an open-loop plot: The gain crossover frequency is about 50 Hz and the PM is 55°.

The second measure of stability is *gain margin*, or GM. GM is defined by the gain at the *phase crossover frequency* (the frequency where phase = $-180°$); GM is the difference between the actual gain and 0 dB. As with PM, were the GM 0 dB, the gain would be 0 dB and the phase $-180°$, so the system would be unstable. Figure 3-11b is the same as Figure 3-11a except that it shows a GM of 15 dB measured at the phase crossover, which is about 240 Hz.

3.4.1 Quantifying GM and PM

There are a number of factors that recommend using PM and GM to measure stability. First, these are among the most intuitive measures of control system stability. Some competing methods suffer from subjective measures, such as the shape of an

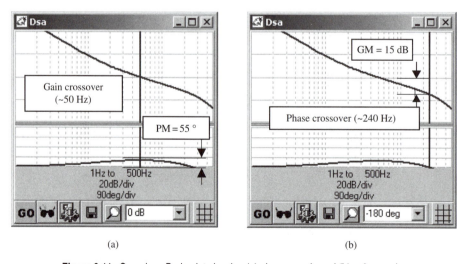

(a)   (b)

Figure 3-11. Open-loop Bode plot showing (a) phase margin and (b) gain margin.

abstract plot. Also, the open-loop method is numeric and thus works well with computers. Finally, the method requires the user to make few simplifying assumptions about the control system.

Although the measurements of PM and GM are objective, determining the desired values for these measures requires judgment. One reason is that applications vary in the amount of margin they require. For example, some applications must follow commands, such as the step, which will generate overshoot in all but the most stable systems. These applications require higher margins of stability than those that respond only to gentler commands. Also, some applications can tolerate more overshoot than others. Finally, some control methods require more PM or GM than others for equivalent response. For example, a PI controller requires typically 55° of PM to achieve 20% overshoot to a step, whereas a PID (proportional-integral-differential) controller might be able to eliminate all overshoot with just 40° PM.

Experience teaches that GM should be between 10 and 25 dB, depending on the application and the controller type; PM should be between 35° and 80°. All things being equal, more PM is better. Since phase lag at and around the gain crossover reduces PM, this teaches one of the most basic rules in controls: *Eliminate unnecessary phase lag*. Every noise filter, feedback device, and power converter contributes to phase lag around the loop, and each erodes the phase margin. Unnecessary lags limit the ultimate performance of the control system.

3.4.2 Experiment 3A: Understanding the Open-Loop Method

The examples in this chapter were generated with Experiment 3A. That experiment allows easy comparison of the open-loop Bode plots (PM, GM), closed-loop Bode plots (bandwidth, peaking), and step responses (overshoot and rise time). Install *Visual ModelQ* according to instructions in Chapter 1. Launch the program. Load Experiment 3A.mqd and press the Run button (black triangle icon, upper left of screen), which compiles the model and starts simulation; the diagram of Figure 3-12 should be displayed.

Experiment 3A is a simple PI controller. Starting at the left, "Wave Gen" is a waveform generator set to produce a square wave command signal for the system. Moving to the right, the Dynamic Signal Analyzer (DSA) normally passes the command signal through to the summing junction. However, each time a Bode plot is requested, the DSA interrupts the waveform generator and produces a specialized command that excites the system over a broad range of frequencies and then measures the results; those results are used to calculate Bode plots. The summing junction compares the command and feedback, which is connected to the bottom, and produces the error signal. That error is fed into a PI controller with two gains, K_P, the proportional gain, and K_I, the integral gain; both gains can be adjusted in *Visual ModelQ* using the *Live Constants* above the PI control law. The output of the control law is fed to the power stage, which here is modeled as a two-pole low-pass filter with a break frequency of 800 Hz. The output of the power stage feeds the plant gain (G) and plant integrator to produce the

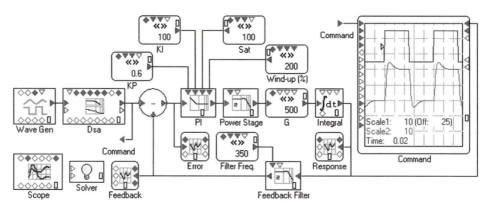

Figure 3-12. Experiment 3A, a simple PI control system.

system response. That response is plotted below the command on the *Live Scope* "Command." The response also connects to the feedback filter, a two-pole low-pass filter with a break frequency of 350 Hz. The output of the feedback filter is the system feedback signal. Four variables are continuously monitored by the main oscilloscope ("Scope") and DSA; three are *Visual ModelQ* variables (Feedback, Response, and Error); the fourth is the channel-one input (left side) of the *Live Scope*.

One of the advantages of *Visual ModelQ* is that it provides numerous on-screen displays (such as the *Live Scope* "Command") and *Live Constants* (K_P, K_I, Wind-up, and G). *Live Constants* can be adjusted while the model runs. After the model is running, double-click on a *Live Constant* and the adjuster box of Figure 3-13 will come into view. Using the six buttons under the numerical display ($<<$, $<$, and so on), the gain can be adjusted up or down in large increments ($<<$, $>>$) or small ($<$, $>$). The results are displayed almost immediately. Note that the scope shots for this chapter are taken from "Scope" rather than the *Live Scope* command, because this display is more detailed; double-click on the Scope icon while the model is running to view the Scope output. To display a Bode plot while the model is running double-click on the Dsa icon to see the DSA display. Click the "GO" button to start the excitation signal; this takes a few seconds. Note that the model must be running. You can hold Bode plots for comparison by right-clicking anywhere in the plot area to bring a popup menu into view and selecting "Save as ... " and choosing a color. For documentation

Figure 3-13. Adjuster box for the *Live Constant* K_P.

on *Visual ModelQ* blocks, hold the cursor over an icon and click the F1 key; this will launch Internet Explorer, which will open the *Visual ModelQ Reference Manual*, and point to documentation for that block.

3.4.3 Open Loop, Closed Loop, and the Step Response

The stability of the system modeled in Experiment 3A can be quantified in four ways: In the closed loop, there is about 1 dB of peaking (Figure 3-10a), the phase margin is 55°, the gain margin is 15 dB (Figure 3-11), and there is about 20% overshoot to a step (Figure 3-14). These measures all describe the stability of the system. This model represents a well-stabilized system. Now, what happens to the stability if the loop gain (K_P) is modified?

The default value of K_P is 0.6. Increase it to 2.4 (multiply by 4, equivalent to adding about 12 dB), and the system will become marginally stable. The effect in the open-loop plot is that the loop gain increases by 12 dB while the phase is unchanged. Compare the gain plots in Figure 3-15 to see this effect. This is the expected result. Referring to Figure 3-4, increasing K_P by 12 dB will raise the loop gain 12 dB and will not change the phase around the loop. Increasing the gain 12 dB moves the gain crossover to the right, from 50 Hz to 200 Hz (again, see Figure 3-15). The PM is just 20°, compared with the 55° PM when the gain crossover was 50 Hz. The GM falls from 15 dB to just 3 dB. Both PM and GM are low, indicating marginal stability.

When running Experiment 3A, note that *Visual ModelQ* simplifies calculation of the GM and PM by providing an *Autofind* button at the bottom center of the DSA. Click the "open" button on the left of the DSA to display the open loop; select "0 dB" from the Autofind combination box and press the magnifying glass icon to move the cursor to the gain crossover frequency. Similarly, select "−180 deg" to move the cursor to the phase crossover frequency. The cursor display window will appear, automatically showing gain, phase, and frequency; using those measures, the PM and GM can be calculated.

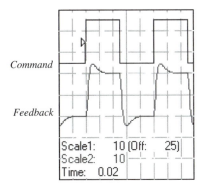

Figure 3-14. Step response of PI system with default values ($K_P = 0.6$, $K_I = 100$).

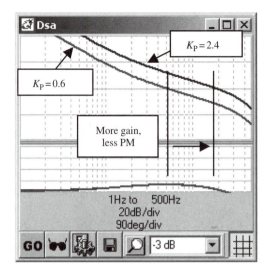

Figure 3-15. Open-loop Bode plot of system with excessive gain.

Raising the gain has reduced both the PM and GM, indicating loss of stability. The step response in Figure 3-16a shows overshoot exceeding 60% and, more important, several cycles of ringing. Figure 3-16b shows the closed-loop Bode plot of the system with the higher gain. The plot indicates peaking of 12 dB, well above the 1 or 2 dB normally considered acceptable. (Incidentally, it is a coincidence that a loop gain

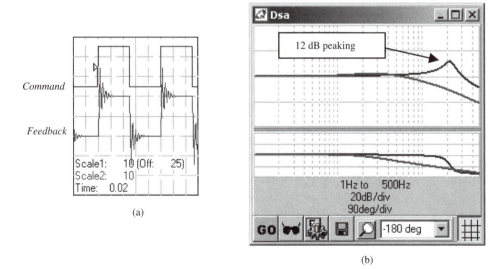

Figure 3-16. (a) Step response and (b) closed-loop Bode plot of system with excessive gain.

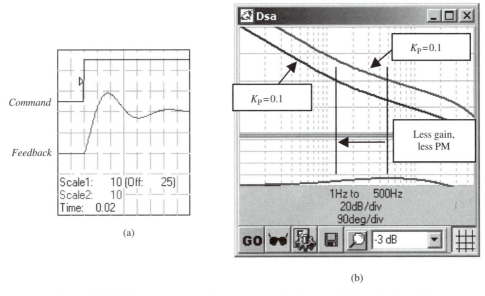

Figure 3-17. (a) Step response and (b) open-loop gain of system with low gain ($K_P = 0.1$).

increase of 12 dB caused 12 dB of peaking.) Almost no application could accept such marginal stability.

Returning to the open-loop plot of Figure 3-11a, notice that in the well-tuned system, the gain crossover of 50 Hz lines up almost directly over the peak of the open-loop phase. This is not a coincidence. Tuning a PI controller usually requires that the gain crossover be located near the peak of the open-loop phase, even if the person tuning is unaware of it. Herein lies another example of the intuition provided by the open-loop method. Observe in Figure 3-11a that substantially raising *or lowering* the loop gain will cause instability, because both reduce PM. For example, if K_P is lowered from 0.6 to 0.1, the step response, as shown in Figure 3-17a, overshoots by 50%. Notice that the gain crossover frequency drops to about 15 Hz, as shown in Figure 3-17b. Here the PM is only about 35°, a loss of 20° PM. Lowering gain further would erode the PM more and reduce stability further.

3.5 A Zone-Based Tuning Procedure

The challenge of tuning is that multiple gains must be varied, and each affects many of the performance measures. The goal of the *zone-based* tuning process is to decouple the multiple tuning gains so that they may be adjusted individually [57]. This can be done by considering the effects of each gain as being dominant over a certain zone of frequency. Developing a tuning gain for a PI controller is simple because there are two gains and, thus, just two zones. As this book progresses, this procedure will be

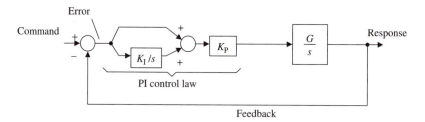

Figure 3-18. Simplified PI controller.

extended to more complex structures. Consider Figure 3-18, which is a simplified version of the PI control loop in Figure 3-4.

By using the $G/(1 + GH)$ rule and applying some algebra, the transfer function for the simplified system can be derived:

$$\frac{\text{Response}(s)}{\text{Command}(s)} = \frac{sK_P + K_I K_P}{s^2/G + sK_P + K_I K_P} \tag{3.2}$$

To consider this transfer function in terms of frequency zones, focus on the denominator. Although the numerator does have an impact on response, the denominator determines the overall stability and response of the system. At high frequencies, s is large and s^2/G dominates the denominator. This is intuitive; at very high frequencies, almost all control systems respond according to the plant gain (inductance, capacitance, inertia, etc.); at those frequencies, the response is above the reach of the controller.

As the frequency declines, the $s \times K_P$ term will become noticeable; as the frequency falls further, it will dominate the denominator. This is the *middle-frequency* range. Here, the proportional gain dominates; the frequency is still too high for the integral gain (K_I) to have much impact. As the frequency approaches zero, both s terms will fall away, leaving only the constant $K_I K_P$ term. This is the *low-frequency* zone.

3.5.1 Zone One: Proportional

To apply the zone-based approach to tuning, tune the highest-frequency terms first. Assuming the plant gain cannot be changed, start with K_P. Set K_I to zero and set K_P low to ensure stability. Apply a square wave and raise K_P until significant overshoot is generated. How much overshoot is "significant" depends on the application. Many applications can tolerate 5% or 10%, while others can tolerate none at all. For demanding applications, it is important to raise K_P as high as possible, because higher K_P will provide more responsive control. For Experiment 3A, K_P can be raised to about 1 before overshoot begins.

Consider the open-loop response as K_P varies (Figure 3-19 shows the system for $K_P = 1$). As K_P increases, PM and GM decline. When K_P becomes large enough that

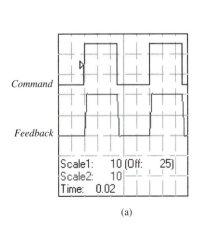

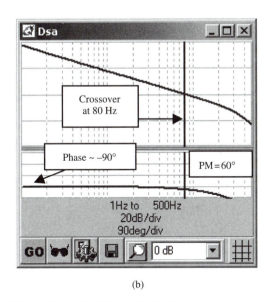

Figure 3-19. (a) Step response and (b) open-loop plot of P-controller for $K_P = 1$.

signs of instability become apparent, it indicates that the GM is too small. When optimizing K_P, you are balancing responsiveness (higher K_P) against stability (smaller K_P).

By looking at the open-loop gain you can see the advantage of setting K_I to 0 when tuning K_P. Notice that the open-loop phase is $-90°$ at low frequencies, compared with the $-180°$ seen in the PI controller of Figure 3-10b. So no matter how low K_P is set, the system will not become unstable. In the PI controller of Figure 3-17b, when the gain crossover was well below the phase "hump" (about 50 Hz), the PM fell so that the system was not sufficiently stable. The P-controller will not be unstable with low K_P. The advantage is that the designer can start with K_P very low and have little risk of instability. With PI controllers, it is necessary to find a good "starting place" for K_P — one that sets the gain crossover somewhere near the "hump" in the open-loop phase.

One question that often arises is: "What causes instability seen with high K_P?" As always, instability is caused by phase lag accumulating to 180° with 0-dB gain. Phase from the plant here is fixed at $-90°$, but phase from the power converter and from the feedback filter fall as frequency rises. In this process, K_P will be raised to provide the maximum responsiveness possible with the given system. If this is insufficient, the designer must look toward reducing phase lag in the components of the loop, for example, increasing the responsiveness of the power converter, the feedback device, or both.

3.5.2 Zone Two: Integral

When you have raised K_P as high as possible, move to the next zone, that served by the integral term. Integral gain is important because it removes all the long-term error.

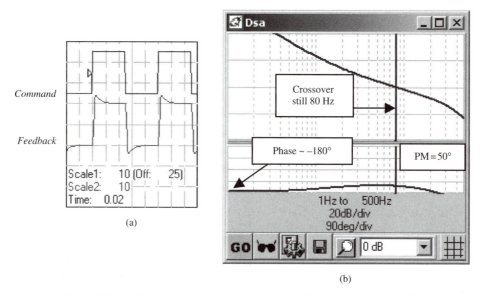

Figure 3-20. (a) Step response and (b) open-loop plot of PI controller with $K_I = 100$.

For example, if a proportional loop were employed to control a motor, disturbances such as friction would slow the system below the commanded value. The integral of a PI controller forces the average value of the feedback to match that of the command. With a PI controller, when 2000 RPM is commanded, 2000 RPM is delivered (at least on average), independent of disturbances such as friction.

When tuning, the value of integral gain will have little bearing on the stability of the next-higher-frequency (K_P) zone. In other words, changing K_I will not require return-ing to change K_P. Continuing from before, raise K_I from zero until overshoot is excessive — usually 5% or 10%. This occurs at $K_I = 100$, as shown in Figure 3-20a; the open-loop Bode plot is shown in Figure 3-20b.

Notice in Figure 3-20 that the gain crossover remained at 80 Hz but that the PM dropped 10°. Both are expected; the integral did not change the gain crossover because its frequency zone is well below 80 Hz. So K_I has little effect on GM, but it still contributes 10° of phase lag at that frequency, reducing the PM by that amount. Notice that the PI controller has overshoot. PI controllers generate overshoot in response to a square wave; it is one of the weaknesses of the method.

▨ 3.6 Variation in Plant Gain

A block diagram is only an approximation of an actual system. For example, the plant gain (*G* in Figure 3-4 and Experiment 3A) is often difficult to characterize. This gain commonly varies both during operation and from one unit to another. Examples of this gain include capacitance in a voltage controller, inductance in a current controller, inertia in a motion controller, and thermal mass in a temperature controller. Capacitance of electrolytic capacitors varies over time and from one unit to another; inductors saturate, effectively reducing the incremental inductance, often by factors of several times; and in motion systems inertia varies as material is added or when mechanism geometries change, such as occurs with robotic arms.

Varying plant gain is similar to changing the proportional gain in Experiment 3A; both vary the loop gain. For example, the fixed gain of 500 in Experiment 3A is modified in Experiment 3B (Figure 3-21) to vary based on the feedback signal, much as the gain of a saturating inductor varies with current. Here, a *Visual ModelQ* programming block is used to implement the following function:

$$Gain = 500/(1 - (Feedback/28)^2)$$

The effective gain increases from 500 to about 2500 as the feedback traverses from 0 to 25 (the program block clamps the formula at Feedback $\leq \pm 25$ to avoid division by zero). The result is that operation will be similar to Experiment 3A when the feedback signal is near zero; however, when the feedback signal approaches 25, where the gain is five times larger, margins of stability will decline. The command of Experiment 3B is modified from Experiment 3A; it now spans from 0 to 20 (not ± 10) to show the effects of gain variation. The *Live Scope* of Figure 3-21 shows signs of marginal stability only on the rising edge, where the feedback signal, and thus the gain, is high.

Variations of $\pm 20\%$ (about 2 dB) in the plant gain usually have little effect on control system performance. Larger variations may cause the system to be either unresponsive or marginally stable at the extremes. In such cases, it is difficult to find a single set of servo gains that will accommodate the range of variation. The reader is encouraged to return to Experiment 3A and attempt to find servo gains that work well when *G* spans from 500 to 2500; it is a difficult task.

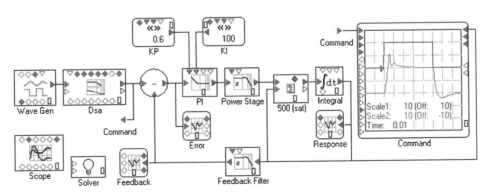

Figure 3-21. Experiment 3B, Experiment 3A with gain variation.

3.6.1 Accommodating Changing Gain

When plant gain varies as it did in the preceding example, the cure is often to adjust the loop gain to accommodate the change. For example, if the loop gain could be reduced by a factor of 5 when G changed from 500 to 2500 in Experiment 3A, the effect of the changing plant on loop dynamics would be removed. Often, changes in gain can be anticipated with reasonable accuracy based on operating conditions. For example, consider an inductor with a transfer function of

$$I(s)/V(s) = 1/Ls \qquad (3.3)$$

In saturation, inductance falls as the inductor current increases. This effect is nearly constant over reasonable operating conditions and from one unit to another. Thus, by characterizing the inductor at design time and knowing only the inductor current during normal operation, changes in plant gain can be predicted. Now, if the loop gain parameter (K_P) can be reduced as inductance falls, the loop dynamics will be maintained. This technique, called *gain scheduling*, is covered in Section 12.3.3. Gain scheduling requires knowledge of the gain in question and a controller with gains that can be modified during operation. This technique is frequently used in robotics, where changing geometries change the apparent inertia in a manner that can be anticipated with accuracy.

If the variations in the plant gain cannot be anticipated, normal gain scheduling will not be effective. To sense gain for scheduling in this condition, a technique known as *adaptive control* is sometimes employed. The inputs and outputs of a control system are compared during normal operation to monitor the behavior of the plant. This technique is often highly noise sensitive, so adaptive control usually must rely on complex statistical methods such as the Kalman filter.

3.7 Multiple (Cascaded) Loops

Multiple loops are common in control systems. For example, in a motion controller, the innermost loop is a current loop that may reside in a velocity loop, which itself may reside within a position or force loop. A typical motion-control loop is shown in Figure 3-22.

When tuning multiloop control systems, follow the procedure developed earlier. The inner loops operate in the next-higher-frequency zone to the outer loop. After an inner loop is tuned, it acts like a low-pass filter within the outer loop. For Figure 3-22, when tuning a velocity loop, the current loop and power stage look so much like a low-pass filter that they can be modeled as such. Similarly, when tuning the position loop, the velocity loop acts like a low-pass filter.

The process of tuning a multiloop controller is an extension of the zone-based method discussed earlier: Turn off all outer loops and tune the inner loop. When the inner loop is tuned, lump all the high-frequency effects together under the moniker

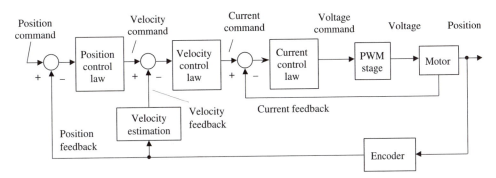

Figure 3-22. Position controller: three loops cascaded.

"power converter" and tune the next outer loop. Continue this process until the outermost loop is tuned. Each loop operates over a different frequency zone, so once the inner loop is tuned, there is little need to return to it. When tuning, expect the bandwidth of the outer loop to be between 20% and 40% of that of the inner loops. Remember to tune each loop to be as responsive as possible, because it becomes the barrier for the next outer loop.

3.8 Saturation and Synchronization

This chapter has focused on tuning, and most of the issues surrounding tuning are linear. However, saturation, the most prevalent nonlinear behavior in industrial controllers, bears discussion in this context. Saturation sometimes causes a poorly clamped system to overshoot excessively when it is actually quite stable. This problem frequently shows itself during tuning, although changes in tuning gains will not cure it properly. The overshoot is not an indicator of instability but rather of an improperly clamped integral.

While you should avoid nonlinear behavior in system design, all systems have limits as to how much energy they can supply to the plant. If a command (for example, a step) is so large that it requires more power than is available from the power converter, the system enters saturation. During saturation, the control system is applying all the power that is available. There is nothing more the control laws can produce. This condition is often called *slewing* or *large-signal response*. Full power is applied, and the plant is making the transition as quickly as is physically possible.

Note that slewing implies a type of saturation that is entered and exited rapidly. If a current controller is scaled to provide a maximum of 10 amps, a command from a velocity loop for 10.5 amps will produce no more current than one for 10 amps; the incremental gain for the last 0.5 amps is zero. This stands apart from the softer saturation of physical elements, such as the inductors discussed earlier, where the inductance fell gradually as current increased (note also that that form of saturation raised the incremental gain at high current rather than forcing it to zero). All systems have power converter limitations, and most will enter saturation immediately when those limits are exceeded.

Slewing is a nonlinear behavior. There are no stability issues. In fact, for simple proportional controllers, there are few concerns at all. Saturation does, however, cause a serious problem for integral controllers called *windup* or *loss of synchronization*. The problem is that while the system is slewing, there is still a considerable difference between command and feedback. In other words, the "error" remains large and the integrator continues charging rapidly. If the slew time is long enough, the value in the integrator can grow very large indeed. When the feedback approaches the command so that the system exits saturation, the integral can be so "wound up" that it causes large overshoot in what may actually be a stable system.

The cure for synchronization problems is to hold the control system just in saturation during slewing. In that way, at the point in time when the system exits saturation, the integral will be small enough that the system will be "synchronized." Note that some loss of synchronization is acceptable. Usually the integral will be allowed to charge up somewhat during slewing so that the system will remain in saturation in the presence of small amounts of noise. In other words, the integral clamp should be large enough to prevent small changes in the error term from bringing the system out of saturation. Remember, you want to apply full power during slewing, and that implies that the system should remain fully saturated. As long as the loss of synchronization is small, no significant ill effects will occur.

One cure for windup is to clamp the integral at a fixed value that prohibits uncontrolled growth. This helps, but it still allows substantial overshoot. Consider the PI control law shown in Figure 3-23: If the integral is limited with a fixed value, the sum of the integral and the error may still be large enough, after being scaled by K_P, to drive the system well into saturation.

Another way to cure saturation is to hold the integral so that the system is just in saturation. This implies that the clamping level will be a function of the error value. The result is to hold the expression (Integral $\times K_I +$ Error) $\times K_P$ to just above the maximum — just in saturation. For analog controllers, a circuit to maintain

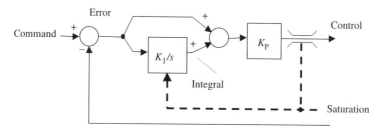

Figure 3-23. Saturation must clamp integral to maintain synchronization.

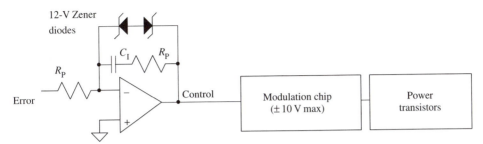

Figure 3-24. Analog circuit with 12-V integral clamp into 10 V saturation.

synchronization is shown in Figure 3-24. For digital controllers, the following pseudocode demonstrates the process:

```
Control = (Error + Integral*KI)*KP; //Remark: run PI control
//Remark: Set saturation and windup levels
WindUp = 2.0; //Remark: WindUp typically ranges from 1.2 to 2.0
ControlMax = WindUp*PowerMax; //Remark: Assume PowerMax is set
//Remark: Check for positive saturation
if (Control > ControlMax) {
  Integral = (ControlMax/KP - Error)/KI;
  Control = ControlMax;
}
//Remark: Check for negative saturation
if (Control < -ControlMax) {
  Integral = (-ControlMax/KP - Error)/KI;
  Control = -ControlMax;
}
```

Normally, the integral is allowed to charge so that the system is more than 100% of the saturation level, usually between 120% and 200%. For the example in Figure 3-24, notice that 12-V Zeners were used to clamp a signal where the power converter saturates at 10 V. For the preceding pseudocode, windup was set to 200%

The problems generated by windup can be seen in Figure 3-25. Figure 3-25a shows a system where the integral is properly limited; "Windup" is set to 200%. Figure 3-25b shows the same controller, but with windup effectively uncontrolled ("Windup" is set to 1000%, making it ineffective). Notice that Figure 3-25b overshoots 50%, indicating that the system is marginally stable; however, this is windup, not true overshoot, and as such does not provide information about loop stability.

To generate Figure 3-25, return to Experiment 3A. First, lower the saturation level ("Sat") from 100 to 3; this better demonstrates problems with windup. Run the model; the results should be equivalent to Figure 3-25a. The system is saturated but windup is limited. Now, set "Windup" to 1000%, effectively removing windup control. The results should be similar to Figure 3-25b.

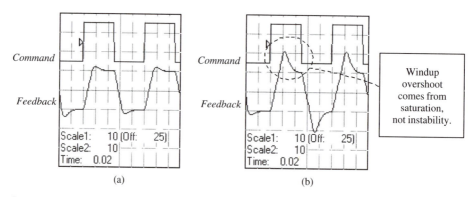

Figure 3-25. From Experiment 3A: Improper windup causes overshoot. Windup = 200% (a) and 1000% (b). (In both cases, Sat = 3.)

3.8.1 Avoid Saturation When Tuning

Finally, when tuning, keep the commands small enough that the "Control" signal does not saturate. This applies to both step response and Bode plots. Notice that in Figure 3-25a and b, the system shows little overshoot and then excessive overshoot. Yet the stability measures of the loop have not changed—there is about 25% overshoot in response to a step in the absence of saturation. Saturation is a nonlinear effect; it does not indicate either stability or responsiveness, because the system is locked fully on. When generating step response or Bode plots, ensure that the system does not enter saturation by monitoring the control signal.

3.9 Questions

1. a. Using Experiment 3A, what is the gain margin with the default settings? What is this on the linear scale?
   b. What gain of K_P would cause instability? (*Hint: Adding gain to K_P directly reduces GM.*)
   c. What gain of K_P will yield 3 dB of GM? Install this gain and comment on the step response. How much peaking does this cause? (*Hint: To measure peaking, run closed-loop Bode plot with DSA.*)
   d. What conclusion does this cause you to draw regarding 3 dB of GM?
2. a. Using Experiment 3A, what is the gain margin with the default settings? What is this on the linear scale? (same as 1A)
   b. What gain of G would cause instability?
   c. What gain of G will yield 3 dB of GM? Install this gain and comment on the step response. How much peaking does this cause?
   d. Compare Questions 1c and 2c and draw a conclusion.

3. Using Experiment 3A, follow the zone-based tuning procedure of Section 3.5 using a square wave command with the following criteria:

 For K_P, allow no overshoot.

 For K_I, allow about 10% overshoot.

 a. What are K_P and K_I?

 b. Repeat Question 1a, but inject extra phase lag in the loop by reducing the feedback filter break frequency to 100 Hz.

 c. Repeat Question 1a, but reduce phase lag in the loop by increasing the feedback filter break frequency to 1000 Hz.

 d. What conclusion does this lead to?

4. a. For Question 3a, what is the phase margin and gain margin?

 b. Repeat for Question 3b.

 c. Repeat for Question 3c.

 d. What conclusion does this lead to?

5. a. For Question 3a, what is the bandwidth?

 b. Repeat for Question 3b.

 c. Repeat for Question 3c.

 d. What conclusion does this lead to?

Chapter 4

Delay in Digital Controllers

The availability of inexpensive processing power has allowed digital control to supplant its analog counterpart in many industries. Most of the advantages of digital control are well known: analog drift and noise are eliminated, and accuracy is often dramatically improved. Tuning gains and other parameters can be changed with ease and accuracy, and advanced functions, such as disturbance decoupling (Chapter 7) and feed-forward (Chapter 8), can be supported with minimal increase in controller complexity.

The advantages of digital control distract attention from its primary weakness: increased delay in the control loop. The delay incurred from sampling feedback at regular intervals generates additional phase lag in the loop. As discussed in Chapter 3, phase lag limits loop gains and ultimately reduces the responsiveness of the system. Too often, engineers develop digital control systems without recognizing that digital controllers support lower loop gains than equivalent analog controllers. In the end, the digital controller may perform below expectations.

Engineers employing digital controllers should comprehend the impact of sampling delay. This allows accurate prediction of the performance of a control system and provides an objective approach to selecting the digital controller's sample time. Fortunately, the effects of delay can be quantified as phase lag, so the methods employed in Chapter 3 are easily extended to digital control systems.

The focus in this chapter will be on the effects of sampling delay. The principles of sampling are discussed and three types of delay are discussed. Experiment 4A is presented; it is a digital PI controller that you can run to observe the effects of sampled systems. The chapter concludes by providing upper boundaries of performance for digital controllers based on sample time.

4.1 How Sampling Works

Digital systems operate in discrete time steps. At regular intervals, the controller reads the feedback and command, executes control algorithms, and outputs a signal to the

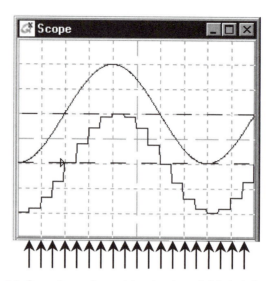

Figure 4-1. Comparing continuous (above) and sampled (below) sine waves.

power converter. Between intervals, the output to the power converter is held at a fixed level. This process is often referred to as *sample-and-hold* or "S/H". The sample-and-hold gives digital control waveforms their characteristic "stair step" appearance. This is shown in Figure 4-1, which compares two waveforms, a continuous sine wave above and its sampled equivalent below; the arrows at the bottom indicate the sampling instants.

Digital systems and analog systems have much in common, and the techniques studied in Chapter 3 can be applied to analyze both. The functions used in analog controllers, such as filters and integrators, have direct counterparts in digital controllers. The delays added by digital control can also be characterized by phase and gain.

4.2 Sources of Delay in Digital Systems

The sampling process adds delay in up to three independent processes: sample-and-hold, calculation time, and, for many motion control systems, velocity estimation.

4.2.1 Sample-and-Hold Delay

The delay from a sample-and-hold results from using stored data. Consider the digitized sine wave of Figure 4-1. At the beginning of the sample (just after the vertical segment), the data is new. But during the sample interval, the data ages. At the end of the sample interval, the data is approximately one sample old. The average age of the data over the interval is half the sample interval. The phase, assuming this half-sample-interval delay and using Equation 2.11, is

$$\text{Phase}_{\text{S/H}} = -360 \times F \times T_{\text{SAMPLE}}/2° = -180 \times F \times T_{\text{SAMPLE}}°$$

where F is the frequency in Hz and T_{SAMPLE} is the sample interval in seconds. The sample-and-hold does little to change the amplitude of the sampled signal; in other words, the gain is approximately 0 dB. So the transfer function of a sample-and-hold is

$$T_{\text{S/H}}(s) \approx 0\,\text{dB}\angle(-180 \times F \times T_{\text{SAMPLE}})° \tag{4.1}$$

Figure 4-2 shows the Bode plot of a sample-and-hold with a sample 250 µsec, which is equivalent to a sample frequency of 4 kHz.

The approximation of Equation 4.1 is nearly exact. In fact, the phase is exact, but the gain is a bit inaccurate at high frequencies. However, Equation 4.1 is still commonly used because the gain inaccuracy is usually insignificant. Below one-fourth the sample frequency the impact on gain is less than 1 dB and so is small enough to ignore. Above one-fourth the sample frequency the impact is still limited to just a few dB and the system rarely needs to be analyzed to accuracy better than that at such high frequencies. So, although the exact gain for the sample-and-hold is available (see Equation 5.33), the simpler form of Equation 4.1 is usually adequate.

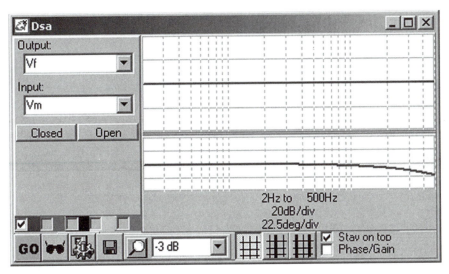

Figure 4-2. Bode plot of a 4-kHz sample-and-hold from 5 to 500 Hz.

4.2.2 Calculation Delay

The second form of delay caused by digital controllers is calculation delay. This delay is generated because of the time required to execute the control laws. Unlike analog systems, which process control laws continuously, digital systems require a finite amount of time after sampling to calculate the control law output. The transfer function of calculation time is simply

$$T_{CALC}(s) = 0 \, dB \angle (-360 \times F \times T_{CALC})° \tag{4.2}$$

where T_{CALC} is in seconds. Calculation delay does not cause attenuation.

Calculation delay can vary from a small portion of the sample interval to nearly the entire interval. It is a function of the complexity of the control laws and the skill with which they are coded. Calculation delay is rarely discussed in university courses on controls, probably because it does not fit well with the z-transform analysis methods ubiquitously taught in digital controls. That is unfortunate because this delay can be large; a portion often comes from careless construction of the control algorithms. Perhaps if engineers were better exposed to this effect, the problem would occur less frequently.

There are well-known techniques for reducing calculation delay, and the subject will be discussed in Section 5.7. However, these practices are often disregarded, so calculation delay needlessly limits performance in many control systems. Engineers applying off-the-shelf controllers will usually need to contact their vendor to obtain estimations for calculation delay. It is often large and should not be ignored. The cautious engineer will consider this delay along with the sample rate when selecting digital controllers.

4.2.3 Velocity Estimation Delay

The third form of delay is caused by estimating velocity from position. Only motion control systems that rely on position sensors are subject to this delay. Most controllers are designed for single-integrating plants, such as those described in Table 2-2. Motion controllers control a double-integrating plant because they apply torque, but they usually measure position rather than velocity. (Note that motion controllers relying on tachometer feedback do not suffer from velocity-estimation delay.) The controller usually forms velocity as the difference of the two most recent positions: $V_N \approx (P_N - P_{N-1})/T$, where V_N and P_N are the current velocity and position and P_{N-1} is the position from the previous sample. Imperfections in this estimation generate additional phase lag equivalent to a sample-and-hold. Consider that the difference is formed by a combination of new data (P_N) and data one sample old (P_{N-1}) so that the average age of the data is half of the sample interval. This delay is identical to that generated by the sample-and-hold (Equation 4.1):

$$T_{VEL \, EST}(s) \approx 0 \, dB \angle - (180 \times F \times T_{SAMPLE})° \tag{4.3}$$

TABLE 4-1 SOURCES OF DELAY

Source	General systems	Position-based motion systems
Sample-and-hold	0.5T	0.5T
Calculation delay (common range)	0.1T–0.9T	0.1T–0.9T
Velocity estimation		0.5T
Total	0.6T–1.4T	1.1T–1.9T

Velocity estimation delay can be reduced by the *inverse trapezoidal method* (Equation 5.28).

4.2.4 The Sum of the Delays

The total delay caused by digital controls is the sum of the sample-and-hold delay, calculation delay, and, for position-based motion systems, velocity estimation delay. These are summed in Table 4-1.

The effect of delay on a digital control system can be large. Consider the control system from Experiment 3A converted to a digital system that sampled at 1000 Hz. Were this a position-based motion system with a large calculation delay (90% of a sample), the total delay according to Table 4-1 would be 1.9 msec; at 50 Hz, the gain crossover frequency, 1.9 msec is equivalent to a loss of $0.0019 \times 360 \times 50 = 34°$ PM. A loss of 34° PM would cause an intolerable degradation of stability in most systems. As a result, the gains in such a system would have to be reduced to achieve stable performance, and this would result in a significant loss of responsiveness.

Digital systems often perform with less responsiveness than their analog counterparts. Still, the advantages of digital systems often outweigh the performance loss, especially if the performance differences are small. A "small" performance difference implies that the sample rate is high enough that the resulting phase lags do not dominate the total phase lag in the loop.

4.3 Experiment 4A: Understanding Delay in Digital Control

This section will present an experiment with the digital controller shown in Figure 4-3; this is a velocity controller. Starting at the upper left of this figure, a waveform generator creates a square wave of ±20 RPM, which flows through the DSA to create the command (Vc). The *Live Scope* shows command vs. feedback (Vf). A summing junction forms the velocity error (Ve), which feeds into the Digital PI control law; the PI gains are set by the *Live Constants* K_{VP} and K_{VI}. The control law output feeds "TCalc," a block that simulates a time delay for computing PI control algorithms as a portion of the sample time. The signal then proceeds to the current loop and plant, much like an analog system would. The feedback can be selected from one of two sources: The velocity feeds

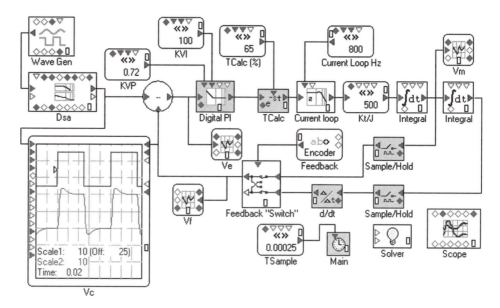

Figure 4-3. Experiment 4A, a digital PI controller.

out of the first integrator through a sample/hold, such as when a tachometer is used. As an alternative, velocity can be calculated as the derivative of position, which feeds out of the second integral; this alternative is commonly used with encoder and resolver feedback. Both signals feed an analog switch, which is controlled with another *Live Constant, Feedback*; feedback can be set to *Encoder* or *Tach*, selecting from the two paths. The *Live Constant* TSample sets the sample time of the digital controller Main. All the digital blocks (Digital PI, TCalc, and d/dt) are sampled according to TSample.

Launch *Visual ModelQ* and load Experiment 4A; click "Run." The time-domain response should be the same as that shown in Figure 4-3. The closed- and open-loop Bode plots are shown in Figure 4-4. Notice that these Bode plots are similar to Bode plots for analog systems. The causes of instability, the tuning methods, and the performance measures are all quite similar for analog and digital controllers.

If you want to use Experiment 4A to study nonmotion systems, set the feedback to "Tach." In this case, velocity is measured directly. This is equivalent to controllers for voltage, current, pressure, flow, and most other topologies, where the feedback variable is sensed directly.

4.3.1 Tuning the Controller

As with analog controllers, digital controllers are tuned in zones. To tune the controller in Experiment 4A, first convert the system to a proportional controller and maximize

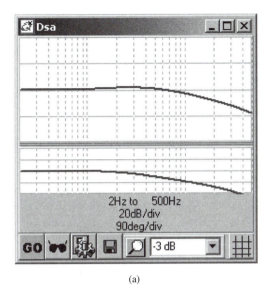

(a)

(b)

Figure 4-4. Closed and open loop for a digital PI controller.

K_{VP} without generating overshoot. For the default settings, the maximum value of K_{VP} is about 1.2; this provides a bandwidth of about 185 Hz. Increase the sample time from 0.00025 sec to 0.001 sec, and notice that almost 60% overshoot is generated along with some ringing. In most systems, this would be unacceptable. To restore stability, the gain must be reduced to about 0.4, resulting in a bandwidth of about 65 Hz; responsiveness is greatly reduced when the sample time is increased. This indicates that in this case the sample time is the determining factor in system performance; in other words, the effects of sampling at 1 kHz dominate the relatively fast current loops. Notice that switching "Feedback" in the Scope constants tab to "Tach," which eliminates velocity estimation delay, allows the gain to increase to 0.55; reducing the calculation delay ("TCalc") also allows higher gain without compromising stability.

Repeat each of the preceding three cases: (1) 2-kHz sampling with encoder, (2) 1-kHz sampling with encoder, and (3) 1-kHz sampling with tachometer. In each case, save the Bode plots (right-click anywhere in the DSA grid to save waveforms). Compare the bandwidth shown in the closed-loop plots. Notice how increasing delay forces bandwidth down in order to hold stability criteria constant. Click the "Open" button on the DSA and measure the phase margins for each of the three cases. Notice that in each case adjusting the gain to eliminate overshoot produces about the same phase margin (65°), providing further evidence that PM and overshoot are closely related measures of stability.

The rest of tuning a digital controller is the same as it was for the analog controller. The integral gain is raised until the overshoot meets the maximum allowable level for the application. You are encouraged to experiment with the digital controller. Change the sample time and change the calculation delay and notice the effects on stability. After each change, retune K_{VP} and note how longer delays force K_{VP} down. Since high proportional gains are required for responsive control systems, the link between reducing phase lag and responsiveness should be clear.

4.4 Selecting the Sample Time

The ultimate performance of a digital system can be estimated based on a combination of the sample interval and assumptions about the elements of the control loop. Aggressive assumptions rely on the sample time's being the limiting factor of performance. This is not always the case; many systems do not achieve the full potential possible with a given sample interval because other elements in the loop may be slow or noisy, limiting the loop gain independently. The following discussion begins with aggressive assumptions; it produces the best outcome at a given sample interval. Later, the results will be extended to cases where assumptions are less aggressive. The guidelines here are split between general and position-based motion systems because the difference between the two (the sample delay incurred by velocity estimation) is significant.

The figures in the next section are generated with *Visual ModelQ* Experiment 4A. In order to repeat these figures, make the following changes: Set the current loop bandwidth to 2 kHz, turn off the control law integrator ($K_{VI} = 0$), and set the calculation delay to a low 5%. These steps reduce phase lag from the other elements of the loop so that the sample-and-hold delay and velocity-estimation delay, if applicable, are the primary generators of phase lag.

4.4.1 Aggressive Assumptions for General Systems

General systems are those that have a single-integrating plant. If such a system is tuned to have the maximum bandwidth, and assuming other elements in the loop do not limit performance, it is possible to tune the system to settle in four or five sample intervals. Equivalently, the bandwidth of the digital controller can be as high as one-sixth the sample rate while still maintaining good stability. Figure 4-5 shows such a system. It settles in four or five samples (Figure 4-5a), and the bandwidth is 319 Hz (Figure 4-5b), or about one-sixth the sample rate (2 kHz). To run the experiment in this section, modify the "Feedback" constant to "Tach" by double-clicking to bring up the adjuster box. Recall that a motion system such as Experiment 4A behaves like standard (single integrating plant) systems when a tachometer is used for feedback.

4.4.2 Aggressive Assumptions for Position-Based Motion Systems

Position-based motion systems are those that have a double-integrating plant. These systems have velocity-estimation delay and so cannot perform as quickly as the general (single-integrating plant) systems. For the aggressive conditions where the dominant sources of phase lag are the sample-and-hold and the velocity-estimation delays, the lag in position-based motion systems is about double that of the general system. To run the experiments in this section, change the "Feedback" constant to "Encoder."

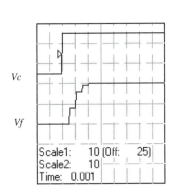

(a)

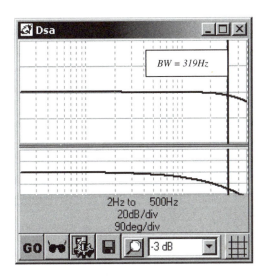

(b)

Figure 4-5. (a) Settling in four samples; (b) Bode plot.

If a position-based motion system is tuned to have the maximum bandwidth, it can settle as fast as 8 or 10 sample intervals. The bandwidth of such a digital controller can be as high as 1/12 the sample rate. Figure 4-6 shows such a system. It settles in eight samples (Figure 4-6a), and the bandwidth is 168 Hz (Figure 4-6b), or about 1/12 the sample rate.

The examples here provide guidance in selecting the sample time based on aggressive assumptions. Settling in four samples (eight samples for position-based systems) or achieving a bandwidth of one-sixth the sample frequency (1/12 for position-based systems) requires very fast calculation time, fast power conversion and feedback, and the near elimination of integral terms in the control law.

4.4.3 Moderate and Conservative Assumptions

More moderate expectations would allow for other elements of the loop to contribute some phase lag. The calculation delay may be well over the 5% assumed earlier, the power converter and feedback elements may add significant phase lag, and the control law may include a large integral term. All of these assumptions would lower the responsiveness that could be achieved from a given sample rate. Conservative assumptions further reduce expectations. Table 4-2 provides a guide with three levels of assumptions: aggressive, moderate, and conservative.

The aggressive assumptions in Table 4-2 are those discussed earlier in this section. The delays from sample-and-hold and velocity estimation dominate all other delays.

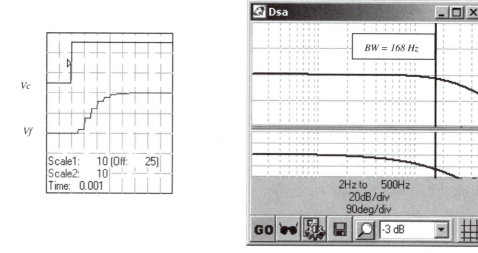

Figure 4-6. (a) Settling in eight samples; (b) Bode plot.

TABLE 4-2 MAXIMUM BANDWIDTH AND MINIMUM SETTLING TIME FOR A GIVEN SAMPLE INTERVAL

Assumptions	Based on target ...	General systems	Position-based motion systems
Aggressive	Max. bandwidth	1/6 sample time	1/12 sample time
	Min. settling time	4–5 sample intervals	8–10 sample intervals
Moderate	Max. bandwidth	1/10 sample time	1/20 sample time
	Min. settling time	6–8 sample intervals	15–20 sample intervals
Conservative	Max. bandwidth	<1/25 sample time	<1/30 sample time
	Min. settling time	>15 sample intervals	>20 sample intervals

The moderate assumptions allow phase lag from other system components to be a factor in determining maximum performance; these assumptions are more realistic for typical well-designed, industrial control systems. The conservative assumptions allow the other elements in the loop to dominate the phase lag so that the lag from the sample interval is less significant. The conservative assumptions only set an upper bound; the lower bound cannot be determined because other elements in the loop may contribute so much phase lag that the delays from sampling do not significantly limit the system. Such a situation indicates that the sample rate may be higher than is necessary; if reducing the sample rate will reduce system cost, it probably makes sense to consider that option. Note that as the assumptions range from aggressive to conservative, the differences between general and position-based motion systems narrow; this is because the difference between the two (the velocity-estimation delay) becomes smaller compared with the total phase lag.

4.5 Questions

1. a. Using the DSA in Experiment 4A to generate open-loop Bode plots, what is the phase margin when the sample time is set to its default (0.00025 sec)?
    b. Repeat with a sample time of 0.0005 sec.
    c. Repeat with a sample time of 0.001 sec.
    d. What conclusion could you draw?
2. a. In Experiment 4A, set the sample frequency to 50 μsec and measure the phase margin.
    b. Compare this to the phase margin with a 1-msec sample time (same as Question 1c).
    c. Assume that a 50-μsec sample time in Experiment 4A is short enough to ignore. What is the measured phase lag due to digital delay when the sample time is 1 msec?
    d. When the sample time is 1 msec, calculate the phase lag due to sample-and-hold. (*Hint: The gain crossover frequency is 57 Hz.*)
    e. When the sample time is 1 msec, calculate the phase lag due to velocity estimation.

  f. When the sample time is 1 msec, calculate the phase lag due to calculation delay of 65% of one sample time, which is the default setting in Experiment 4A.

  g. Compare the total measured phase lag (Question 2c) to the three calculated components of phase lag (Questions 2d–2f).

3. The first row of Table 4-2 provides "aggressive" assumptions. Using Experiment 4A, examine the meaning of "aggressive" in this context.

  a. Validate the entry in Table 4-2 for "aggressive, position-based motion systems." Use parameter settings from Section 4.4.2.

  b. What is the PM?

  c. Aside from the 90° lag from the integrator, what portion of the phase lag in the loop comes from sample delay? Comment.

Chapter 5

The z-Domain

The z-domain is used to analyze digital controllers in much the same way that the s-domain is used to analyze analog controllers. The most important feature of the z-domain is that it allows you to account easily for the sampling process, which is inherent in computer-based systems. Actually, s and z are so closely related that it may be a misnomer to refer to them as being different domains. The basic principles — transfer functions, phase and gain, block diagrams, and Bode plots — are the same for both.

5.1 Introduction to the z-Domain

Digital controllers process data in regular intervals. At least three things occur during each cycle: The input is sampled, the control algorithm is executed, and the output is stored. The length of the intervals is referred to as T, the cycle time, or sample time. The sample rate, or sample frequency, is $1/T$.

The z-domain is an extension of the s-domain [32]. It is based on the s-domain delay operation, which was shown in Equation 2.6 to be e^{-sT}. That is, if $f(t)$ and $F(s)$ are a transform pair, then so are $f(t-T)$ and $e^{-sT}F(s)$. If $f(t)$ is delayed $N \times T$ seconds, then its Laplace transform is $e^{-sNT}F(s)$, or $(e^{-sT})^N F(s)$.

5.1.1 Definition of z

The term z is defined as e^{+sT} [41], which implies that $1/z$ is e^{-sT}, the delay operation. Developing the z-domain based on this simple equality may appear to require unwarranted effort. However, digital systems need to include the effects of delays so frequently that the effort is justified. Most control algorithms process data that is either new (just sampled) or delayed an integer number of cycles (sampled earlier). With z, delays of an integer number of cycles can be represented easily.

In the strictest sense, the *s*-domain is for continuous (not analog) systems and *z* is for sampled (not digital) systems. *Sampled* is synonymous with *digital* because digital systems normally are sampled; computers, the core of most digital controllers, cannot process data continuously. On the other hand, most analog systems are continuous. Recognizing this, in this book *digital* will imply *sampled* and *analog* will imply *continuous*.

5.1.2 z-Domain Transfer Functions

Transfer functions of *z* are similar to those of *s*, in that both are usually ratios of polynomials. Several *z*-domain transfer functions are provided in Table 5-1. For

TABLE 5-1 UNITY-DC-GAIN S-DOMAIN AND Z-DOMAIN FUNCTIONS

Operation	s-Domain Transfer Function	z-Domain Transfer Function
Integration (accumulation)	$1/s$	$Tz/(z-1)$
Trapezoidal integration	$1/s$	$\dfrac{T}{2}\left(\dfrac{z+1}{z-1}\right)$
Differentiation (simple difference)	s	$(z-1)/Tz$
Inverse trapezoidal differentiation	s	$\dfrac{1+a}{T}\left(\dfrac{z-1}{z+a}\right),\ 0<a<1$
Delay T seconds	e^{-sT}	$1/z$

Simple filters ($\omega = 2\pi F$)
  Single-pole low-pass
  Two-pole low-pass*
	
	$\omega/(s+\omega)$
	$\omega_N^2/(s^2+2\zeta\omega_N s+\omega_N^2)$
		
		$z(1-e^{-\omega T})/(z-e^{-\omega T})$
		$Az^2/(z^2+B_1z+B_2)$
		$B_1=-2e^{-\zeta\omega_N T}\cos(\omega_N T\sqrt{1-\zeta^2})$
		$B_2=e^{-2\zeta\omega_N T}$
		$A=1+B_1+B_2$
		$\zeta=\text{Damping}$

Two-pole notch*	$(s^2+\omega_N^2)/(s^2+2\zeta\omega_N s+\omega_N^2)$	$K(z^2+A_1z+A_2)/(z^2+B_1z+B_2)$ $B_1=-2e^{-\zeta\omega_N T}\cos(\omega_N T\sqrt{1-\zeta^2})$ $B_2=e^{-2\zeta\omega_N T}$ $A_1=-2\cos(\omega_N T)$ $A_2=1$ $K=(1+B_1+B_2)/(1+A_1+A_2)$ $\zeta=\text{Damping}$

Compensators
  PI
  Lead
	
	$(K_1/s+1)K_P$
	$1+K_D s\times\omega/(s+\omega)$
		
		$(K_1 Tz/(z-1)+1)K_P$
		$1+K_D(z-1)/Tz\times$
		$z(1-e^{-\omega T})/(z-e^{-\omega T})$

*If $\zeta>1$, negate the term under the radical and substitute hyperbolic cosine for cosine.

example, consider a single-pole low-pass filter with a break frequency of 100 Hz (628 rad/sec) and a sample time (T) of 0.001 sec. From Table 5-1, the transfer function would be

$$\frac{C(z)}{R(z)} = T(z) = \frac{z(1 - e^{-0.628})}{z - e^{-0.628}} = \frac{0.4663z}{z - 0.5337} \tag{5.1}$$

If you are familiar with the s-domain, the presence of a minus sign in the denominator of z transfer functions may seem odd. It rarely occurs in the s-domain, and when it does it usually indicates instability. However, in the z-domain, it is common to have negative constants in the denominator.

5.1.3 Bilinear Transform

An alternative to Table 5-1 for determining the z-domain equivalent of an s-domain transfer function is to approximate s as a function of z:

$$s \cong \frac{2}{T}\left(\frac{z - 1}{z + 1}\right) \tag{5.2}$$

This is called the *bilinear transform*, and it is developed in Appendix D. This text will rely on Table 5-1 because it is usually less tedious than use of the bilinear transform.

5.2 z Phasors

Phasors in z are also similar to phasors in s. Again, the transfer function is evaluated with complex (versus real) math to determine the phase and gain at one frequency. The resulting complex number represents gain and phase, as it did in the s-domain; the only difference is that z must be evaluated instead of s.

Evaluating z requires the use of the following identity:

$$e^{jx} = \cos(x) + j \times \sin(x), \quad \text{where } j = \sqrt{-1}$$

Substituting z at steady state ($s = j\omega$), we get

$$z = e^{sT} = e^{j\omega T} = \cos(\omega T) + j \times \sin(\omega T) \tag{5.3}$$

The magnitude of z is

$$|z| = \sqrt{\cos^2(\omega T) + \sin^2(\omega T)} = 1$$

and the angle is

$$\angle(z) = \tan^{-1}\left(\frac{\sin(\omega T)}{\cos(\omega T)}\right)$$
$$= \tan^{-1}(\tan(\omega T))$$
$$= \omega T$$

This implies that the phasor representation for z is

$$z = e^{+sT}\big|_{s=j\omega} = 1\angle(+\omega T) \tag{5.4}$$

For example, Equation 5.1 can be evaluated at its break frequency, 100 Hz. In this case, $s = 628.3j$, $T = 0.001$, and $z = e^{j\omega T} = 1\angle\omega T = 1\angle 0.6283$ radians $= 1\angle 36°$. The response at 100 Hz is

$$T(z) = \frac{0.4663z}{z - 0.5337}, z = 1\angle 36° = 0.8090 + j0.5878$$

$$= \frac{0.46637 \times 1\angle 36°}{(0.8090 + j0.5878) - 0.5337}$$

$$= \frac{0.4663\angle 36°}{0.2753 + j0.5878}$$

$$= \frac{0.4663\angle 36°}{0.6492\angle 64.9°}$$

$$= 0.7182\angle - 28.9°$$

The equivalent analog low-pass filter from Table 5-1 with a break frequency (ω in the table) of 100 Hz, or 628.3 rad/sec, also evaluated at 100 Hz ($s = j628.3$), is

$$T(s) = \frac{628.3}{s + 628.3}, s = j628.3$$

$$= \frac{628.3}{j628.3 + 628.3} \tag{5.5}$$

$$= \frac{628.2}{888.6\angle 45°}$$

$$= 0.707\angle - 45°$$

The results of the equivalent filters in the s- and z-domains are similar but not identical. Digital functions are never identical to their analog counterparts, but they can be designed to be equivalent in a single facet of operation. For example, Table 5-1 derives digital filters assuming they should have the same response to a step command as their analog equivalents. On the other hand, the bilinear transform can be used to ensure phase and gain equivalence at one frequency (see "Prewarping" in Appendix D).

5.3 Aliasing

Aliasing is an undesirable effect that is seen in sampled systems. When the input frequency is greater than half the sample frequency, the sampled points do not adequately represent the input signal. Inputs at these higher frequencies are observed at a lower, *aliased* frequency.

A simple thought experiment demonstrates the problem. Imagine observing the motion of a pendulum with a frequency of around 1 Hz. Sample the position of the pendulum: Once every cycle, blink your eyes open just long enough to see the pendulum. If you blink at the same frequency as the pendulum is swinging, it will appear to stand still. Here, the actual input frequency is about 1 Hz, but the observed (or aliased) frequency is DC (0 Hz).

Figure 5-1 shows actual input frequency versus observed (aliased) frequency. Frequencies above one-half the sampling rate alias to lower frequencies. If an input is constructed of many frequencies, then the components below half the sample rate are unaffected and those above half the sample rate alias to lower frequencies. The frequency equal to one-half the sample rate is called the *Nyquist frequency*.

A computer cannot correct aliasing after it has occurred because there is no way to differentiate between legitimate and aliased inputs. Aliasing must be corrected *before* the data enters the computer. The simplest remedy is an analog prefilter that conditions the input signal. The filter should have enough attenuation so that the amplitude of components with frequencies above half the sample rate is insignificant.

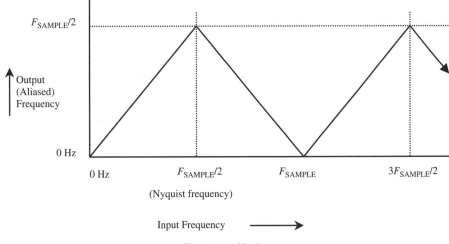

Figure 5-1. Aliasing.

Antialiasing can also be carried out with digital filters using a technique called *oversampling*. Here, a wave form is sampled at a higher frequency than the frequency where the control laws are evaluated. For example, you could sample an input at 10 kHz for a 1-kHz control law. Aliasing in the input would begin at 5 kHz rather than 500 Hz as it would for the non-oversampled system. The input could then be digitally filtered to remove frequency components above 500 Hz before the signal was shuttled to the control laws. There would still probably be a need for an analog prefilter to remove components above 5 kHz to prevent aliasing in the input.

Fortunately, few digital control systems have problems with aliasing, and often you will be able to ignore the effect. High-frequency signals in analog control systems, if they are present, are normally generated by the controller. This does not occur in digital systems because the microprocessor cannot normally generate components with frequencies above half the sample rate. However, there are at least two possible sources of high frequencies: the command and resonances generated by the plant.

High-frequency components may be added to the command with electrical noise or interference (EMI) coupled from other equipment or from the controller's own power converter. Electronic power conversion often relies on modulating (rapidly turning on and off) transistors. This process produces large voltage and current swings that can couple into the command signals. Since these frequencies are almost always much higher than the sample rate, they can alias down to cause low-frequency problems. For example, a 2-MHz noise source in a 1000-Hz sample system would alias to a DC offset. High-frequency components in the command often can be corrected with an antialiasing, low-pass filter.

Resonances in the plant, such as an L-C (inductance-capacitance) network or mass-spring combination, are more difficult to correct because the filter must be inside the loop and thus will be more likely to degrade stability. Again, a low-pass filter will often correct the problem. Always set the filter break frequency as high as possible to reduce phase lag, which destabilizes the loop.

5.4 Experiment 5A: Aliasing

Experiment 5A is a *Visual ModelQ* model that demonstrates aliasing. Starting from the upper left in Figure 5-2, time is scaled by frequency (note that the *Live Constant* "Freq (Hz)" has an implicit multiplier of 2π and so provides units of hertz). This feeds a sine wave block, which outputs a 100-Hz signal. This signal feeds a sample/hold that samples at 1 kHz, as specified by the *Live Constant* "TSample." The *Live Scope* displays the sine wave and its sampled counterpart.

The results of Experiment 5A are shown in Figures 5-2 and 5-3. In Figure 5-2 the scope trace shows a continuous sine wave above and the same wave sampled at 100 Hz below. Aliasing is avoided because the sample frequency is 1000 Hz so that input signals below 500 Hz (the Nyquist frequency) do not alias.

If the input frequency is increased to 900 Hz, the signal should alias; according to Figure 5-1, the aliased frequency should be 100 Hz. In Experiment 5A, increase the

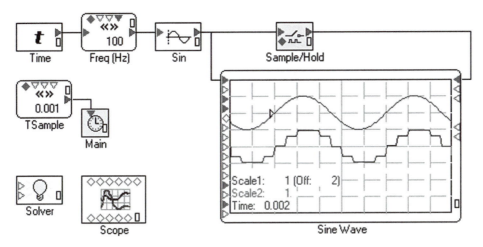

Figure 5-2. Experiment 5A: Sample-and-hold.

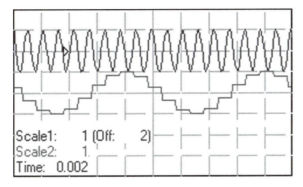

Figure 5-3. Experiment 5A: A 900-Hz signal (above) aliases to 100 Hz (below) when sampled at 1 kHz.

frequency of the input signal by changing the Live Constant "Freq (Hz)". Figure 5-3 shows the expected results.

Spend some time experimenting with aliasing. Notice that as you raise the frequency of the waveform generator starting from 100 Hz, both signals have the same frequency, although the sampling process can induce considerable distortion as you near the Nyquist frequency. When the signal generator frequency reaches half the sample rate, or about 500 Hz, aliasing begins; when the input frequency is well above the Nyquist frequency, the distortion is reduced and the effect of aliasing is obvious. As the input frequency increases, the observed frequency aliases even lower. When the input frequency is equal to the sampling frequency (1000 Hz), the observed frequency aliases down to DC. Raising the frequency further starts the process over again. A sampled sine wave of 1100 Hz is indistinguishable from one of 100 Hz, as it is from those of 2100 Hz, 3100 Hz, and so on.

5.4.1 Bode Plots and Block Diagrams in z

As discussed in Chapter 4, Bode plots in digital systems (that is, the z-domain) are the same as those in the s-domain. Block diagrams in the z-domain and s-domain are also the same. They can be combined with the $G/(1 + GH)$ rule or Mason's signal flow graphs. It is commonly thought that block diagrams must be either all in z or all in s. This implies that functions of s (such as the plant) must be approximated by functions of z to be included in a block diagram with other functions of z. Actually, this is not the case when using the open-loop method; this method relies on phase and gain, which can be provided equally well by functions of s or z or any mixture of the two. A block diagram can contain functions of both, even in the same block.

5.4.2 DC Gain

The DC gain of a transfer function of z is evaluated by setting z to 1, which is equivalent to setting s to 0. Returning to the definition of z for $s = 0$, $z \equiv e^{sT} = e^{0T} = 1$. For example, the low-pass filter of Equation 5.1 has a DC gain of unity, as is shown in Equation 5.6:

$$\left.\frac{C(z)}{R(z)}\right|_{z=1} = \left.\frac{z(1 - e^{-0.628})}{z - e^{-0.628}}\right|_{z=1} = \left.\frac{1 - e^{-0.628}}{1 - e^{-0.628}}\right|_{z=1} = 1 \qquad (5.6)$$

All the filters of Table 5.1 are adjusted for unity DC gain.

5.5 From Transfer Function to Algorithm

This section will present a procedure to convert z-domain transfer functions to time-based algorithms. z-Domain functions are implemented with computer algorithms just as s-domain functions are implemented with circuits. These algorithms are executed once each cycle of the controller after the input is sampled. The algorithm calculates the new output based on the new input and on inputs and outputs from previous cycles.

A five-step procedure to convert z-domain transfer functions to time-based algorithms follows. Note that digital controllers do not have access to an entire function, $f(t)$, but only to discrete samples: $f(0), f(T), \ldots, f(NT)$. This procedure will use a more compact notation for the same values, $f_0, f_1, \ldots, f_N$.

1. Write the transfer function in the z-domain as a ratio of output to input. For the filter of Equation 5.1, if $R(z)$ is the input and $C(z)$ is the output, then

$$\frac{C(z)}{R(z)} = \frac{0.4663z}{z - 0.5337} \qquad (5.7)$$

2. Multiply out the equation so that there is no z in any denominator:

$$C(z) \times (z - 0.5337) = R(z) \times 0.4663 \times z \qquad (5.8)$$

3. Divide both sides by the largest power of z that occurs on either side. For the example of Equation 5.8, the largest power of z is 1:

$$C(z) \times (1 - 0.5337/z) = R(z) \times 0.4663 \qquad (5.9)$$

4. Transform to the time domain. This is done while recognizing that a factor of $1/z$ in the z-domain implies that the time-domain signal is delayed by one sample. The expression $1/z^N$ indicates a delay of N cycles. For example, $C(z)$ transforms to C_N and $C(z)/z$ transforms to C_{N-1}. For the example filter,

$$C_N - 0.5337C_{N-1} = 0.4663 \times R_N \qquad (5.10)$$

5. Rewrite the equation to calculate C_N, the newest value of the output. Rewriting Equation 5.10 produces

$$C_N = 0.4663 \times R_N + 0.5337C_{N-1} \qquad (5.11)$$

You can use a spreadsheet program to demonstrate a digital low-pass filter, as shown in Microsoft Excel format in Table 5-2. The formula in block C8 is C7*B$3 + B8*B$4. That formula is copied to all the cells below C8 to calculate the filter.

TABLE 5-2 EXCEL FILE SIMULATING STEP RESPONSE OF DIGITAL FILTER

	A	B	C
1	T	0.001	Seconds
2	Break Frequency	100	Hz
3	exp$(-2{*}PI(){*}T)$	0.533488	
4	$1 - \exp(-2{*}PI(){*}T)$	0.466512	
5			
6	Time	Rn	Cn
7	0.000	0	0
8	0.001	1	0.466512
9	0.002	1	0.71539
10	0.003	1	0.848164
11	0.004	1	0.918997
12	0.005	1	0.956786
13	0.006	1	0.976946
14	0.007	1	0.987701
15	0.008	1	0.993439
16	0.009	1	0.9965
17	0.010	1	0.998133

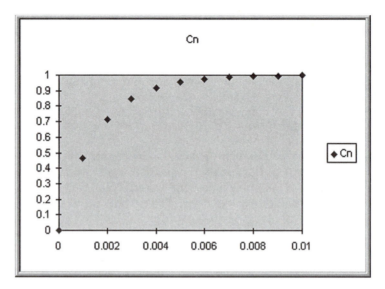

Figure 5-4. Microsoft Excel plot of low-pass filter step response.

Figure 5-4 shows a plot (from Excel) of the step response for this filter, which, at the sample points, acts very much like an analog low-pass filter.

This method of deriving the algorithm from the transfer function results in what is called the *controllable form*. An alternative means of converting transfer functions creates the *parallel form*, which is presented in Appendix E. The controllable form is presented here because it is the easiest to use and generally works well.

5.6 Functions for Digital Systems

The following section reviews several functions used in digital systems.

5.6.1 Digital Integrals and Derivatives

Digital systems cannot integrate or differentiate exactly because signals are sampled. Several useful approximations have been developed, such as trapezoidal integration and Simpson's method. This section discusses a few types of integration and differentiation.

5.6.1.1 Simple Integration

The most common digital integration method is Euler's integration, or accumulation. Each time step, the input (R_N) is scaled by the sample time (T) and accumulated in the integrator output (C_N). The algorithm is

$$C_N = C_{N-1} + TR_N \tag{5.12}$$

Figure 5-5 shows a block diagram of Euler's integration.

Euler's method can be translated to the z-domain to allow some observations. From Equation 5.12,

$$C(z) = \frac{C(z)}{z} + TR(z) \tag{5.13}$$

$$z \times C(z) = C(z) + Tz \times R(z) \tag{5.14}$$

$$\frac{C(z)}{R(z)} = \frac{Tz}{z-1} \tag{5.15}$$

which is the integration shown in Table 5-1. The phase and gain of Euler's integration is

$$\frac{Tz}{z-1} = \frac{T \times 1 \angle \omega T}{1 \angle \omega T - 1} = \frac{T[\cos(\omega T) + j \times \sin(\omega T)]}{\cos(\omega T) + j \times \sin(\omega T) - 1} \tag{5.16}$$

The Bode plot of Equation 5.16 is shown in Figure 5-6 from 2 to 500 Hz, where T is 0.001 sec. Note that 500 Hz is the Nyquist frequency, so there is no need to measure the Bode plot at higher frequencies. Ideal integration, $1/s$, always has a phase of $-90°$ and a gain of $1/(2\pi f)$, or $-20 \log_{10} (2\pi f)$ dB. The gain of simple integration is similar to true (continuous) integration at low frequencies. The phase is less accurate. At first this would seem a disadvantage for Euler's method. However, a second look shows that the phase always leads the ideal $-90°$. As discussed in Chapter 3, phase lead normally increases stability; in that sense, Euler's method can be better than ideal integration for controls applications, since phase lag is normally detrimental.

Euler's method is the most common integration method for control systems. Control methods such as proportional-integral-differential (PID) do not need exact integration to work well. This is because the function of the integral is usually to force the average error to zero to ensure that the controlled signal matches the command signal over long periods of time. Only rarely is there a need to control the integral of a signal. In those cases you may need a more accurate integration method. The next section provides two alternatives to Euler's method.

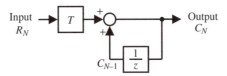

Figure 5-5. Simple (Euler's) integration.

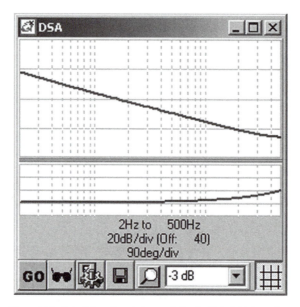

Figure 5-6. Gain (above) and phase of simple (Euler's) integration.

5.6.1.2 Alternative Methods of Integration

If you need more accurate integration, you can use trapezoidal integration, where the present and previous input values are averaged to approximate the area under the curve. This is shown in Figure 5-7. In the time domain, the formula for trapezoidal integration is

$$C_N = C_{N-1} + \frac{T}{2}(R_N + R_{N-1}) \tag{5.17}$$

which is converted to the z-domain as

$$T_{\text{TRAPEZOIDAL}}(z) = \frac{T}{2}\left(\frac{z+1}{z-1}\right) \tag{5.18}$$

A second alternative method of integration, Simpson's rule, often produces more accurate results than trapezoidal integration because the sampled points are connected in parabolas rather than the straight lines of the trapezoidal method. Simpson's rule is expressed in the time domain as

$$C_N = C_{N-2} + \frac{T}{3}(R_N + 4R_{N-1} + R_{N-2}) \tag{5.19}$$

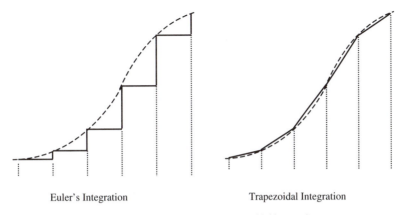

Euler's Integration   Trapezoidal Integration

Figure 5-7. Euler's (left) versus trapezoidal integration.

and converts to the z-domain as

$$T_{\text{SIMPSON'S}}(z) = \frac{T}{3}\left(\frac{z^2 + 4z + 1}{z^2 - 1}\right) \tag{5.20}$$

Simpson's rule has one clear disadvantage compared with trapezoidal integration: It can be calculated at most every other cycle because the parabola requires three points — two new points in addition to a third, old point. This further increases the phase lag induced by Simpson's method.

5.6.2 Digital Derivatives

Digital differentiation, like integration, is only an approximation. As you might expect, the most common method is simple differences, as shown in Table 5-1, the inverse of Euler's integration:

$$\frac{C(z)}{R(z)} = \frac{z - 1}{Tz} = \frac{1 - 1/z}{T} \tag{5.21}$$

When this function is transformed to the time domain, it shows that the derivative is calculated as the slope of a straight line formed by the new and the most recent input values:

$$C_N = \frac{R_N - R_{N-1}}{T} \tag{5.22}$$

As discussed with Figure 5-6, the phase of Euler's integration was in one way superior to exact integration: It had less phase lag. The simple difference is just the opposite — the phase lags the ideal +90°. (The phase is the negative of the phase in Figure 5-8 because the 5.21 is just 5.15 inverted.) As discussed in Chapter 4, the simple difference method (Euler's differentiation) in the form of velocity estimation increases phase lag and thus decreases the stability of the system.

5.6.2.1 Inverse Trapezoidal Differentiation

The delay from velocity estimation can be reduced in several ways [8,43,63]. One alternative that the author has used with success is a modification of trapezoidal integration. Trapezoidal integration draws a straight line to connect successive input values rather than using the stair-step approach of Euler's method:

$$C_N = C_{N-1} + T/2 \times (R_N + R_{N-1}) \tag{5.23}$$

The z-domain transfer function for trapezoidal integration is

$$\frac{C(z)}{R(z)} = \frac{T}{2}\frac{z+1}{z-1} \tag{5.24}$$

Trapezoidal integration is rarely used in control systems. Although it is more accurate, it has the 90° phase lag of true integration; as discussed earlier, this is not

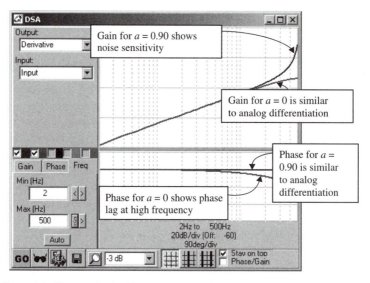

Figure 5-8. Phase and gain of inverse trapezoidal differentiation for $a = 0$ and $a = 0.90$.

normally desirable. However, if trapezoidal integration (Equation 5.24) were inverted to approximate differentiation, it would produce the 90° phase lead of ideal differentiation, which is indeed an advantage:

$$\frac{C(z)}{R(z)} = \frac{2}{T}\frac{z-1}{z+1} \tag{5.25}$$

If the function is transformed to the time domain, the result is

$$C_N = -C_{N-1} + T/2 \times (R_N - R_{N-1}) \tag{5.26}$$

Unfortunately, this algorithm is not stable. This can be seen by observing that if the inputs (R_N and R_{N-1}) are zero, C_N changes sign on each successive sample. A modification eliminates this problem: Add a parameter a, which scales between Euler's differentiation ($a = 0$) and the impractical inverse trapezoidal ($a = 1$). Then, by setting a close to but less than 1, the majority of benefits of inverse trapezoidal can be enjoyed while maintaining stability in the algorithm. The modification is shown in Equation 5.27:

$$C_N = -a \times C_{N-1} + \frac{1+a}{T}(R_N - R_{N-1}), a < 1 \tag{5.27}$$

and in the z-domain,

$$T(z) = \frac{1+a}{T}\left(\frac{z-1}{z+a}\right), a < 1 \tag{5.28}$$

If the microprocessor that implements the control algorithm cannot execute multiplications rapidly, Equation 5.27 can be rewritten to be implemented with arithmetic shifts for special cases of a:

$$C_N = 2 \times \left(\frac{R_N - R_{N-1}}{T}\right) - C_{N-1} - (1-a) \times \left(\frac{R_N - R_{N-1}}{T} - C_{N-1}\right) \tag{5.29}$$

For Equation 5.29, if a is greater than or equal to 1/2 (1/2, 3/4, 7/8) so that $(1 - a)$ is a binary fraction (1/2, 1/4, or 1/8), $(1 - a)$ can be calculated with a shift right (1, 2, or 3 bits, respectively) rather than a multiplication. Typical values of a are 1/2 and 3/4. The phase lag recovered from inverse trapezoidal is proportional to a; $a = 0.75$ removes 75% of the phase lag generated by simple differences.

Figure 5-8 compares the phase lag of inverse trapezoidal differentiation for two values of a: 0 and 0.90. When a is zero (simple differences), the delay from differentiation is $1/2T$, the sample time. When a is 0.90, only one-tenth of the error from simple differences remains. Figure 5-8 shows the gain and phase; note that ideal (analog) differentiation has a gain that increases in a straight line at 20 dB/decade and a fixed phase of +90°. By moving a to 0.90, the phase lag improves to be nearly equal to the ideal, $+90°$. However, an increased noise sensitivity also appears, as indicated by the gain increase of about 20 dB at 500 Hz.

So when a is larger, the phase advance added by differentiation increases; unfortunately, the function is also more noise sensitive. However, this noise susceptibility can be offset by using more highly resolved feedback or by otherwise reducing noise sources in the system.

5.6.2.2 Experiment 5B: Inverse Trapezoidal Differentiation

Experiment 5B, shown in Figure 5-9, demonstrates the frequency-domain performance of inverse trapezoidal differentiation. An adjustable-frequency sine wave similar to that in Experiment 5A is provided. The signal proceeds through the DSA and then is sampled and held. Inverse trapezoidal differentiation is provided through the block "d/dt." Note that this block is similar in appearance to simple differences (see Experiment 4A) except that the symbol Δ is replaced with a rhombus and the parameter a is accepted along the top edge. In Experiment 5B, a is set with a *Live Constant*. The *Live Scope* displays the input (amplitude ± 1) against the differentiated signal (amplitude $\sim \pm 600$); the differentiated signal leads the input by $90°$. Note that the input scale is 200 units/div to account for the gain of differentiation at 100 Hz, which is $2\pi \times 100 \approx 628$.

Load Experiment 5B and start the model running. Double-click on the DSA to get the DSA output screen to appear. Press GO on the DSA to produce a Bode plot. Adjust the *Live Constant a* to different values and take new Bode plots. Notice that as a increases from 0 (Euler's method) to 1, the phase lag in the higher frequencies improves but the noise sensitivity increases. This is the behavior of the transfer function $T(z)$ from Equation 5.28.

Correctly applied, inverse trapezoidal can significantly reduce phase lag in the loop. For example, return to Experiment 4A and replace the simple difference block with inverse trapezoidal differentiation and notice that the proportional gain, K_P, can be

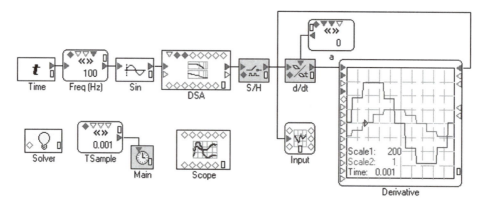

Figure 5-9. Experiment 5B: Inverse trapezoidal differentiation.

increased by about 25% when a is set to 0.90. Be aware, however, that this model is essentially noiseless and that noise is the primary disadvantage of inverse trapezoidal differentiation. If the signal being differentiated is noisy enough to require a filter to be placed in the loop, the benefit of inverse trapezoidal differentiation may be lost. For this reason, the benefits of the method will be more easily gained with low-noise input signals.

5.6.3 Sample-and-Hold

Digital controllers calculate outputs once each cycle. That output is stored, often in a D/A (digital-to-analog) converter, and then held constant until the next cycle. This function, sample (output to D/A)-and-hold (keep constant for one cycle), is rare in analog systems but present in virtually all digital systems.

The effect of holding the output constant introduces phase lag because the output is getting old from the time it is stored. By the end of the cycle, the output is a full cycle old. As discussed in Section 4.2.1, since the data is, on average, one-half cycle old, the sample-and-hold (S/H) acts like a delay of a half-cycle. That is, in fact, an excellent approximation:

$$T_{S/H}(z) \approx e^{-sT/2} = 1\angle(-\omega T/2) \text{ rad} \tag{5.30}$$

or, in degrees and Hz,

$$T_{S/H}(s) \approx 1\angle(-180 \times F \times T)° \tag{5.31}$$

At higher frequencies, the S/H also begins attenuating the input somewhat. The more exact transfer function for S/H is

$$T_{S/H}(z) = \frac{z-1}{Tz}\left(\frac{1}{s}\right) \tag{5.32}$$

which turns out to be digital differentiation cascaded with analog integration. This form is shown as a zero-order hold in Ref. 33, although the T is not included. Few textbooks include the T, although it is required to reflect the sample-and-hold's intrinsic unity DC gain.

Recognizing that $z = e^{sT}$, some algebra can provide Equation 5.32 in a simpler form for sinusoidal excitation:

$$T_{S/H}(z) = \frac{z-1}{Tz}\left(\frac{1}{s}\right)$$

$$= \frac{e^{sT}-1}{e^{sT}}\left(\frac{1}{Ts}\right)$$

$$= \frac{e^{sT/2}-e^{-sT/2}}{e^{sT/2}}\left(\frac{1}{Ts}\right)$$

To apply steady-state sinusoids, set $s = j\omega$:

$$T_{S/H}(z) = \frac{e^{j\omega T/2} - e^{-j\omega T/2}}{e^{j\omega T/2}}\left(\frac{1}{Tj\omega}\right)$$

$$= e^{-j\omega T/2} \times \frac{e^{j\omega T/2} - e^{-j\omega T/2}}{2j(\omega T/2)}$$

$$= e^{-j\omega T/2} \times \frac{\sin(\omega T/2)}{\omega T/2}$$

$$= \left(\frac{\sin(\omega T/2)}{\omega T/2}\right)\angle -\omega T/2 \text{ rad}$$

(5.33)

So the precise sample-and-hold (Equation 5.33) and the approximations presented in Equations 5.30 and 5.31 (and Equation 4.1) have the same phase lag ($-\omega T/2$ radians) but different gains. The gain, $\sin(\omega T/2)/(\omega T/2)$, also known as the *sync* function, is nearly unity for most frequencies of interest. For example, at one-fourth the sample frequency ($\omega = 2\pi/4T$), the sync function evaluates to 0.9, or about -1 dB, which is a value so close to 0 dB that the difference can usually be ignored. And recognizing that the system bandwidth will usually be at much lower frequencies, say, 1/10 the sample frequency, there is rarely interest in the precise gain at so high a frequency. Even at the Nyquist frequency, the sync function evaluates to 0.637, or -4 dB, a value that can often be ignored, considering that this is the highest frequency the system can process. This is why the simpler forms (Equations 4.1, 5.30, and 5.31) are sufficiently accurate to use in most controls problems.

5.6.4 DAC/ADC: Converting to and from Analog

Computers process digital values. An analog-to-digital converter (ADC) converts a voltage to an integer, where the value of the integer is proportional to the amount of voltage. For example, suppose you are using a 12-bit ADC that converts up to 8 V. The range of voltages is 0 to 8 V and the range of integers is 0 to 4095 ($2^{12} - 1$) counts. The simple model of the ADC is a constant of proportionality:

$$T_{ADC}(z) = \frac{4095\,\text{counts}}{8.0\,\text{V}} = 511.9\frac{\text{counts}}{\text{V}}$$

(5.34)

Similarly, a digital-to-analog converter (DAC) converts integers to voltages. For example, the constant of proportionality for a 0 to 10 V, 10-bit DAC is

$$T_{DAC}(z) = \frac{10.0\,\text{V}}{1023\,\text{counts}} = 0.0978\frac{\text{V}}{\text{count}}$$

(5.35)

A sample-and-hold is an implicit part of the output DAC. That is, a DAC can be modeled with two sections: a constant in volts per count and an S/H. So a common model for a DAC is

$$T_{DAC}(z) = 0.0978\angle\left(\frac{-\omega T}{2}\right)\frac{V}{count} \tag{5.36}$$

DACs and ADCs also can convert to and from signed voltages. For example, an 8-bit DAC may output $\pm 5\,V$. The model for this DAC is

$$T_{DAC}(z) = \frac{5.0 - (-5.0)\,V}{255\,counts}\angle\left(\frac{-\omega T}{2}\right) = 0.0392\left(\frac{-\omega T}{2}\right)\frac{V}{count} \tag{5.37}$$

In general, the model for DACs and ADCs is the ratio of the integer range to the voltage range. In this text, the sample-and-hold is shown as part of the DAC. However, when studying the effects of phase lag on stability, the sample-and-hold can be placed elsewhere in the loop, with little impact.

5.7 Reducing the Calculation Delay

As discussed in Chapter 4, digital systems have up to three types of delays: sample-and-hold, calculation delay, and, for position-based motion systems, velocity estimation delay. The delay for calculations can be greatly reduced by careful construction of the control laws. After the input is sampled, there are usually many calculations that must be performed to arrive at the new output. The time to execute calculations often exceeds half of the cycle. Since delay within the loop decreases stability, it is desirable to reduce this delay by redesigning the control algorithms. Consider this function:

$$T(z) = 1.4\left(\frac{z - 0.8}{z - 0.6}\right)\left(\frac{z}{z - 1}\right) \tag{5.38}$$

Using the five-step method from the beginning of this chapter, the algorithm for this transfer function is

$$C_N = 1.4R_N - 1.12R_{N-1} + 1.6C_{N-1} - 0.6C_{N-2} \tag{5.39}$$

Notice that the algorithm has the form

$$C_N = K \times R_N + history \tag{5.40}$$

where K is 1.4 and *history* is a combination of old inputs and outputs. Generally, control algorithms can be written in this manner. The advantage is that *history* can be calculated before R_N is read, because it is based on old data. This reduces the number of calculations between the time when R_N is read and C_N is output to one addition and one multiplication. The procedure for each cycle that is required to take advantage of this modification is as follows:

1. Read R_N.
2. Calculate $C_N = KR_N + history$.
3. Output C_N.
4. Calculate *history* for the next cycle.

5.8 Selecting a Processor

Selecting a processor is a challenging task. The designer must choose a processor, either directly or indirectly when choosing a vendor's product, that meets all the application needs for the current product and has an upgrade path that will allow future product derivatives. The processor must have the resources to execute the control algorithms with enough margin to allow execution of higher-level logic. The challenge is compounded because of the great difficulty of changing the processor during the development project or at any time during the life of the product or its derivatives. The control system designer must start by accurately estimating the processing requirements of the product and then match the processor to the application.

The first step is to determine which control algorithms will be run and how often. Knowing the performance requirements and using the techniques discussed in Section 4.4, a designer should be able to estimate the sample rate with reasonable accuracy. As shown in Table 4-2, the sample rate should be at least 6–10 times higher than the bandwidth of the control system and 12–20 times higher for position-based motion systems. Estimating the complexity of the control algorithm requires that the designer choose between floating-point and fixed-point math.

5.8.1 Fixed- and Floating-Point Math

Floating-point math represents numbers in two parts: a mantissa and an exponent. The mantissa is scaled to hold the maximum precision possible in the given word space. The exponent scales the value of the mantissa up or down. Floating-point math offers high precision and a large dynamic range with little attention required from the programmer.

Fixed-point math performs all operations as if the numbers were integers. However, in almost all control problems, most of the math must include fractions. The programmer must scale constants and results to achieve the appropriate resolution while ensuring that individual calculations do not overflow. Microprocessors execute integer math in fixed lengths — bytes (8 bits), words (16 bits), or long words (32 bits) — allowing dynamic ranges for parameters of 256:1, 65,536:1, and about 4×10^9: 1, respectively.

Floating-point math is much easier to code than is fixed point. In the past, the cost of floating-point processors was prohibitive for many control applications. In recent years, this cost has fallen so that in all personal computers (PCs) and in many digital signal processors (DSPs) floating-point and fixed-point arithmetic come at equal cost. However, in the most cost-sensitive applications, floating-point math still comes at a premium. For applications that can afford it, floating-point math is usually preferred. Algorithms are coded faster and without the puzzling programming bugs that are common when coding fixed-point math.

When fixed-point math is appropriate, consider that z-domain transfer functions use constants that require subinteger resolution; the designer must scale constants into integers. For example, the algorithm from Equation 5.39 had four constants that required subinteger resolution:

$$C_N = 1.4R_N - 1.12R_{N-1} + 1.6C_{N-1} - 0.6C_{N-2} \tag{5.41}$$

In order to represent 1.4 or 1.12, you must scale integers and use fixed-point math. Since these constants are less than 2 (as most z-function constants are), scale the maximum value of the microprocessor constant to 2. If the microprocessor uses signed 16-bit words with a range of -2^{15} to $2^{15} - 1$, then 1.4 scales to

$$\frac{\text{Maximum positive word}}{\text{Maximum positive constant}} \times \text{constant} = \frac{\approx 2^{15}}{2} \times 1.4 = 22{,}938 \tag{5.42}$$

The first multiplication of Equation 5.41 can be written as

$$1.4R_N = 22{,}938 \frac{R_N}{2^{14}} \tag{5.43}$$

Since the divisor is an even power of 2, the division can be executed with a shift right of 14 places. However, to accelerate execution, you can also shift it left 2 places (multiply by 4) and take the most significant word:

$$1.4R_N = 22{,}938 \frac{R_N 2^2}{2^{16}} \tag{5.44}$$

Executing a two-word multiply and ignoring the least significant word is the same as shifting right 16 places.

5.8.2 Overrunning the Sample Time

Another factor in processor selection is the evaluation of the worst-case execution time for the real-time algorithms. When estimating the processing resources to run the control algorithms, consider that there is often some variation in the amount of time it takes to execute the control algorithms. For example, some subroutines may be executed only when a variable is outside a certain range or there may be extensive setup required when the controller changes operating modes. The designer must be certain that the time to process real-time algorithms never exceeds the sample time. The results of overrunning the sample interval can be serious; in the worst case, it may generate a catastrophic fault, such as overwriting the program counter, a fault that usually can be cured only with a hard reset.

5.8.3 Other Algorithms

Most machines require more algorithms than just control laws. Many systems execute customer programs or may need to respond to queries or commands from a higher-level controller. The processor must be able to execute the control laws with enough time left over for other processes.

In many systems, the control algorithms can require more than half of the total computing time. This implies that higher sampling rates, which improve system performance from a control designer's viewpoint, may significantly increase the elapsed time for the system to execute other algorithms.

5.8.4 Ease of Programming

Ease of programming is a subjective feature. It depends on the tools, on the processor structure, and on the experience of the programming team. Chips belonging to families that are broadly used may be better supported than those that are in a limited market. For example, Texas Instruments, Motorola, and Analog Devices produce several popular families and they can leverage revenue from many markets to develop the programming tools. Vendors with narrow markets may not be able to invest as much in tools.

5.8.5 The Processor's Future

Longevity of software is another factor in processor selection. Some companies are more careful to keep future processor families backward compatible. For example, Analog Devices and Motorola enjoy a good reputation for creating backward-compatible product families. Other companies are less concerned with this area. If you want to reuse software in future generations of products, choose chips from vendors whose history and plans include backward compatibility.

5.8.6 Making the Selection

Before you select a processor, ask yourself the following questions:

- Does the processor have enough power to execute the necessary control laws with enough margin for other algorithms?

- Does the processor have a clear upgrade path in case more processing resources are required in the future?
- Does the vendor have a history of keeping new processors backward compatible with earlier products?
- Is the complexity of programming in the processor's language reasonable for my organization?
- Does the processor vendor provide the high-level language support and development tools that the project will require?

Processor selection is a task that may affect product development for many years. Be certain the choice is wisely made.

5.9 Quantization

Quantization is a nonlinear effect of digital control systems. It can come from the resolution limitations of transducers or of internal calculations. One example, as shown in Figure 5-10, is a 12-bit ADC that converts a continuous range of 10 V to 4096 different values. Quantization also occurs in integer multiplications because the result of a multiplication is usually rounded.

If the resolution of sensors and the control system mathematics is fine enough, quantization can be ignored. Otherwise it must be taken into account either in modeling or by use of statistical methods. Quantization is nonlinear and cannot be represented in the z-domain.

5.9.1 Limit Cycles and Dither

One effect of quantization is called *limit cycles* [1, p. 367; 33]. Limit cycles are low-level oscillations that occur because of quantization error in digital mathematics. Limit

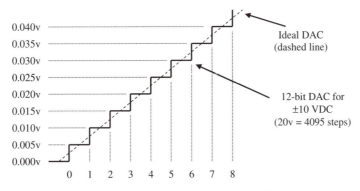

Figure 5-10. Quantization in an ideal DAC.

cycles can produce sustained oscillations that are low in frequency and many times larger than the quantization level. For example, consider placing in series three single-pole low-pass filters. Here, assume the break frequency is 50 Hz and the sample time is 0.001 second; using the "single-pole low-pass filter" entry from Table 5-1, and noting that $e^{-2\pi f T} \approx 0.76$, each of the three low-pass filters would have the form

$$T(z) = \frac{z(1 - \text{Pole})}{(z - \text{Pole})}$$

where Pole $= 0.76$. Three such low-pass filters could be put in series as

$$T(z) = \frac{z^3(1 - \text{Pole})^3}{(z - \text{Pole})^3}, \ \text{Pole} = 0.76 \tag{5.45}$$

Equation 5.45 can be converted to a sampled algorithm by expanding the denominator and applying the techniques presented earlier (see Equation 5.7 and following) to derive Equation 5.46:

$$C_N = (1 - \text{Pole}^3)R_N + 3C_{N-1} \times \text{Pole} - 3C_{N-2} \times \text{Pole}^2 + C_{N-3} \times \text{Pole}^3 \tag{5.46}$$

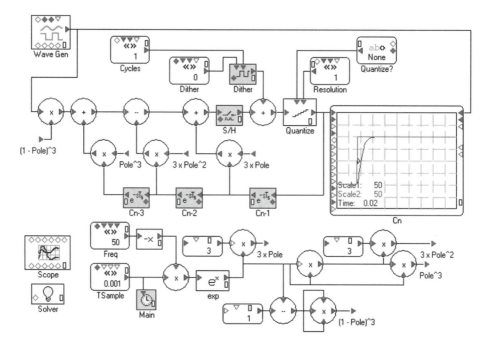

Figure 5-11. Experiment 5C: quantization and dither.

Experiment 5C, shown in Figure 5-11, displays the time-domain response of Equation 5.46. This system is modeled in two parts; above, the filter is executed based on the filter constants, which are calculated below. The pole for this filter is calculated as Pole $= \exp(-2\pi fT)$ (note that the *Live Constant* Freq has an implicit multiplier of 2π to convert hertz to radians/second). The math blocks in the bottom of the model calculate the following filter constants for Equation 5.46: $3 \times$ Pole, $3 \times$ Pole^2, Pole^3, and $(1 - \text{Pole})^3$. The filter is implemented with these constants using the delay function (e^{-sT}) three times to create C_{N-1}, C_{N-2}, and C_{N-3}. The filter output is optionally quantized using the *Live Constants* Resolution and Quantize?, where Quantize is either "None," "Round," or "Truncate." The default condition is for "None," or no quantization. Additionally, an optional dither signal can be added; this signal is zero by default. Dither will be discussed in the succeeding paragraphs. The default (without quantization) response is shown in Figure 5-11. Note that the step response is typical of a low-pass filter.

Now use Experiment 5C to study the effects of quantization. Click on the "Constants" tab and change "Quantize?" from "None" to "Round." This will cause each of the four quantities on the right side of Equation 5.46 to be rounded to $\pm 1/2$ the value specified in "Resolution," another *Live Constant* in the model. This process simulates quantization in the final calculations. With quantization the wave has a self-sustaining oscillation. This oscillation, shown in Figure 5-12a, is called a *limit cycle*.

5.9.2 Offset and Limit Cycles

Quantization causes offset and limit cycles in digital control systems. Offset can be improved by rounding results of multiplications instead of truncating. For example, Experiment 5C allows you to specify quantization as "Truncate" instead of "Round." Truncation sets the fractional part of a resolution to 0; rounding sets fractions less than 0.5 to 0 and those greater than 0.5 to 1. Rounding reduces the tendency of quantization to induce offset. Return to Experiment 5C and specify truncation; notice that the DC offset increases. Rounding helps offset but provides little benefit for the limit cycle.

Another step is to increase the resolution of sensors when practical. Increasing the resolution of the arithmetic reduces limit cycles that come from math quantization. You can increase the resolution of the arithmetic by choosing a more powerful computer, switching to floating-point operations, or increasing the word length of integer math operations.

A third step available is to add dither [21]. Dither is high-frequency noise intentionally added to remove the low-frequency limit cycles. While this adds noise, it is at a sufficiently high frequency that the system will not respond substantially. Dither can reduce the amplitude of limit cycle oscillation dramatically. Experiment 5C allows you to specify "Dither." Here, the controller adds dither every other cycle and subtracts it in the remaining cycles. Notice how using dither of just 1/2 count nearly removes the limit cycle, as shown in Figure 5-12b.

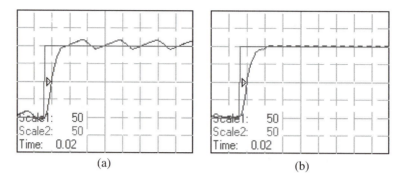

Figure 5-12. Experiment 5C: response to quantization without (a) and with (b) dither.

Because limit cycles are highly nonlinear, they can present perplexing problems. In Experiment 5C using 2 counts of dither eliminates limit cycles but 4 counts does not. However, changing the pole frequency from 50 to 70.7 Hz changes the results. Once you find a value of dither that almost removes the limit cycle, changing one constant in the algorithm may bring it back. According to Ref. 33, "Unfortunately, the selection of the amplitude and signal shape of effective dither remains more a matter of experimentation than theory."

Even getting dither to show up can be challenging. If the dither does not show up in Experiment 5C try changing "Quantize?" from "float" to "round" once or twice. Note also that the construction of the algorithm affects dither. In the preceding example, when the filter was reconstructed so that three single-pole low-pass filters were cascaded and processed in series rather than being combined together in a single polynomial as they were in Equation 5.46, there were no limit cycles.

5.10 Questions

1. a. Provide the transfer function for an accumulator (Euler's integrator) assuming a sample time of 0.01s.
   b. What is the time-domain function?
   c. Repeat 1a) with trapezoidal integration.
   d. Repeat 1b) with trapezoidal integration.
2. For a sample time of 0.005 s,
   a. evaluate z at 10 Hz.
   b. Repeat for 20 Hz.
   c. Repeat for 50 Hz.
3. a. For a sample time of 0.001 s, what is the transfer function of a single-pole low-pass filter with a bandwidth of 20 Hz?
   b. What is the time-domain function?
   c. Run the first four samples for $R_N = 1$, N > 0.
   d. What is the DC gain of this filter?

4. a. For a sample time of 0.00025 s, what is the transfer function of a two-pole
      notch filter with a bandwidth of 200 Hz and a $\zeta = 0.2$?
   b. Evaluate this transfer function at the following frequencies: 150 Hz, 180 Hz,
      195 Hz, 200 Hz, and 205 Hz.
5. a. For a system sampling at 0.05 s, where does aliasing start?
   b. What frequency components are indistinguishable from DC?
6. a. What is the DC gain of a 14-bit A/D converter with a span of ±10 V?
   b. Repeat for a 16-bit ADC.
   c. Repeat for a 14-bit D/A converter with a span of ±5 V.

Chapter 6

Six Types of Controllers

The proportional-integral-differential (PID) controller [32,47] is perhaps the most common controller in general use. Most programmable logic controllers (PLCs) support a variety of processes with this structure; for example, many temperature, pressure, and force loops are implemented with PID control. PID is a structure that can be simplified by setting one or two of the three gains to zero. For example, a PID controller with the differential ("D") gain set to zero reduces to a PI controller. This chapter will explore the use of six variations of P, I, and D gains.

When choosing the controller for an application, the designer must weigh complexity against performance. PID +, the most complex of the six controllers in this chapter, can accomplish anything the simpler systems can do, but there is a cost. More complex controllers require more capability to process, in the form of either faster processors for digital controllers or more components for analog controllers. Beyond that, more complex controllers are more difficult to tune. The designer must decide how much performance is worth paying for.

The focus in this chapter will be on digital controls, although issues specific to analog controls are covered throughout. As discussed in Chapter 4, the basic issues in control systems vary little between digital and analog controllers. For all control systems, gain and phase margins must be maintained, and phase loss around the loop should be minimized. The significant differences between the two controller types relate to which schemes are easiest to implement in analog or digital components.

The controllers here are all aimed at controlling a single-integrating plant. Note especially that the PID controller discussed in this chapter is for a single-integrating plant, unlike a PID position loop, which is for a double-integrating plant. As will be shown in Chapter 17, a PID position loop is fundamentally different from the classic PID loops discussed here.

6.1   Tuning in This Chapter

Throughout this chapter, a single tuning procedure will be applied to multiple controllers. The main goal is to provide a side-by-side comparison of these methods. A consistent set of stability requirements is placed on all of the controllers. Of course, in industry, requirements for controllers vary from one application to another. The requirements used here are representative of industrial controllers, but designers will need to modify these requirements for different applications. The specific criteria for tuning will be as follows: In response to a square wave command, the high-frequency zone (P and D) can overshoot very little (less than 2%), and the low-frequency zone can overshoot up to 15%. Recognizing that few people have laboratory instrumentation that can produce Bode plots, these tuning methods will be based on time-domain measures of stability, chiefly overshoot in response to a square wave. This selection was made even though it is understood that few control systems need to respond to such a waveform. However, square waves are the signals of choice in many cases for exposing marginal stability; testing with gentler signals may allow marginal stability to pass undetected.

This chapter will apply the zone-based tuning method of Chapter 4. Each of the six controllers has either one or two zones. The proportional and differential gains combine to determine behavior in the higher zone and thus will be set first. So the P and D gains must be tuned simultaneously. The integral gain and a command filter, which will be presented in due course, determine behavior in the lower zone.

The higher zone is limited by the control loop outside the control law: the plant, the power converter, and the feedback filter. The lower zone is limited primarily by the higher zone. Note that sampling delays can be thought of as parts of these processes; calculation delay and sample-and-hold delay (see Section 4.2) can be thought of as part of the plant and feedback delay as part of the feedback filter.

The tuning in this chapter will set the loop gains by optimizing the response to the command. Higher loop gains will improve command response and they will also improve the disturbance response. Depending on the application, command or disturbance response may be more important. However, command response is usually preferred for determining stability, for a practical reason: Commands are easier to generate in most control systems. Disturbance response is also an important measure, as will be discussed in detail in Chapter 7.

When tuning, the command should be as large as possible to maximize the signal-to-noise ratio. This supports accurate measurements. However, the power converter must remain out of saturation during these tests. For this chapter, the example systems are exposed only to the relative quiet of numerical noise in the model; in real applications, noise can be far more damaging to accurate measurements.

6.2   Using the Proportional Gain

Each of the six controllers in this chapter is based on a combination of proportional, integral, and differential gains. Whereas the latter two gains may be optionally zeroed, virtually all controllers have a proportional gain. Proportional gains set the boundaries of performance for the controller. Differential gains can provide incremental improvements at higher frequencies, and integral gains improve performance in the lower frequencies. However, the proportional gain is the primary actor across the entire range of operation.

6.2.1 P Control

The proportional, or "P," controller is the most basic controller. The control law is simple: Control $\propto$ Error. It is simple to implement and easy to tune. A P-control system is provided in Experiment 6A and is shown in Figure 6-1. The command is provided by a square wave feeding a DSA. The error is formed as the difference between command and feedback. That error is scaled by the single control law gain K_P to create the command to the power converter. The command is clamped (here, to ±20) and then fed to a power converter modeled by a 500-Hz, two-pole low-pass filter with a damping ratio of 0.7. The plant is a single integrator with a gain of 500. The feedback must also pass through a sample-and-hold. The sample time for the digital

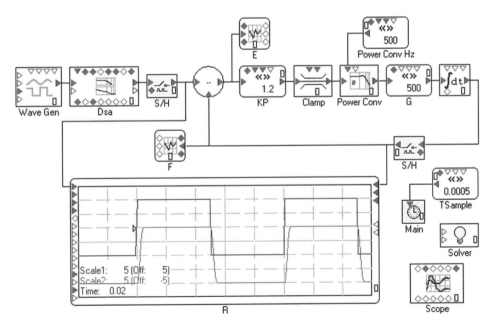

Figure 6-1. Experiment 6A, a P controller.

controller, set by the *Live Constant* "TSample," is 0.0005 seconds. The response vs. command is shown on the *Live Scope* at the bottom left.

The chief shortcoming of the P-control law is that it allows DC error; it droops in the presence of fixed disturbances. Such disturbances are ubiquitous in controls: Ambient temperature drains heat, power supply loads draw DC current, and friction slows motion. DC error cannot be tolerated in many systems, but where it can, the modest P controller can suffice.

6.2.1.1 *How to Tune a Proportional Controller*

Tuning a proportional controller is straightforward: Raise the gain until instability appears. The flowchart in Figure 6-2 shows just that. Raise the gain until the system begins to overshoot. The loss of stability is a consequence of phase loss in the loop, and the proportional gain will rise to press that limit. Be aware, however, that other factors, primarily noise, often ultimately limit the proportional gain below what the stability criterion demands.

Noise in a control system may come from many sources. In analog controllers, it is often from electromagnetic interference (EMI), such as radio frequency interference (RFI) and ground loops, which affects signals being connected from one device to another. Noise is common in digital systems in the form of limited resolution, which acts like random noise with an amplitude of the resolution of the sensor. Independent

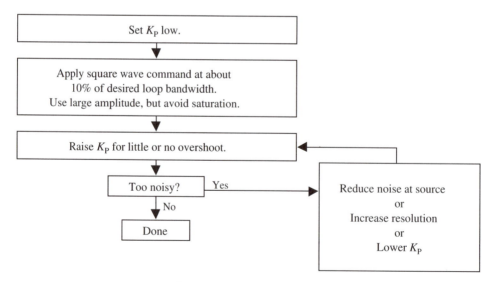

Figure 6-2. Tuning a P controller.

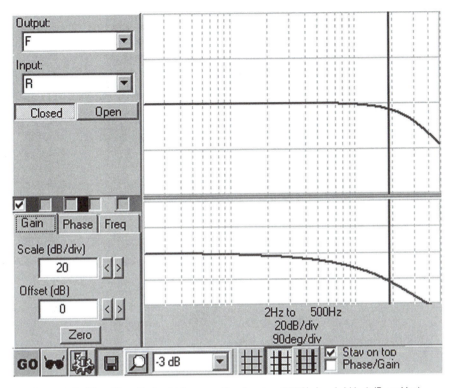

Figure 6-3. Closed-loop Bode plot for proportional system (186 Hz bandwidth, 0 dB peaking).

of its source, noise will be amplified by the high-frequency gains in the controller, such as the proportional gain.

Noise is a nonlinear effect and one that is generally difficult to characterize mathematically. Usually, the person tuning the system must rely on experience to know how much noise can be tolerated. Noise at some level is acceptable in every control system. Higher gain amplifies noise, so setting the gain low will relieve the noise problem but at the expense of degrading the control system performance. In cases of substantial noise, setting the proportional gain requires balancing the need for performance and the elimination of noise. Things are simpler for tuning the examples in this chapter; these systems deal only with the small numerical noise in the model.

Figure 6-1 shows the step response of the P controller tuned according to the procedure of Figure 6-2. The result was $K_P = 1.2$. The step response has almost no overshoot. Using Experiment 6A, the closed- and open-loop responses can be measured. As shown in Figure 6-3, the closed-loop response has a comparatively high bandwidth (186 Hz) without peaking. The open-loop plot shows 65° PM and 12 dB GM (Figure 6-4).

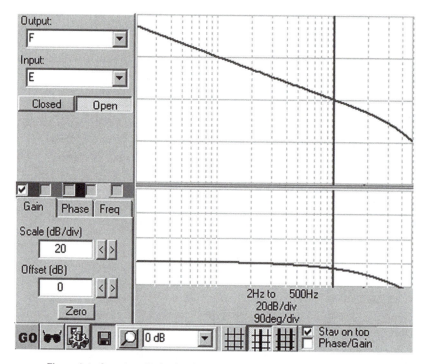

Figure 6-4. Open-loop Bode plot of proportional system (65° PM, 12.1 dB GM).

▌ 6.3 ▐ Using the Integral Gain

The primary shortcoming of the P controller, tolerance of DC error, is readily corrected by adding an integral gain to the control law. Because the integral will grow ever larger with even small DC error, any integral gain (other than zero) will eliminate DC droop. This single advantage is why PI is so often preferred over P control.

Integral gain provides DC and low-frequency stiffness. When a DC error occurs, the integral gain will move to correct it. The higher the gain, the faster the correction. Fast correction implies a "stiffer" system. In other words, higher integral gain translates to higher DC stiffness. Don't confuse DC stiffness with dynamic stiffness. A system can be at once quite stiff at DC and not stiff at all at high frequencies. These concepts are discussed in detail in Chapter 7. For the present, be aware that higher integral gains will provide higher DC stiffness but will not substantially improve stiffness near or above the system bandwidth.

Integral gain does bring a certain amount of baggage. PI controllers are more complicated to implement; the addition of a second gain is part of the reason. Also, saturation becomes more complicated. In analog controllers, clamping diodes must be added; in digital controllers, saturation algorithms must be coded. The reason is that the integral must be clamped during saturation to avoid the problem of "windup," as discussed in Section 3.8. Integral gain also causes instability. In the open loop, the integral, with its 90° phase lag, reduces phase margin. In the time domain, the common result of adding integral gain is overshoot and ringing.

6.3.1 PI Control

With PI control, the P gain provides similar operation to that in the P controller, and the I gain provides DC stiffness. Larger I gain provides more stiffness and, unfortunately, more overshoot. The controller is shown in Figure 6-5. Note that the K_I is in series with K_P; this is common, although it's also common to place the two gains in parallel.

It should be noted that the implementation of Figure 6-5 is for illustrative purposes. The PI controller lacks a windup function to control the integral value during saturation. The standard control laws supported by *Visual ModelQ* provide windup control and so would normally be preferred. (In addition, they take less space on the screen.) However, Experiment 6B and other experiments in this chapter break out the control law gains to make clear their functions. Because the purpose of this section is to compare similar control laws, the clarity provided by explicitly constructed control laws outweighs the need for wind-up control or compact representation.

6.3.1.1 How to Tune a PI Controller

PI controllers have two zones: high and low. The high zone is served by K_P and the low by K_I. As Figure 6-6 shows, the process for setting the proportional gain is the same as

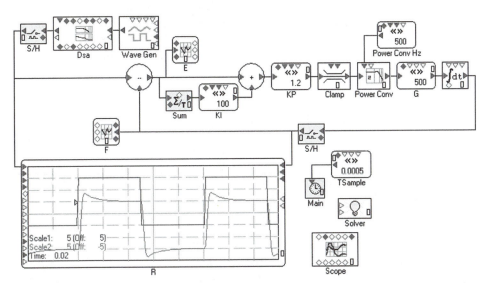

Figure 6-5. Experiment 6B, a PI Controller.

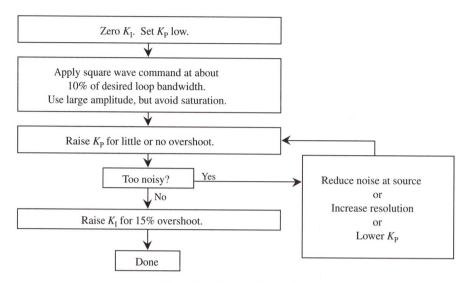

Figure 6-6. Tuning a PI controller.

it was in the P controller. After the higher zone is complete, K_I can be tuned. Here it is raised for 15% overshoot to a square wave. Again, a square wave is an unreasonably harsh command to follow perfectly; a modest amount of overshoot to a square wave is tolerable in most applications.

As Figures 6-5, 6-7, and 6-8 show, the PI controller is similar to the P controller, but with slightly poorer stability measures. The integral gain is high enough to cause a 15% overshoot to a step. The bandwidth has gone up a bit (from 186 Hz to 206 Hz), but the peaking is about 1.3 dB. The PM has fallen 9°, and the GM is nearly unchanged, just down 0.4 dB to 11.7 dB.

6.3.1.2 Analog PI Control

A simple analog circuit can be used to implement PI control. As shown in the schematic of Figure 6-9, a series resistor and capacitor are connected across the feedback path of an op-amp to form the proportional (R_L) and integral (C_L) gains. Clamping diodes clamp the op-amp and prevent the capacitor from charging much beyond the saturation level. A small leakage path due to the diodes is shown as a resistor. The input-scaling resistors are assumed here to be equal ($R_C = R_F$).

The control block diagram for Figure 6-9 is shown in Figure 6-10. Note that the gains in this figure are constructed to parallel those of the general PI controller in Figure 6-5. Tuning the analog controller is similar to tuning the general controller. Short (remove) the capacitor to convert the system to a P controller, and determine the

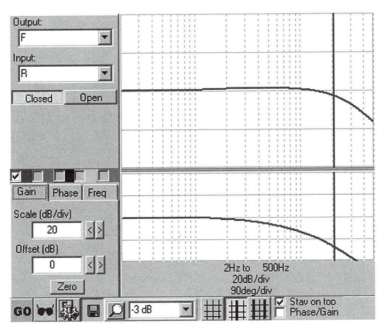

Figure 6-7. Closed-loop Bode plot for a PI controller (206-Hz bandwidth, 1.3 dB of peaking).

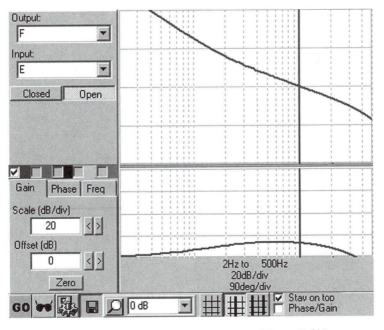

Figure 6-8. Open-loop plot of PI controller (56°, PM 11.7 dB GM).

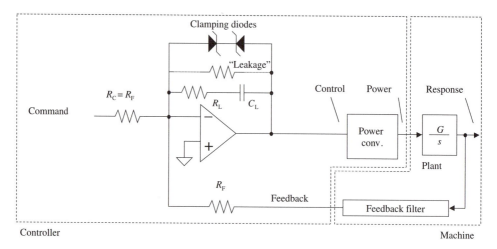

Figure 6-9. Schematic for analog PI controller.

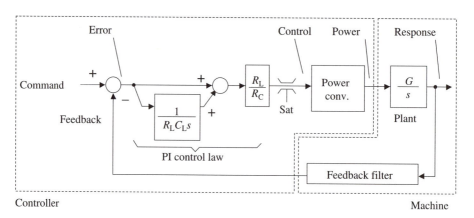

Figure 6-10. Block diagram of analog PI controller.

appropriate setting of R_L, as was done for K_P. Then adjust C_L for 15% overshoot. The analog controller will behave much like the digital controller.

One compromise that must be made for analog PI control is that op-amps cannot form true integrators. The diodes and capacitor will have some leakage, and, unlike a true integrator, the op-amp has limited gains at low frequency. Often, the PI controller is modeled as a lag network, with a large resistor across the op-amp feedback path, as shown in Figure 6-9. This "leaky" integrator is sometimes called a lag circuit. In some cases a discrete resistor is used to cause leakage intentionally. This is useful to keep the integral from charging when the control system is disabled. Although the presence of the resistor does have some effect on the control system, it is usually small enough and at low enough frequency not to be of much concern.

6.3.2 PI+ Control

PI+ control, as the name indicates, is an enhancement to PI. Because of the overshoot, the integral gain in PI controllers is limited in magnitude. PI+ control uses a low-pass filter on the command signal to remove overshoot. In this way, the integral gain can be raised to higher values. PI+ is useful in applications where the rejection of DC disturbances is paramount, for example, in a motion controller driving a high-friction mechanism such as a worm gear. The primary shortcoming of PI+ is that the command filter also reduces the controller's command response.

The PI+ controller is shown in Figure 6-11. The system is the PI controller of Figure 6-5 with a command filter added. The degree to which a PI+ controller filters the command signal is determined by the gain K_{FR}. As can be seen in Figure 6-11, when K_{FR} is 1, all filtering is removed and the controller is identical to a PI controller. Filtering is most severe when K_{FR} is zero. As can also be seen in Figure 6-11, when K_{FR} is zero, command is filtered by $K_I/(s + K_I)$, which is a single-pole low-pass filter at the frequency K_I (in rad/sec). This case will allow the highest integral gain but also will most severely limit the controller command response. Typically, $K_{FR} = 0$ will allow an increase of almost three times in the integral gain but will reduce the bandwidth by about one-half when compared with $K_{FR} = 1$ (PI control). Finding the optimal value of K_{FR} depends on the application, but a value of 0.65 has been found to work in many applications. This value typically allows the integral gain to more than double while reducing the bandwidth by only 15%–20%. References 22 and 23 provide detailed discussions of the performance of motion control systems using a range of values for K_{FR}.

One question about PI+ that naturally arises is why to select K_I as the frequency of the command low-pass filter. Why not set that frequency either higher or lower? The reason is that this frequency is excellent at canceling the peaking caused by the integral gain. One way to look at PI+ control is that it uses the command filter to attenuate the peaking caused by PI. The peaking caused by K_I can be canceled by the attenuation of a low-pass filter with a break of K_I.

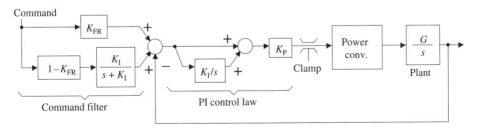

Figure 6-11. Block diagram for PI+ control.

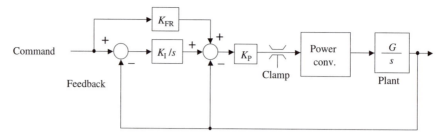

Figure 6-12. Alternative implementation for PI+ control, a PDFF controller.

6.3.2.1 Comparing PI+ and PDFF

PI+ is often referred to as pseudo-derivative feedback with feed-forward, or PDFF [22,23,74,75]. This method is shown in Figure 6-12. Although the equivalence between Figures 6-11 and 6-12 is not obvious upon inspection, construction of the control law for Figure 6-11 is:

$$\text{Control} = K_P \left(\text{Command} \left(K_{FR} + (1 - K_{FR}) \frac{K_I}{s + K_I} \right) - \text{Feedback} \right) \left(1 + \frac{K_I}{s} \right) \quad (6.1)$$

And of the control law for Figure 6-12 is:

$$\text{Control} = K_P \left((\text{Command} - \text{Feedback}) \frac{K_I}{s} + K_{FR} \text{Command} - \text{Feedback} \right) \quad (6.2)$$

With some algebra, Equation 6.1 reduces to Equation 6.2.

PDFF is an extension of a control method developed by Phelan [81] called PDF, which is equivalent to PDFF with K_{FR} set to 0. PDFF is an alternative way to implement PI+; it is useful in digital systems because there are no multiplications before the integral. Multiplication, when not carefully constructed, causes numerical noise. That noise prior to the integrator may cause drift in the control loop as the round-off error accumulates in the integrator. PDFF has a single operation, a subtraction, which is usually noiseless, before the integration and thus easily avoids such noise.

6.3.2.2 How to Tune a PI+ Controller

Tuning a PI+ controller (Figure 6-13) is similar to tuning a PI controller except that you must choose the amount of filtering (K_{FR}) before tuning the integral gain. For the highest possible stiffness, K_{FR} should be set to zero; here the PDFF controller reduces to PDF. However, for most applications K_{FR} should be at least 0.4; there is a substantial loss of response for setting K_{FR} lower, and the stiffness is improved only marginally. For applications that require the highest response to command, select

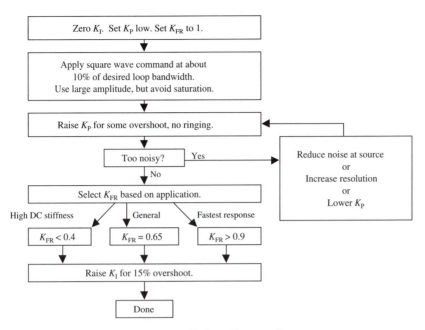

Figure 6-13. Tuning a PI+ controller.

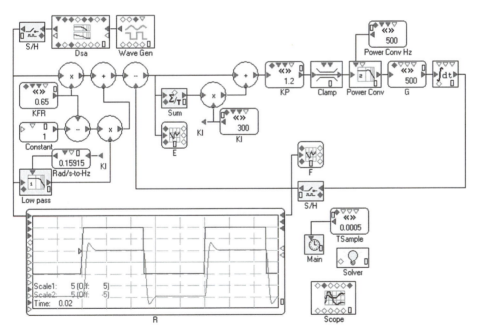

Figure 6-14. Experiment 6C, a PI+ controller.

$K_{FR} = 1$ (equivalent to PI) or at least above 0.9. Setting $K_{FR} = 0.65$ is a good compromise for many applications.

Experiment 6C, shown in Figure 6-14, will be used to demonstrate the PI+ system. This is similar to the PI controller of Experiment 6B (Figure 6-5) except that a command filter has been added between the sample-and-hold ("S/H") at top left and the summing junction just under the waveform generator ("Wave Gen"). This filter implements Equation 6.1. The *Live Constant* K_{FR} scales the command directly; "$1 - K_{FR}$" scales the command passing through the low-pass filter. The low-pass break frequency is set by K_I, although K_I must be scaled by 0.159 to convert K_I (which is in rad/sec) to Hz, the scaling for filter break frequency. One other minor change was required; the *Live Constant* K_I was converted from a "Scale-by" constant to a standard *Live Constant*, with scaling accomplished by a multiplication just above the block. This was necessary because K_I had to be provided explicitly because it is used in two places: as the integral gain and as the break frequency for the command filter.

The results of tuning a PI+ system are shown in Figures 6-14 through 6-16. The setting for K_{FR} was 0.65, the compromise value. This allowed the integral gain to increase from 100 in PI (Figure 6-5) to 300 while maintaining the same overshoot. The closed-loop plot shows a decline in bandwidth, from 206 Hz in the PI system (Figure 6-7) to 180 Hz (Figure 6-15). The open-loop plot (Figure 6-16) shows a PM of 40°, a decline of 15° compared with PI. Since K_{FR} is outside the loop, it has no direct impact

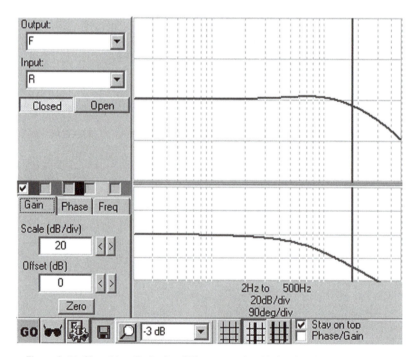

Figure 6-15. Closed-loop Bode plot of PI+ system (180 Hz bandwidth, 1.5 dB peaking).

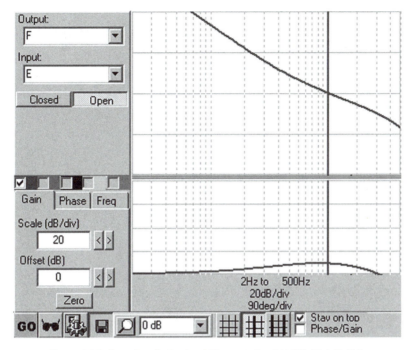

Figure 6-16. Open-loop Bode plot of PI + controller (40° PM, 10.4 dB GM).

on PM. However, because it filters high-frequency components in the command signal, lower K_{FR} allows higher integral gains, which, in turn, reduce the PM.

6.4 Using the Differential Gain

The third gain that can be used for controllers is the differential, or "D," gain. The D gain advances the phase of the loop by virtue of the 90° phase lead of a derivative. Using D gain will usually allow the system responsiveness to increase, for example, allowing the bandwidth to nearly double in some cases.

Differential gain has shortcomings. Derivatives have high gain at high frequencies. So while some D does help the phase margin, too much hurts the gain margin by adding gain at the phase crossover, typically a high frequency. This makes the D gain difficult to tune. The designer sees overshoot improve because of increased PM, but a high-frequency oscillation, which comes from reduced GM, becomes apparent. The high-frequency problem is often hard to see in the time domain because high-frequency ringing can be hard to distinguish from normal system noise. So a control system may be accepted at installation but have marginal stability and thus lack the robust performance expected for factory equipment. This problem is much easier to see using Bode plots measured on the working system.

Another problem with derivative gain is that derivatives are sensitive to noise. Even small amounts of noise from wiring or resolution limitations may render the D gain useless. In most cases, the D gain needs to be followed by a low-pass filter to reduce the noise content. The experiments in this section assume a near-noiseless system, so the D filter is set high (2000 Hz). In many systems, especially in analog controllers, such a value would be unrealistic.

6.4.1 PID Control

The PID controller adds differential gain to the PI controller. The most common use of differential gain is adding it in parallel with the PI controller shown in Figure 6-17. Here, a low-pass filter with a break frequency (2000 Hz by default) is added to the derivative path. As with the PI controller, the differential and integral gains will be in line with the proportional gain; note that many controllers place all three gains in parallel.

6.4.1.1 How to Tune a PID Controller

A PID controller is a two-zone controller. The P and D gains jointly form the higher-frequency zone. The I gain forms the low-frequency zone. The benefit of the D gain is that it allows the P gain to be set higher than it could be otherwise. The first step is to

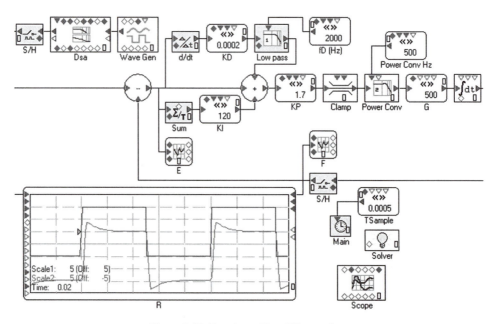

Figure 6-17. Experiment 6D, a PID controller.

tune the controller as if it were a P controller, but to allow more overshoot than normal (perhaps 10%), understanding that the D gain will cure the problem. Typically, the P gain can be raised 25%–50% over the value from the P and PI controllers. The next step is to add a little D gain to cure the overshoot induced by the higher-than-normal P gain. The P and D gains together form the high-frequency zone. Next, the integral gain is tuned, much as it was in the PI controller. The expectation is that the P and I gains will be about 20–40% higher than they were in the PI controller.

The results of the tuning procedure in Figure 6-18 are shown in Figures 6-17, 6-19, and 6-20. The PID controller allowed the proportional gain to increase to 1.7, about 40% more than in the PI controller (Figure 6-5), and the integral gain to increase to 120, about 20% more than the PI. However, the PID controller overshoots no more than the PI controller.

The closed-loop Bode plot of Figure 6-19 shows a dramatic increase in bandwidth; the PID controller provides 359 Hz, about 70% more than the 210 Hz provided by PI (Figure 6-7). Notice, though, that the phase lag of the closed-loop system is 170°, which is about 45° more than the PI. That makes this PID system more difficult to control as an inner loop than the PI controller would be. More phase lag at the bandwidth means an outside loop (such as a position loop surrounding this PID velocity controller) would have to deal with greater lag within its loop and thus have more stability problems.

The open-loop plot of the PID controller in Figure 6-20 shows a PM of 55°, about the same as the PI controller. However, the GM is about 8.5 dB, 3 dB less than the PI controller. Less GM is expected because the high-frequency zone of the PID controller is so much higher than that of the PI controller, as evidenced by the higher bandwidth.

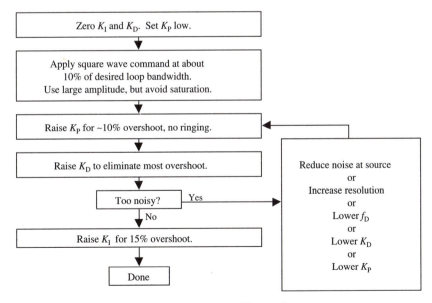

Figure 6-18. Tuning a PID controller.

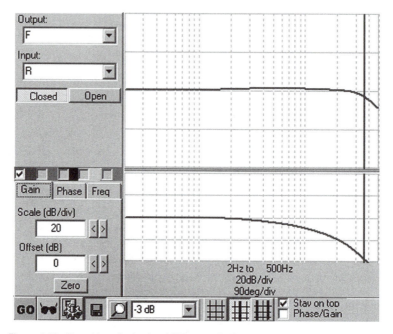

Figure 6-19. Closed-loop Bode plot of PID controller (359-Hz bandwidth, 1.0 dB peaking).

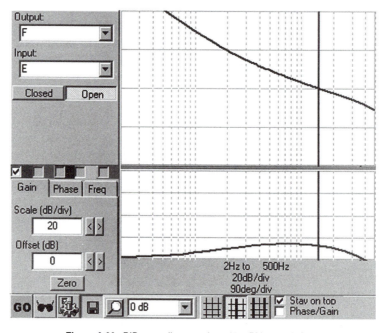

Figure 6-20. PID controller open loop (55° PM, 8.5 dB GM).

Reduced GM is a concern because the gains of plants often change during normal operation. This is of particular concern in systems where the gain can increase, such as saturation of an inductor (which lowers the inductance) in a current controller, declining inertia in a motion system, or declining thermal mass in a temperature controller; these effects all raise the gain of the plant and chip away at the GM. Given the same plant and power converter, a PID controller will provide faster response than a PI controller but will often be harder to control and more sensitive to changes in the plant.

6.4.1.2 Noise and the Differential Gain

The problems with noise in the PI controller are exacerbated by the use of a differential gain. The gain of a true derivative increases without bound as the frequency increases. In most working systems, a low-pass filter is placed in series with the derivative to limit gain at the highest frequencies. If the noise content of the feedback or command signals is high, the best cure is to reduce the noise at its source. Beyond that, lowering the frequency of the derivative's low-pass filter will help, but it will also limit the effectiveness of the D gain. Noise can also be reduced by reducing the differential gain directly, but this is usually a poorer alternative than lowering the low-pass filter frequency. If the signal is too noisy, the D gain may need to be abandoned altogether.

6.4.1.3 The Ziegler–Nichols Method

A popular method for tuning P, PI, and PID controllers is the Ziegler–Nichols method. This method starts by zeroing the integral and differential gains and then raising the proportional gain until the system is unstable. The value of K_P at the point of instability is called K_{MAX}; the frequency of oscillation is f_0. The method then backs off the proportional gain a predetermined amount and sets the integral and differential gains as a function of f_0. The P, I, and D gains are set according to Table 6-1 [32].

If a dynamic signal analyzer is available to measure the GM and phase crossover frequency, there is no need to raise the gain all the way to instability. Instead, raise the gain until the system is near instability, measure the GM, and add the GM to the gain. For example, if a gain of 2 had a GM of 12 dB (a factor of 4), K_{MAX} would be 2 plus 12 dB, or 2 times 4, or 8. Use the phase crossover frequency for f_0. A flowchart for the Ziegler–Nichols method is shown in Figure 6-21.

TABLE 6-1 SETTINGS FOR P, I, AND D GAINS ACCORDING TO THE ZIEGLER–NICHOLS METHOD

	K_P	K_I	K_D
P controller	$0.5\,K_{MAX}$	0	0
PI controller	$0.45\,K_{MAX}$	$1.2\,f_0$	0
PID controller	$0.6\,K_{MAX}$	$2.0\,f_0$	$0.125/f_0$

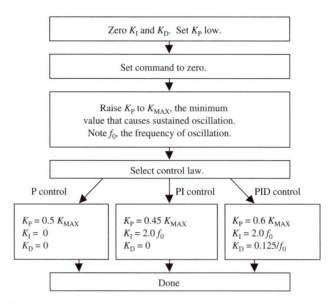

Figure 6-21. Ziegler–Nichols method for tuning P, PI, and PID controllers.

Note that the form shown here assumes K_P is in series with K_I and K_D. For cases where the three paths are in parallel, be sure to add a factor of K_P to the formulas for K_I and K_D in Table 6-1 and Figure 6-21. Note, also, that these formulas make no assumption about the units of K_P, but K_I and K_D must be in SI units (rad/sec and sec/rad, respectively). This is the case for the experimental model but often is not the case for industrial controllers. Finally, the Ziegler–Nichols method is frequently shown using T_0, the period of oscillation when $K_P = K_{MAX}$; of course, $T_0 = 1/f_0$.

The Ziegler–Nichols method is too aggressive for many industrial control systems. For example, for a proportional controller, the method specifies a GM of just 6 dB, compared with the 12 dB in the P controller tuned earlier in this chapter (Figure 6-5). In general, the gains from Ziegler–Nichols will be higher than from the methods presented here. Table 6-2 shows a comparison of tuning the P, PI, and PID controllers

TABLE 6-2 COMPARISON OF RESULTS FROM TUNING METHOD IN THIS CHAPTER AND THE ZIEGLER–NICHOLS METHOD

	Method of Chapter 6	Ziegler–Nichols method $K_{MAX} = 4.8$ and $f_0 = 311$ Hz
P controller	$K_P = 1.2$	$K_P = 2.4$
PI controller	$K_P = 1.2$	$K_P = 2.2$
	$K_I = 100$	$K_I = 373$
PID controller	$K_P = 1.7$	$K_P = 2.9$
	$K_I = 120$	$K_I = 622$
	$K_D = 0.0002$	$K_D = 0.0004$

according to the method in this chapter and to the Ziegler–Nichols method. (The terms $K_{MAX} = 4.8$ and $f_0 = 311$ Hz were found experimentally.) Both sets of gains are stable, but the Ziegler–Nichols method provides smaller stability margins.

6.4.1.4 Popular Terminology for PID Control

Often PID controllers involve terminology that is unique within controls. The three gains, proportional, integral, and differential, are called *modes* and PID is referred to as *three-mode control*. Error is sometimes called *offset*. The integral gain is called *reset* and the differential gain is called *rate*. The condition where the error is large enough to saturate the loop and continue ramping up the integral is called *reset windup*. Synchronization, the process of controlling the integral during saturation, is called *anti-reset wind-up*. You can get more information from PID controller manufacturers, such as the Foxboro Company (www.foxboro.com).

6.4.1.5. Analog Alternative to PID: Lead-Lag

PID presents difficulties for analog circuits, especially since extra op-amps may be required for discrete differentiation. The lead-lag circuit of Figure 6-22 provides performance similar to that of a PID controller but does so with a single op-amp. The differentiation is performed only on the feedback with the capacitor C_A. The resistor, R_A, forms a low-pass filter on the derivative with break frequency of $R_A \times C_A/2\pi$ Hz. Because the differential gain is only in the feedback path, it does

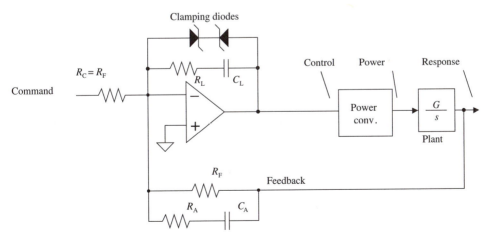

Figure 6-22. Lead-lag schematic.

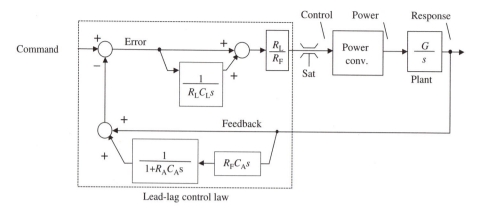

Figure 6-23. Alternative controller 4, a lead-lag controller.

not operate on the command; this eliminates some of the overshoot generated by a fast changing command.

Tuning a lead-lag circuit is difficult because the tuning gains are coupled. For example, raising C_A increases the effective differential gain but also increases the proportional gain; the derivative from C_A is integrated through C_L to form a proportional term, although the main proportional term is the signal that flows through R_F to R_L. Lead-lag is often not used in digital controls because numerical noise caused by the lead circuit (here, R_A and C_A) is fed to the integral (here, C_L); such noise can induce DC drift in digital systems, which could be avoided with the standard PID controller. On the other hand, lead circuits are sometimes used by digital designers to a larger extent than is practical in analog lead circuits. For example, multiple digital lead circuits can be placed in series to advance the phase of the feedback to increase the phase margin; this is usually impractical in analog circuits because of noise considerations.

Tuning a lead-lag controller (Figure 6-23) is similar to tuning a PID controller. Set R_A as low as possible without generating excessive noise. Often, R_A will be limited to a minimum value based on experience with noise; a typical value might be $R_A \geq R_F/3$. When tuning, start with a proportional controller: Short C_L and open C_A, raise R_L until the system just overshoots, and then raise it, perhaps 30% (how much depends on R_A, because lower R_A will allow C_A to cancel more overshoot from R_L). Start with low C_A and raise it to cancel overshoot. Then set C_L to a high value and reduce it to provide a predetermined amount of overshoot.

6.5 PID+ Control

The fifth controller is PID+: a PID controller modified with the command filter (Figure 6-24). As with PI+, the goal for PID+ is to allow higher integral gains for improved DC

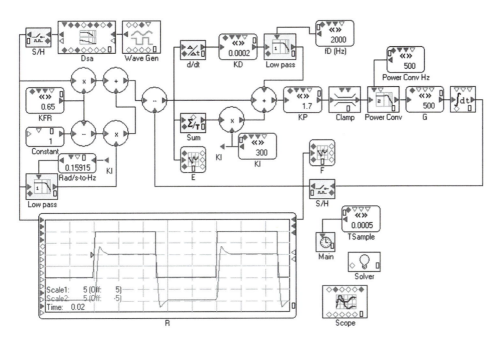

Figure 6-24. Experiment 6E, a PID+ controller.

stiffness. Again, the input filter cancels peaking caused by high integral gains; as with PI+, the command response suffers as the stiffness improves.

6.5.1 How to Tune a PID+ Controller

Tuning a PID+ controller is the same as tuning a PID controller except the value of K_{FR} must be selected before tuning the integral gain (similar to PI+). The process is shown in Figure 6-25.

The results of the tuning process of Figure 6-25 are shown in Figures 6-24, 6-26, and 6-27. The integral gain increased to 300, up from 120 in the PID controller. The closed-loop Bode plot shows the bandwidth fell to 282 Hz, down from 359 Hz in the PID controller. However, the PID+ controller has only 140° phase lag, which is superior to the 170° phase lag of the closed-loop PID controller.

Comparing the PID+ and PI+ controllers, introduction of D gain allows the PID+ controller to have higher bandwidth (282 Hz compared to 180 Hz) and similar DC stiffness, as indicated by the integral gain (300). As shown in Figure 6-27, the GM for the PID+ controller is similar to that of the PID controller but with 45° PM, 10° less than the PID controller. This is expected; as with the PI+ controller, the command filter allows the controller to work with a lower PM.

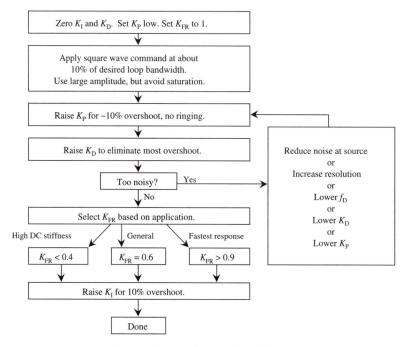

Figure 6-25. Tuning a PID+ controller.

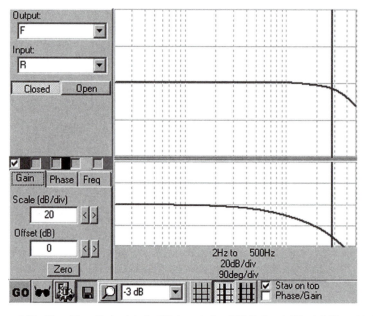

Figure 6-26. Closed-loop Bode plot of a PID+ controller (282-Hz bandwidth, 0.4 dB peaking).

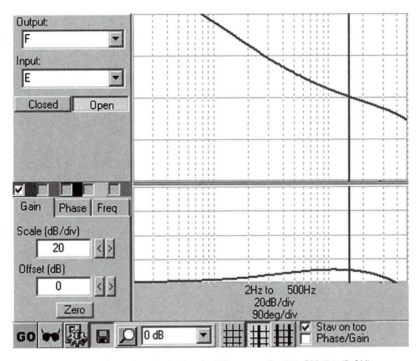

Figure 6-27. Open-loop Bode plot of a PID + controller (45° PM, 7.9 dB GM).

6.6 PD Control

The sixth controller covered in this chapter is a PD controller; the P controller is augmented with a D term to allow the higher proportional gain. The controller is shown in Figure 6-28. It is identical to the PID controller with a zero I gain.

6.6.1 How to Tune a PD Controller

Tuning a PD controller (Figure 6-29) is the same as tuning a PID controller, but assume K_I is zero. The effects of noise are the same as those experienced with the PID controller.

The results of tuning are shown in Figures 6-28, 6-30, and 6-31. The step response is square. The introduction of the D gain allowed the P gain to be raised from 1.2 to 1.7. This allows much higher bandwidth (353 Hz for the PD controller compared to 186 Hz for the P controller), although the phase lag at that bandwidth is much higher (162° for the PD controller compared to 110° for the P controller). As with the PID controller, the PD controller is fast but more susceptible to stability problems.

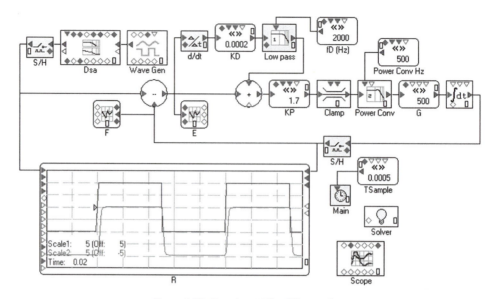

Figure 6-28. Experiment 6F, a PD controller.

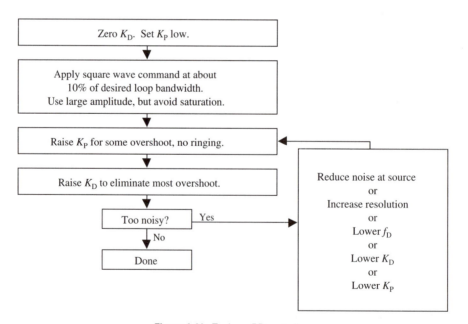

Figure 6-29. Tuning a PD controller.

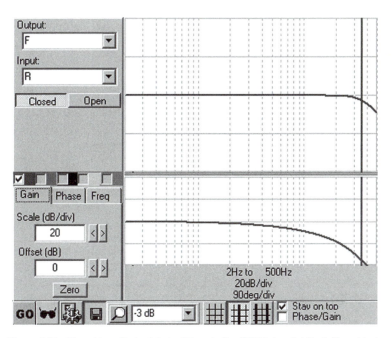

Figure 6-30. Closed-loop Bode plot of a PD controller (353 Hz bandwidth, 0 dB peaking).

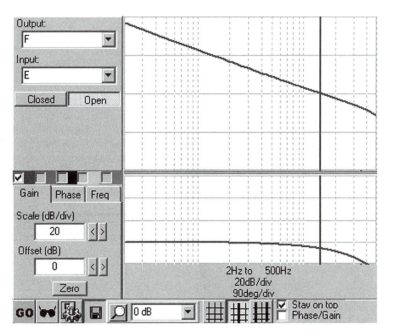

Figure 6-31. Open-loop Bode plot of a PD controller (63° PM, 8.8 dB GM).

Also, the GM is smaller (8.8 dB, 3 dB lower than for the P controller). The PD controller is useful in the cases where the fastest response is required.

6.7 Choosing the Controller

The results of tuning each of the six controllers in this chapter are tabulated in Table 6-3. Each has its strengths and weaknesses. The simple P controller provides performance suitable for many applications. The introduction of the I term provides DC stiffness but reduces PM. The command filter in PI+ and PID+ allows even higher DC stiffness but reduces bandwidth. The D term provides higher responsiveness but erodes gain margin and adds phase shift, which is a disadvantage if this loop is to be enclosed in an outer loop.

The chart in Figure 6-32 provides a procedure for selecting a controller. First determine whether the application needs a D gain; if not, avoid it, because it adds complexity, increases noise susceptibility, and steals gain margin. Next, make sure the application can support D gains; systems that are noisy may not work well with a differential gain. After that, examine the application for the needed DC stiffness. If none is required, avoid the integral gain. If some is needed, use the standard form (PI or PID); if maximum DC stiffness is required, add the input filter by using PI+ or PID+ control.

6.8 Experiments 6A–6F

All the examples in this chapter were run on *Visual ModelQ*. Each of the six experiments, 6A–6F, models one of the six methods, P, PI, PI+, PID, PID+, and PD, respectively. These are models of digital systems, with sample frequency defaulting to

TABLE 6-3 COMPARISON OF THE SIX CONTROLLERS

	P	PI	PI+	PID	PID+	PD
Overshoot	0%	15%	15%	15%	15%	0%
Bandwidth	186 Hz	206 Hz	180 Hz	359 Hz	282 Hz	353 Hz
Phase lag at BW	110°	126°	127°	169°	140°	162°
Peaking	0 dB	1.3 dB	1.5 dB	1.0 dB	0.4 dB	0 dB
PM	65°	56°	40°	55°	45°	63°
GM	12.1 dB	11.7 dB	10.4 dB	8.5 dB	7.9 dB	8.8 dB
K_P	1.2	1.2	1.2	1.7	1.7	1.7
K_I	—	100	300	120	300	—
K_D	—	—	—	0.0002	0.0002	0.0002
K_{FR}	—	—	0.65	—	0.65	—
Visual ModelQ Experiment	6A	6B	6C	6D	6E	6F

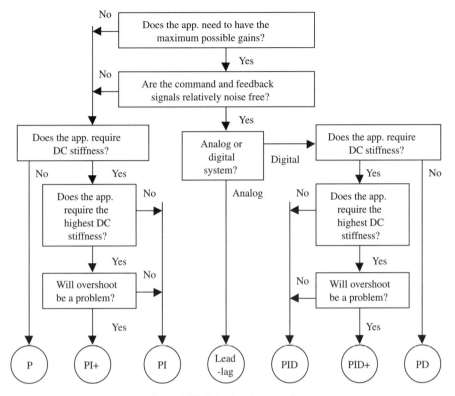

Figure 6-32. Selecting the controller.

2 kHz. If you prefer experimenting with an analog controller, set the sample time to 0.0001 second, which is so much faster than the power converter that the power converter dominates the system, causing it to behave like an analog controller.

The default gains reproduce the results shown in this chapter, but you can go further. Change the power converter bandwidth and investigate the effect on the different controllers. Assume noise is a problem, reduce the low-pass filter on the D gain (f_D), and observe how this reduces the benefit available from the derivative-based controllers (PID, PID+, and PD). Adjust the power converter bandwidth and the sample time, and observe the results.

6.9 Questions

1. Retune the proportional controller of Experiment 6A with the power converter bandwidth set to 100 Hz. Use the criteria of Chapter 6 (no overshoot for proportional gain, etc.). Measure the overshoot, peaking, and bandwidth of the closed-loop system; measure the PM and GM of the open-loop system.

2. Repeat Question 1 for the PI controller of Experiment 6B.
3. Repeat Question 1 for the PI+ controller of Experiment 6C; assume $K_{FR} = 0.65$.
4. Repeat Question 1 for the PID controller of Experiment 6D.
5. Repeat Question 1 for the PID+ controller of Experiment 6E.
6. Repeat Question 1 for the PD controller of Experiment 6F.
7. For the tuning procedures used in Questions 1–6, list two steps you could take to make the tuning more aggressive (that is, producing smaller margins of stability).

Chapter 7

Disturbance Response

Most writing about controls focuses on how well the system responds to commands. Commands are system inputs that should be followed as well as possible. Conversely, disturbances are those inputs that should be overcome. Disturbance response is usually important, and in some applications it is more important than command response. Most power supplies are commanded to an unvarying DC voltage at power-up. Holding the voltage constant in the presence of load current matters much more than rapid command response.

Subjecting systems to various command waveforms is usually straightforward. Command response is relatively easy to specify and measure: bandwidth, step response, and so on. Disturbance response is more difficult to measure because disturbances are more difficult to produce than are commands. This probably explains, at least in part, why command response is discussed more than is disturbance response.

Although command and disturbance responses are different, they share an important characteristic: Both are improved by high loop gains. A high proportional gain provides a higher bandwidth and better ability to reject disturbances with high-frequency content. A high integral gain helps the control system reject lower-frequency disturbances.

You may have noticed in earlier chapters that setting the integral gains in a PI controller had only a modest effect on the command response Bode plots. This is because integral gains are aimed at improving response to disturbances, not to commands. In fact, the process of tuning the integral gain is essentially to raise it as high as possible without significant impact on the command response. (Integral gains may not be noticeable in the command response until they are high enough to cause peaking and instability.) As you will see in this chapter, high integral gains provide unmistakable benefit in disturbance response.

Whereas high gains aid both command and disturbance response, another type of gain, called *disturbance decoupling*, aids disturbance response alone. This type of gain uses measured or estimated disturbances to improve disturbance response. Conversely, *feed-forward* gains, which will be discussed in Chapter 8, are used to improve command response but do not benefit disturbance response.

Sometimes disturbance response is referred to by its inverse, disturbance rejection, or the equivalent, dynamic stiffness. These two alternatives measure the same quality of a control system, providing the reciprocal of disturbance response. Control system designers aspire to create systems with high dynamic stiffness, high disturbance rejection, but low disturbance response. This chapter will use the terms *disturbance response* and *disturbance rejection* in discussion. However, the transfer functions and Bode plots will always show disturbance response.

7.1 Disturbances

Disturbances are undesirable inputs. They are common in control systems; examples include cogging in a motor, changes in ambient temperature for a temperature controller, and 60-Hz noise in a power supply. In each case, you are concerned about the response to an input other than the command. The system should be designed to reject (i.e., the output should not respond to) these inputs. A properly placed integrator will totally reject DC disturbances. High tuning gains will help the system reject dynamic disturbance inputs, but those inputs cannot be rejected entirely.

A typical home heating and air conditioning system, as shown in Figure 7-1, illustrates disturbances. The command temperature for a room is usually fixed, but thermal disturbances move the actual temperature. In the winter, cold wind cools an outside wall and seeps in through loose windowpanes, driving the temperature down, demanding hot air from the furnace. In the summer, the sun beats down on the roof and shines in windows, raising the room temperature and demanding cold air from the air conditioner. Each person in a room radiates tens of watts; many people in a small room can drive up the temperature as well.

Consider a second example: web-handling applications. Web-handling machines process material such as film or paper that moves from one roll to another. The goal of the motion controller is to keep the web (the material between the two rolls) moving at a constant speed or under constant tension. As with the power supply, the ability to

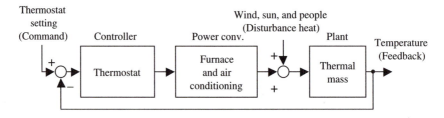

Figure 7-1. Home heating and air conditioning control system.

follow a changing command is much less important than the ability to hold a fixed output in the presence of disturbances.

Figure 7-2 shows a general control system. As in Figure 7-1, the disturbances are applied just before the plant. This is the typical placement of a disturbance. While the command enters the system through the controller ($G_C(s)$) and then through the power converter ($G_{PC}(s)$), the disturbance usually enters just before the plant ($G_P(s)$). The control system cannot reject the disturbance perfectly because the disturbance is detected only after it moves the output; in other words, the controller cannot react until its output has been disturbed.

Disturbance response is evaluated independent of command response using the principle of superposition from Section 2.2.2: Set the command to zero and use the disturbance ($D(s)$) as the input. The command response loop from Figure 7-2 can be redrawn as in Figure 7-3 to emphasize the fact that the concern here is response to disturbance.

Disturbance response is defined as the response of the system output ($C(s)$) to the disturbance force ($D(s)$):

$$T_{DIST}(s) \equiv C(s)/D(s) \tag{7.1}$$

Using the $G/(1 + GH)$ rule, $T_{DIST}(s)$ can be evaluated as

$$T_{DIST}(s) = \frac{G_P(s)}{1 + G_P(s) \times G_C(s) \times G_{PC}(s)} \tag{7.2}$$

The ideal disturbance response is $-\infty$ dB; that is, $T_{DIST}(s) = 0$. One way to improve (that is, reduce) disturbance response is intuitive: Use slow-moving plants, such as large inertia and high capacitance, to provide low plant gains. For example, the model for a capacitor charged by current is $G_P(s) = 1/Cs$. Make C larger, and $G_P(s)$ shrinks. Focusing on the numerator of Equation 7.2, reducing $G_P(s)$ reduces $T_{DIST}(s)$. In fact, the use of slow plants is a time-proven way to improve disturbance response. Large flywheels smooth motion; large capacitors smooth voltage output.

A second way to improve disturbance response is to increase the gains of the controller ($G_C(s)$). Focusing on the denominator of Equation 7.2, larger $G_C(s)$ will improve (reduce) the magnitude of the disturbance response. This is how integral gains grant systems perfect response to DC inputs: The gain of the ideal integral at 0 Hz is

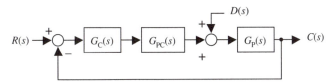

Figure 7-2. Simple loop with disturbance.

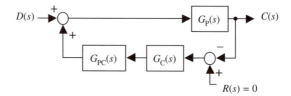

Figure 7-3. Simple loop (Figure 7-2) drawn to emphasize disturbance response.

infinite, driving up the magnitude of the denominator in Equation 7.2 and, thus, driving down the disturbance response. At other frequencies, unbounded gain is impractical, so AC disturbance response is improved with high gains but not cured entirely.

Sometimes dynamic stiffness is substituted for disturbance response. The concept is similar, but the formula for dynamic stiffness is the geometric inverse of disturbance response:

$$T_{\text{STIFFNESS}}(s) \equiv D(s)/C(s) \tag{7.3}$$

Stiffness is a measure of how much force is required to move the system, as opposed to disturbance response, which is a measure of how much the system moves in the presence of a force. Dynamic stiffness parallels mechanical stiffness in a material. A system that is very stiff responds little to disturbances.

7.1.1 Disturbance Response of a Power Supply

As an example of dynamic stiffness, consider the power supply from Figure 7-4. This system is a PI voltage controller feeding a current modulator that creates current to power the load. The modulator is a subsystem that pulses voltage to an inductor to create current. Current from the modulator is fed to the main capacitor, which

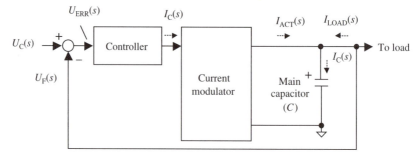

Figure 7-4. Power supply.

smooths the output voltage. The current from the modulator is summed with the load current to feed the main capacitor. With ideal operation, the current from the modulator (I_F) would be equal to and opposite in sign to the load current (I_{LOAD}) to maintain the voltage across the capacitor at a fixed level. Unfortunately, the load current is a disturbance to the voltage controller, and, as we have discussed, perfect disturbance response is impractical; the feedback current cannot exactly offset dynamic changes in the load current.

The model in Figure 7-5 represents the power supply from Figure 7-4. The model for the current modulators is assumed to be an ideal low-pass filter for this discussion. The direction of the load current is shown pointing into the summing junction so that when the load draws power, the load current is negative. This sign allows the tradi-tional sign (positive) for a disturbance.

Using the $G/(1 + GH)$ rule, the disturbance response is the forward path from the disturbance to the voltage output ($G_P(s)$ or $1/Cs$) divided by 1 plus the loop gain $(1 + G_P(s) \times G_C(s) \times G_{PC}(s))$:

$$T_{DIST}(s) = \frac{U_F(s)}{I_{LOAD}(s)} = \frac{1/Cs}{1 + (1/Cs) \times (1 + K_I/s)K_P \times LPF(s)} \qquad (7.4)$$

Equation 7.4 can be multiplied, numerator and denominator, by Cs^2. To make the form more recognizable, ideal power conversion ($G_{PC}(s) = 1$) is assumed. These steps are combined in Equation 7.5:

$$T_{DIST}(s) = \frac{U_F(s)}{I_{LOAD}(s)} = \frac{s}{Cs^2 + K_P s + K_I K_P} \qquad (7.5)$$

Equation 7.5 can be plotted for a typical system as shown in Figure 7-6 [57, 58, 70, 87, 99]. Note that disturbance response is very good (low) at high frequencies because of the main capacitor. The response is also very good at low frequencies; the integrator will remove all the disturbance at DC. In the midrange, the disturbance response is at its worst; this is typical.

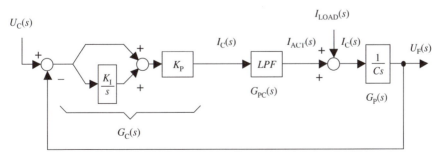

Figure 7-5. Model of power supply.

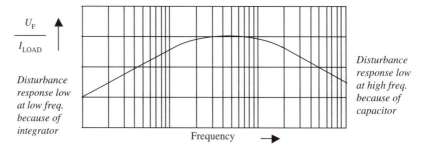

$\dfrac{U_F}{I_{LOAD}}$

Disturbance response low at low freq. because of integrator

Disturbance response low at high freq. because of capacitor

Frequency

Figure 7-6. Typical disturbance rejection (Equation 7.5).

Larger capacitance will provide better (lower) disturbance response at high frequencies. This can be seen by noticing that the Cs^2 term will dominate the denominator when the frequency is high (i.e., when s is large). So

$$T_{\text{DIST–HIGH FREQ}}(s) \approx \frac{s}{Cs^2} = \frac{1}{Cs} \tag{7.6}$$

Figure 7-7 shows a plot of Equation 7.6 against the disturbance response shown in Figure 7-6 in dashed lines. Notice that increasing the value of capacitance improves (reduces) the disturbance response in the higher frequencies; this is intuitive. Notice also that the disturbance response from the capacitance improves as frequency increases. The s in the denominator of Equation 7.6 implies that the disturbance response improves with higher frequency, as the -20 dB/decade slope of $1/Cs$ in Figure 7-7 indicates.

In the medium-frequency range, the $K_P s$ term dominates the numerator of Equation 7.5. Equation 7.5 can be rewritten as

$$T_{\text{DIST–MED FREQ}}(s) \approx \frac{s}{K_P s} = \frac{1}{K_P} \tag{7.7}$$

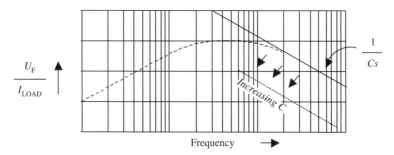

$\dfrac{U_F}{I_{LOAD}}$

Increasing C

$\dfrac{1}{Cs}$

Frequency

Figure 7-7. Effect of C on disturbance response is in the higher frequencies.

As Equation 7.7 illustrates, a larger proportional gain helps in the medium frequencies. Equation 7.7 is plotted in Figure 7-8.

In the lowest-frequency range, s becomes very small so that the term $K_I K_P$ dominates Equation 7.5, which reduces to

$$T_{\text{DIST−LOW FREQ}}(s) \approx \frac{s}{K_I K_P} \tag{7.8}$$

Equation 7.8 is plotted in Figure 7-9.

Larger proportional gain improves the low-frequency disturbance response, as does larger integral gain. Recall also from Section 3.5 that the zone-based tuning approach showed that larger proportional gain allows larger integral gains. So increasing K_P improves medium- and low-frequency disturbance response directly (Equations 7.7 and 7.8) and also indirectly helps low-frequency disturbance response by allowing larger K_I (Equation 7.8).

Figure 7-10 combines all three approximations (Equations 7.6–7.8) across the valid frequency ranges of each. Notice that the straight-line approximations combine to closely parallel the actual disturbance response.

Recall that a smaller plant gain (here, a larger capacitor) requires proportionally larger K_P to maintain the same loop gain. In the case of tuning where the primary limitation on raising K_P is the phase lag of the loop, raising C allows K_P to be raised proportionally. So raising C improves the high-frequency disturbance response directly but improves the rest of the frequency spectrum indirectly by allowing a higher value of K_P. This does not always apply, because in some systems phase lag may not be the primary limitation on K_P. K_P can also be limited by noise generated by the feedback, by resolution, which is a form of noise, or by resonance. In these cases, K_P cannot be raised in proportion to the falling plant gain, so raising capacitance will not always allow an improvement in the disturbance response in the medium and low frequencies.

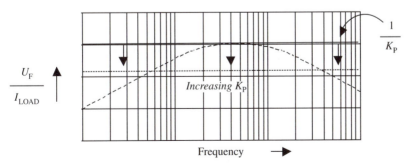

Figure 7-8. Disturbance response is improved in the medium frequencies by higher K_P.

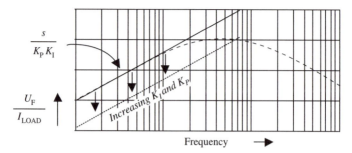

Figure 7-9. Disturbance response is improved in the lower frequencies by higher K_P and K_I.

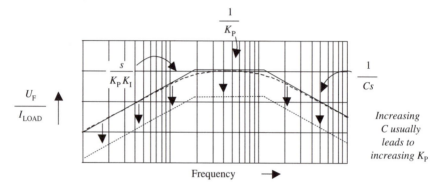

Figure 7-10. Increasing C often improves disturbance response across the entire frequency range.

7.2 Disturbance Response of a Velocity Controller

Consider the velocity controller from Chapter 4, shown in Figure 7-11, as another example of disturbance response. This system is modeled in *Visual ModelQ*, Experiment 7A; displays from this model are used in the discussion that follows. This model is based on the model in Experiment 4A but with three modifications:

- A waveform generator has been added to inject torque disturbances.
- The DSA has been added in line with the disturbance to measure disturbance response.
- The K_T/J multiplier has been separated into two terms, since torque disturbances are injected between K_T and J.

This system is similar to the power supply discussed earlier except that the controller is digital, and the feedback is from the position sensor and thus is delayed a half sample interval compared to measuring velocity directly (Section 4.2.3). Finally, because the velocity controller requires the conversion of electrical to mechanical

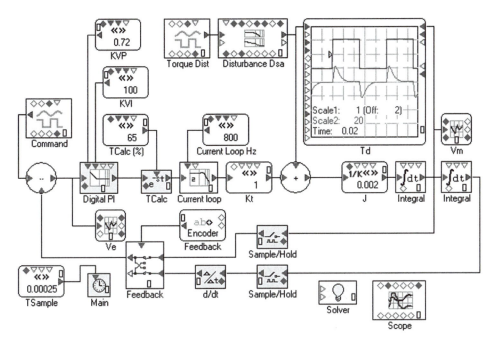

Figure 7-11. Experiment 7A, velocity controller with a disturbance.

energy, the torque conversion term (K_T) must be included. This example uses $K_T = 1$ Nm/amp (as have similar earlier examples) so that the conversion does not add confusion. Finally, motion variables (J, torque, velocity) replace electrical parameters (C, current, voltage). Those differences are all small, and the two controllers perform about the same.

The model of Figure 7-11 will allow more thorough study than was done with the power supply. First, response in either the time or frequency domain can be investigated. Also, the imperfections of phase delay from the digital controller and a realistically limited current loop are included in the simulation.

The transfer function of disturbance response for the model of Figure 7-11 can be approximated, as was the power supply in Equation 7.5. In order to make Equation 7.9 clearer, the current loop dynamics and sampling delays have been ignored; however, those effects are included in all simulations that follow.

$$T_{\text{DIST}}(s) = \frac{V_{\text{M}}(s)}{T_{\text{D}}(s)} \approx \frac{s}{Js^2 + K_{\text{VP}}K_{\text{T}}s + K_{\text{VI}}K_{\text{VP}}K_{\text{T}}} \qquad (7.9)$$

As with the power supply, the motion controller disturbance response can be approximated over ranges of frequency, as shown in Equations 7.10 through 7.12.

$$T_{\text{DIST-HIGH FREQ}}(s) \approx \frac{s}{Js^2} = \frac{1}{Js} \tag{7.10}$$

$$T_{\text{DIST-MED FREQ}}(s) \approx \frac{s}{K_{\text{VP}}K_{\text{T}}s} = \frac{1}{K_{\text{VP}}K_{\text{T}}} \tag{7.11}$$

$$T_{\text{DIST-LOW FREQ}}(s) \approx \frac{s}{K_{\text{VI}}K_{\text{VP}}K_{\text{T}}} \tag{7.12}$$

7.2.1 Time Domain

A time-domain disturbance response is shown in Figure 7-12. The disturbance waveform, ±1 Nm at 10 Hz, is shown above and the velocity response below. At the vertical edges of the disturbance, the motor moves rapidly. The peak velocity of the excursion is bounded by the proportional gain (K_{VP}); the integrator term (K_{VI}) requires more time to provide aid, although given long enough, K_{VI} will eliminate all DC error. In Figure 7-12, after about 3 msec, the integral has returned the system to near ideal (i.e., zero speed) response.

The most immediate way to improve disturbance response is to increase controller gains. Experiment 4A showed that K_{VP} can be increased to about 1.2 while remaining stable. Increasing K_{VP} from 0.72 (the default) to 1.2 cuts the peak response almost proportionally: from about 21 RPM (Figure 7-12) to 14 RPM (Figure 7-13a). Restoring the K_{VP} to 0.72 and increasing K_{VI} to 150 improves the response time of the integral, again approximately in proportion, from about 1.5 time divisions in Figure 7-12 to 1 division in Figure 7-13b. However, changing the integral gain does not improve the peak of the disturbance response, which is a high-frequency effect, out of reach of the integrator; both Figures 7-12 and 7-13b show excursions of about 21 RPM.

Note that for motion systems, the area under the velocity response is often more important than the peak excursion. Area under the velocity curve (i.e., the integral of the velocity) represents position error. You could just as easily monitor the position

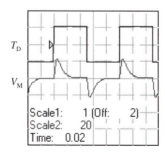

Figure 7-12. Dynamic stiffness: velocity in reaction to a disturbance torque.

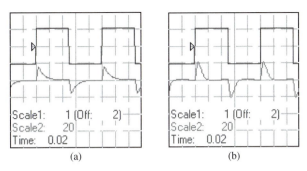

(a)   (b)

Figure 7-13. Disturbance response improves compared to Figure 7-12. (a) K_{VP} is raised to 1.2. (b) K_{VP} is restored to 0.72 and K_{VI} is raised to 150.

deviation from a torque pulse. In that case, the disturbance response would be P_M/T_D rather than V_M/T_D.

7.2.1.1 Proportional Controller

The controller from Experiment 7A can be modified to become a proportional controller by setting $K_{VI} = 0$. Removing the integral term demonstrates its benefit. The time-domain response of the proportional controller is shown in Figure 7-14. Notice that the lack of an integral term implies that the controller does not provide ideal DC disturbance response. Unlike that in Figures 7-12 and 7-13, the velocity response does not decay to the ideal value of zero.

7.2.2 Frequency Domain

Experiment 7A can be used to generate Bode plots of frequency. The concern here will be with gain; the phase of the disturbance response matters less. The system is configured

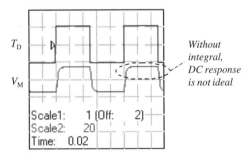

Figure 7-14. Disturbance response of proportional controller ($K_{VP} = 0.72$).

to display disturbance response. The result of the default system is shown in Figure 7-15, with the approximations for the disturbance response (Equations 7.10 through 7.12) drawn in the appropriate zones.

The effect of higher K_{VP} and higher K_{VI} can also be seen in the frequency domain. The effects demonstrated in the time domain in Figure 7-13 are shown in the Bode plot of Figure 7-16. In Figure 7-16a, K_{VP} has been raised from 0.72 to 1.2; the result, with the plot for higher K_{VP} being a bit lower, shows that higher K_{VP} improves (lowers) the disturbance response in both the medium and low frequencies. In Figure 7-16b, K_{VP} is restored to 0.72 and K_{VI} is raised from 100 to 150; note that only the low-frequency disturbance response is improved. In both cases, the high-frequency response, which is governed by J, is unchanged.

The effect of removing the integral controller shows up in the low frequencies. Figure 7-17 shows the disturbance response with K_{VI} set to 100 (PI control) and to 0 (P control). The response of the two controllers is similar in the medium and high frequencies; in the low frequencies, the difference is clear. The response of the PI controller is superior; it improves (declines) as the frequency drops. The P controller does not improve with lower frequency.

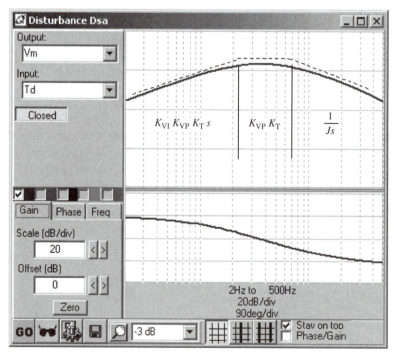

Figure 7-15. Frequency-domain plot of velocity controller.

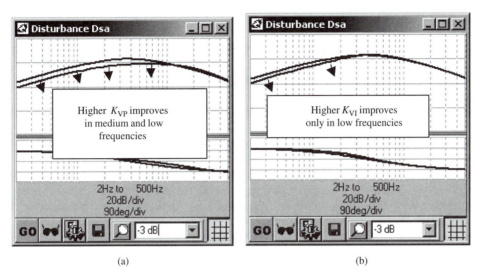

Figure 7-16. Disturbance response with (a) $K_{VP} = 1.2$ and $K_{VI} = 100$ and (b) $K_{VP} = 0.72$ and $K_{VI} = 150$.

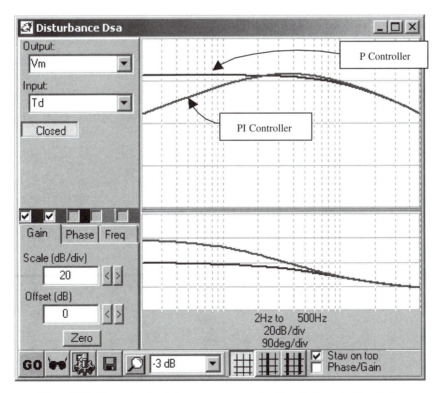

Figure 7-17. Comparing the disturbance response of a PI controller and a P controller.

7.3 Disturbance Decoupling

As discussed earlier, disturbance response can be improved by raising loop gains. However, all systems have stability limitations that cap the value a loop gain can take on. Another method of improving disturbance response is called *disturbance decoupling* [57, 58]. Here, the disturbance is measured or estimated and then an equivalent force is subtracted before power conversion to nearly cancel the disturbance. The advantage of disturbance decoupling is that it can improve disturbance response when low margins of stability prohibit increasing in the loop gain. Disturbance decoupling does suffer from the disadvantage that the disturbance must be measured or estimated. Disturbance decoupling and high loop gains are not exclusive; they can be employed together to obtain disturbance response superior to what either of the other two methods could offer alone.

Figure 7-18 shows the general form of disturbance decoupling in a single loop. The disturbance, which is summed with the output of the power converter, is also measured to a specified accuracy (K_{DD}) and with a limited bandwidth ($G_D(s)$). Note that in Figure 7-18, DC accuracy and dynamic response of the disturbance measurement have been separated into K_{DD} and $G_D(s)$, respectively. The accuracy and speed of the measurement depend on the quality of the sensor and on the quantity being measured. Voltage and current can often be measured for a few dollars to within a few percent and with a bandwidth of several hundred or even several thousand hertz. Measurement of a torque disturbance on a motion system may cost hundreds of dollars and have a bandwidth of 20 Hz and an accuracy of ±10%.

As an alternative to direct measurement, disturbances can be estimated or observed [28,44,54,58,70,87]. Observers are typically fast and inexpensive to implement, but the accuracy is usually low, perhaps ±20%. Whether the disturbance signal is measured or observed, decoupling is carried out in the same manner.

The transfer function of disturbance response as shown in Figure 7-18 can be derived using Mason's signal flow rule (Section 2.4.2). Upon inspection of Equation 7.13, if the disturbance measurement and power converter were ideal, the response to the disturbance would also be ideal and independent of the loop gain. This can be seen

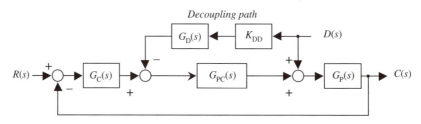

Figure 7-18. Disturbance decoupling.

by assuming K_{DD}, $G_{\text{D}}(s)$, and $G_{\text{PC}}(s)$ are all 1 and noticing that the numerator becomes zero, the ideal response to a disturbance.

$$T_{\text{DIST}}(s) = \frac{C(s)}{D(s)} = \frac{1 - K_{\text{DD}}G_{\text{D}}(s)G_{\text{PC}}(s)}{1 + G_{\text{C}}(s)G_{\text{PC}}(s)G_{\text{P}}(s)} \tag{7.13}$$

Of course, the assumption of perfect measurement and power conversion is unrealistic. The study of disturbance decoupling focuses on understanding the implications of imperfections in the disturbance measurement and, to a lesser extent, in the power converter.

7.3.1 Applications for Disturbance Decoupling

How well disturbance decoupling can be used in an application depends, for the most part, on how practical it is to measure or estimate the disturbance. This section discusses several applications that can be improved with disturbance decoupling.

7.3.1.1 Power Supplies

Power supplies benefit greatly from disturbance decoupling because the disturbance (the load current) is usually sensed and available to the control loop. Figure 7-19 shows the power supply from Figure 7-5 redrawn to include disturbance decoupling. The load current is sensed and deducted from the output of the voltage loop. So rather than waiting for the load current to disturb the feedback voltage ($U_{\text{F}}(s)$) and then having the voltage loop react to the change, the disturbance is shuttled directly to the commanded current. In that way, the controller can react immediately to the disturbance within the accuracy of the current sensor (K_{DD}), which is typically a few percent, and within the response rates of the sensor and power converter, which are faster than the voltage loop.

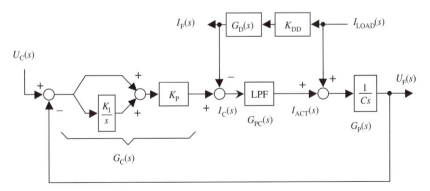

Figure 7-19. Model of power supply (Figure 7-5) with disturbance decoupling.

7.3.1.2 Multizone Temperature Controller

Part of the processing of semiconductor wafers includes heating the wafer to a well-controlled temperature [57]. Often, many wafers are processed simultaneously in a single oven by having the wafers move through multiple thermal zones; the temperature of each zone is controlled independently. The zones are thermally separated, but the separation is incomplete; each zone disturbs its neighbors. This is illustrated with a three-zone oven in Figure 7-20, with one of the disturbance paths (Zone 1 to Zone 2) labeled.

The wafer in Zone 2 is heated by the Zone 2 heater and, to a lesser extent, by the heaters in Zones 1 and 3. This is described in Equation 7.14:

$$J_2 = \frac{1}{C_1} \int (h_{22}q_2 + h_{12}q_1 + h_{32}q_3)dt \qquad (7.14)$$

where h_{22} is the coupling of the heater in Zone 2 to the wafer in Zone 2, h_{12} is the coupling from the heater in Zone 1 to the wafer in Zone 2, and so on. This model is a simplification of the actual process; it ignores heating from outside the oven (ambient) and also from the air in each zone. The controller for the oven in Figure 7-20 is shown in Figure 7-21.

The controller in Figure 7-21 shows a temperature command, j_{2C}, which is compared to the actual temperature, j_2. The temperature controller, which is commonly a PI or PID controller, provides the commanded heat (q_{2C}). The heat command passes through the power converter to produce the actual heat (q_2), which, when scaled by h_{22}, heats the Zone 2 wafer through its thermal mass (C_2).

The "h" constants provide the ratio of heat from the heaters that goes into the wafers. For example, h_{22} provides the ratio of heat generated by q_2 that heats the Zone 2 wafer; h_{22} is less than 1 because some of the heat q_2 is lost to heating the oven, the chamber air, and the ambient. The constant h_{12} provides the ratio of the heat from heater 1 that goes to the Zone 2 wafer. Of course, h_{12} would be much smaller than h_{22} because it corresponds to leakage from the first chamber to the second.

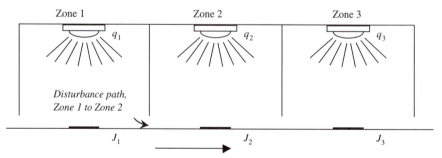

Transport moves wafers through the zones

Figure 7-20. Multizone oven.

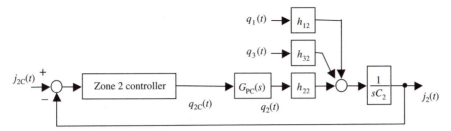

Figure 7-21. System model of Zone 2 in multizone oven.

Figure 7-22 shows the control system of Figure 7-21 augmented with disturbance decoupling. The estimated heat from Zone 1 is scaled by h_{12}, the amount of heat that comes from Zone 1 to heat the wafer in Zone 2, but must be divided down by h_{22}, the amount of heat coming from the heater in Zone 2 to heat the wafer in Zone 2. Each of the controllers can be augmented with a decoupling path, allowing the different zones to operate with more independence. For ovens with more zones, this scheme can be extended to as many zones as necessary; however, decoupling the nearest neighbors provides the majority of the benefits.

One question that arises when designing disturbance decoupling is how well the disturbance path can be measured or estimated. With thermal control systems, it can often be measured quite well. The estimates for the heat ($q_{1\text{-EST}}$ and $q_{3\text{-EST}}$) can be calculated by monitoring the output of their respective power converters. The constants for the decoupling paths ($h_{12\text{-EST}}$ etc.) can be determined by experimentation. The dynamics of the thermal sensors or power converters are of less concern if they are much faster than the thermal controllers, as is often the case.

7.3.1.3 Web Handling

In web-handling applications, material is taken from a roll, processed, and usually put back on another roll. The processing may be printing, the application of a coating layer, or slitting the material into multiple strips. Generally, many processes are performed on

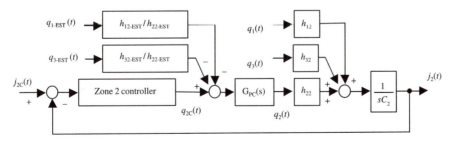

Figure 7-22. Disturbance decoupling in system model of Zone 2 in multizone oven.

stations or stands, machine segments set end to end to form a *web line*. Web lines can include over a dozen stations and can extend to over 50 meters long.

A simple web line is shown in Figure 7-23. Here, material is taken from a feed roll, threaded through the rollers in the coating station, and routed to the take-up roll. The coating station has four rollers, two servo driven and two idle (that is, unpowered). The servos are used to hold tension constant in the coating area.

Disturbances come from rolls that feed material to the station unevenly. A major cause of disturbances is that the material on the roll is not perfectly circular but rather will normally go on in a slightly elliptical shape so that each revolution of the roll produces a perturbation on the web; those perturbations translate as torque disturbances to the servos. The primary function of the servos in web-handling equipment is rejecting such disturbances. Unlike typical servo applications, where servomotors are chosen for the ability to accelerate quickly, normal operation in web applications usually involves moving the web at a constant speed. Quick acceleration is a secondary consideration.

For applications where the tension must be regulated, a force transducer is often fitted to one of the idlers. Because the web pulls at right angles on the idler (e.g., down and to the left for Idler 2), the total force on the idler is proportional to the tension across the web. Strain gauges on the idler can measure the force exerted by the web with accuracy. The constant of proportionality between force and tension is based on the mechanical dimensions of the idler roll.

The two servos are commanded to travel at a constant speed and a third loop, the force loop, is closed around the outside of one or the other servo to maintain the tension in the coating region. The three controllers are shown in Figure 7-24. The path for disturbance decoupling is shown with dashed lines.

The benefit of disturbance decoupling is evident in Figure 7-24. Without the decoupling path, force disturbances sensed by the transducer are delayed, first by the transducer ($G_D(s)$), then by the force controller, and finally by the velocity controller. Feeding the signal ahead to the velocity controller output causes the signal to bypass the delays in the force and velocity controllers. Of course, it must still endure the delay from the sensor, but that will usually be much faster than the force loop. Thus, by decoupling the disturbance signal, the dynamic reaction is improved greatly.

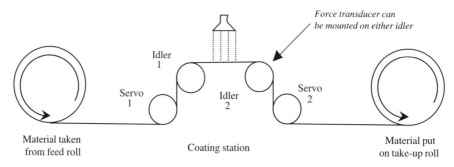

Figure 7-23. A simple web-handling line.

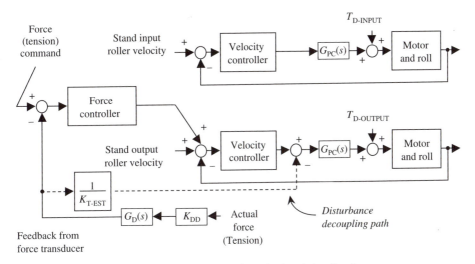

Figure 7-24. Control diagram for a simple web-handling line.

Notice that for decoupling, the force output must be divided by the estimated K_T of the motor. That is because the force feedback must be converted to a current command, and that requires converting the force to a torque and then to a current. Conversion to a torque is straightforward; it requires knowledge of the dimensions of the machine, which are customarily known to great accuracy. However, the conversion to a current command requires knowledge of the motor K_T. Because K_T is known only to within about ±15%, the overall DC accuracy of the disturbance decoupling path (including torque sensor accuracy) is normally ±20% or so.

7.3.2 Experiment 7B: Disturbance Decoupling

Experiment 7B is similar to Experiment 7A, except that a path for disturbance decoupling is added. The model for Experiment 7B is shown in Figure 7-25. This model assumes the disturbance is measured with an accuracy of K_{DD} and passes through a single-pole low-pass filter ($G_{PC}(s)$). As with the web-handling machine, decoupling requires an estimate of K_T ($K_{T\text{-EST}}$) by which to scale the torque disturbance to create a compensating current command. The low-frequency accuracy of the decoupling path is a combination of the sensor accuracy and the accuracy of the estimated K_T. Here, the assumption is that the inaccuracy of K_T dominates and K_{DD} will be set to 1.

The time-domain plots in Figure 7-26 show the improvement offered by nearly ideal disturbance decoupling. Figure 7-26a shows the disturbance response without decoupling; it is equivalent to Figure 7-12. Figure 7-26b shows the dramatic improvement when an accurate, fast disturbance signal is available for decoupling. The signal is assumed

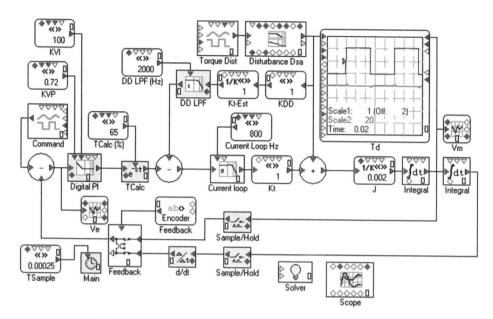

Figure 7-25. Experiment 7B, control loop with disturbance decoupling.

perfectly accurate and has a bandwidth of 2000 Hz. Unfortunately, such a signal is currently unrealistic for motion control systems.

A more realistic sensed signal for a torque disturbance would have 20% error at DC and a bandwidth of perhaps 20–200 Hz. The limitation in bandwidth comes from the force sensor's strain-gauge amplifier, which is typically a sluggish device. The inaccuracy at DC comes from the limitation in accurately estimating K_T. Figure 7-27 shows the effects of more realistic (slower) amplifiers. Figure 7-27a shows disturbance response with a 200-Hz bandwidth disturbance sensor and Figure 7-27b with a 20-Hz sensor.

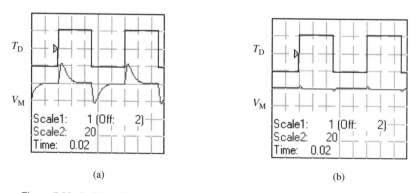

(a)   (b)

Figure 7-26. An ideal disturbance sensor can provide dramatic improvement for disturbance decoupling. (a) without decoupling; (b) with near-ideal decoupling.

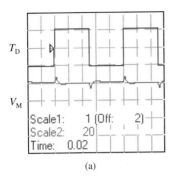
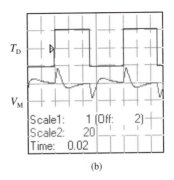

(a)　　　　　　　　　　　　　　　　　(b)

Figure 7-27. The effect of slowing the bandwidth of the disturbance sensor. (a) with a 200-Hz sensor and (b) with a 20-Hz sensor.

Figure 7-27b shows little improvement over no decoupling; this is because the sensor is so much slower than the velocity loop bandwidth (about 100 Hz) that the velocity loop responds reasonably well to the disturbance before the signal passes through the disturbance sensor. These curves are produced by setting *DD_LPF* to 200 for (a) and to 20 for (b).

The impact of DC inaccuracy is shown in Figure 7-28. Figure 7-28a shows the 200-Hz sensor with DC accuracy of 80%; Figure 7-28b shows the 20-Hz sensor with 80% accuracy. The loss of accuracy degrades the performance but in this case, where the accuracy is ±20%, the improvement is still evident, especially for the fast sensor. These curves are produced by setting $K_{T\text{-EST}}$ to 1.2 and again setting *DD_LPF* to 200 and 20 for (a) and (b), respectively.

Disturbance response can also be analyzed in the frequency domain. The results shown in Figures 7-26 through 7-28 can be demonstrated in the frequency domain with Bode plots. Figure 7-29 shows the disturbance response under three conditions: no decoupling, decoupling with a 20-Hz sensor, and decoupling with a 200-Hz sensor. The higher-bandwidth sensor makes a considerable improvement (recall that lower is

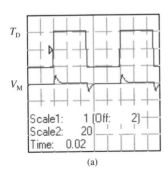
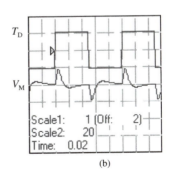

(a)　　　　　　　　　　　　　　　　　(b)

Figure 7-28. The effect of DC inaccuracy in the disturbance decoupling signal with 80% ($K_{T\text{-EST}} = 1.2$) accuracy with (a) a 200-Hz sensor and (b) with a 20-Hz sensor.

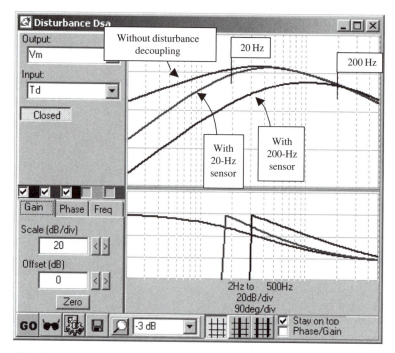

Figure 7-29. Disturbance response without decoupling and with 20- and 200-Hz disturbance sensors.

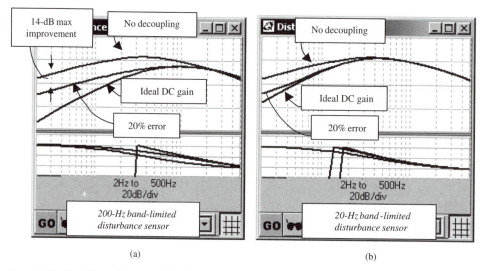

Figure 7-30. The effect of imperfect DC gain on disturbance decoupling. (a) with a 200-Hz sensor and (b) with a 20-Hz filter.

better). As expected, the 200-Hz sensor makes improvement as compared to the nondecoupled response below about 200 Hz (the third vertical line from the right). The 20-Hz sensor makes improvement below about 20 Hz.

The impact of imperfect DC gain can be seen directly in the disturbance response Bode plots. The error is 20%, 0.2, or about $-14\,\mathrm{dB}$. Such an error limits the improvement offered by decoupling to 14 dB. Note that in Figure 7-30a, the disturbance response of the imperfect-DC-gain sensor follows the ideal (unity DC gain) except that it cannot improve the nondecoupled curve by more than 14 dB. This behavior can also be seen with the slower sensor (Figure 7-30), but it is harder to see only because the frequency of the Bode plot does not extend very low compared to the sensor bandwidth.

7.4 Questions

1. What is the primary disturbance to a voltage supply?
2. In the power supply of Figure 7-5, what is the primary means of improving disturbance response at very high frequency? What would the equivalent means be for a motor controller? for a temperature controller?
3. What is the most common way to achieve ideal disturbance response to DC disturbances? What is the most common problem encountered when trying to implement disturbance decoupling?
4. For the system of Experiment 7A:
   a. Slow the sample rate to 1 kHz (TSample = 0.001). Retune the system for max K_{VP} causing no overshoot and K_{VI} causing 10% overshoot. What are the new values of K_{VI} and $K_{VP}t$? *Hint: When tuning, the disturbance waveform generator should be set very low, say, 0.001 Nm, just enough to trigger the scope. Set the velocity command to 40 RPM. Since both waveform generators are set to 10 Hz, the scope will trigger synchronously with the velocity command.*
   b. Restore the command waveform generator back to zero and restore the disturbance to 10 Nm to evaluate disturbance response. What velocity excursion does the new set of gains allow?
   c. Repeat part a with 4-kHz sampling (TSample = 0.00025).
   d. Repeat part b with 4-kHz sampling.
   e. Run a Bode plot with both gain sets. Be sure to set the velocity command to zero. Explain the effects of the higher gains by frequency zone.
   f. What is the relationship between sample time and disturbance response?
5. Run Experiment 7B using the gains of Question 4a. Set the disturbance decoupling term (K_{DD}) to zero. Confirm the velocity excursions are the same as in Question 4b.
   a. Set for full disturbance decoupling ($K_{DD} = 1$). What is the velocity excursion?
   b. How does disturbance decoupling (Question 5a) compare to speeding the sample rate by 4 times (Question 4d)?

Chapter 8

Feed-Forward

Feed-forward uses knowledge of the plant to improve the command response. The command signal is processed and routed ahead of the control law, directly to the power converter. Feed-forward calculates a best guess; it predicts the signal that should be sent to the power converter to produce the ideal response from the plant. For command response, this unburdens the control loop because most of the power converter excitation can be generated in the feed-forward path; the control law is required to provide only corrections and to respond to disturbances.

Feed-forward requires knowledge of the plant and, to a lesser extent, the power converter. Perfect knowledge of both allows for ideal command response. While perfect knowledge is unattainable, it is often practical to predict the operation of the plant and power converter well enough to allow substantial improvement in command response. For example, knowing the operation of the plant and power converter to within ±20% can support dramatic improvement.

With feed-forward, the command response becomes less dependent on the control loop bandwidth. By using an aggressive feed-forward design, the command response of a sluggish controller can be improved by a factor of several times. The use of feed-forward to improve command response is analogous to the use of disturbance decoupling (Chapter 7) to improve disturbance response.

8.1 Plant-Based Feed-Forward

The standard, plant-based feed-forward controller is shown in Figure 8-1. The $G_C/G_{PC}/G_P$ control loop of Chapter 6 is augmented with a feed-forward path: $G_F(s)$ and K_F. In most control systems, $G_F(s)$ is chosen as the inverse of the plant: $G_P^{-1}(s)$. Usually, $G_P(s)$ is an integrator (K/s), so $G_F(s)$ is usually a derivative: s/K. Even though derivatives often generate too much noise to be used in feedback loops, the use of derivatives in the feed-forward path is common. Noise sensitivity is less of a concern in

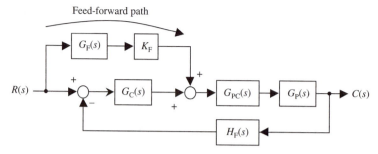

Figure 8-1. Plant-based feed-forward.

the command path, where noise is often minimal. K_F is a gain that is generally equal to or less than 1.

Notice that the command is processed (by $G_F(s)$ and K_F) and fed directly to the power converter ($G_{PC}(s)$). This direct access to the power converter command is what allows feed-forward to enhance command response. With feed-forward, command response is not limited by the controller gains. And because feed-forward gains do not form a loop, they do not impair system stability.

By using Mason's signal flow graphs (Section 2.4.2), the command response transfer function of the system from Figure 8-1 can be written

$$T(s) = \frac{C(s)}{R(s)} = \frac{(K_F G_F(s) + G_C(s))G_{PC}(s)G_P(s)}{1 + G_C(s)G_{PC}(s)G_P(s)H_F(s)} \tag{8.1}$$

The plant-based approach to analyzing feed-forward assumes that the power converter ($G_{PC}(s)$) and feedback ($H_F(s)$) are ideal (i.e., unity). Although these assumptions will be accepted for the present, they will be challenged later in this chapter. The ideal command response is unity (actual = command), so the ideal $T(s)$ in Equation 8.1 is 1. Rewriting Equation 8.1 with these assumptions yields

$$T_{IDEAL}(s) \approx \frac{(K_F G_F(s) + G_C(s))G_P(s)}{1 + G_C(s)G_P(s)} = 1 \tag{8.2}$$

So

$$(K_F G_F(s) + G_C(s))G_P(s) = 1 + G_C(s)G_P(s) \tag{8.3}$$

which, upon inspection, can be satisfied if $K_F = 1$ and $G_F(s) = G_P^{-1}(s)$.

8.1.1 Experiment 8A: Plant-Based Feed-Forward

The *Visual ModelQ* model Experiment 8A (Figure 8-2) will be used to demonstrate several aspects of feed-forward. This is a PI control system with a single-integrating plant. There is a command, R, a feedback signal, F, and an error, E. The power

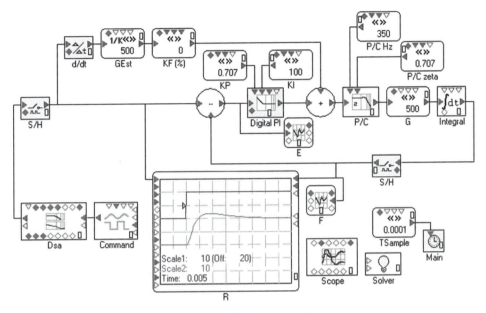

Figure 8-2. Experiment 8A: Plant-based feed-forward system.

converter is modeled as a 350-Hz two-pole low-pass filter with a damping ratio of 0.707. The sample time is fast: 0.0001 seconds. The feed-forward path is $G_P^{-1}(s)$: a digital derivative and a division by 500. Note the block *GEst* is a *Visual ModelQ* "1/K" scaling block, which produces the effect of dividing by 500. The *Live Scope* plots command (above) and feedback; in Figure 8-2, feed-forward is turned off ($K_F = 0$), so the response is that of the control loop gains, which are tuned for a 100-Hz bandwidth, a reasonable level of response considering the 350-Hz power converter.

Using Experiment 8A, the system with varying levels of feed-forward can be compared. Figure 8-3 compares the *Live Scope* plot ($R(t)$ and $F(t)$) system with 0% and 65% feed-forward gains in the time domain. The system with feed-forward responds faster and with less overshoot than the non-feed-forward system.

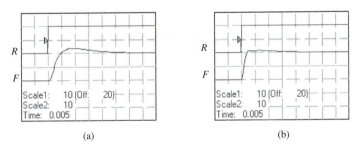

Figure 8-3. Step response (command, above, and feedback) of system with (a) 0% and (b) 65% feed-forward.

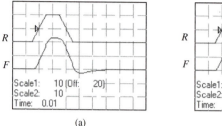
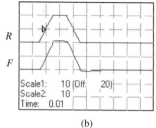

(a)   (b)

Figure 8-4. Trapezoid response (command, above, and feedback) of system with (a) 0% and (b) 65% feed-forward.

The feed-forward level of 65% for Figure 8-3b was determined by finding the value that minimized overshoot. Though Equation 8.3 indicates $K_F = 100\%$ would provide the ideal response, this is not the case. Were K_F set to 100%, the system would overshoot excessively (about 35%) in response to a square wave command. The assumptions that produced Equation 8.2, that the power converter and feedback functions were ideal, are not valid in Experiment 8A, chiefly because of the delay of the 350-Hz power converter. We will discuss how to deal with this problem in due course. For the present, be aware that the feed-forward gain often is set below the ideal 100%.

When evaluating feed-forward, it may be desirable to evaluate the response to a trapezoidal waveform, as shown in Figure 8-4. The shortcoming of relying wholly on a square wave is that an integrating plant cannot follow such a wave perfectly because instantaneous changes in state are impractical. A trapezoid can be followed much more closely. Figure 8-4 compares the settings of Figure 8-3, but against a trapezoidal command. The system with $K_F = 65\%$ follows the command better.

Figure 8-5 is the Bode plot of the command response for the system with 0% and 65% feed-forward gains. This figure also demonstrates dramatic improvement: feed-forward increases the bandwidth of the command response and does so while reducing peaking slightly. A command response bandwidth of 275 Hz in a system with a 350-Hz power converter is impressive. There is no set of loop gains in this example that can approach this level of responsiveness without feed-forward.

8.2 Feed-Forward and the Power Converter

The previous section calculated a feed-forward signal based on Equation 8.2, which takes into account only plant behavior. However, in Experiment 8A, the power converter bandwidth of 350 Hz presented a significant delay in the loop, so the assumption of ideal power conversion was not valid. This led to the problem where using full feed-forward ($K_F = 1$) would have caused excessive overshoot to aggressive commands. This was dealt with by limiting the feed-forward gain to 65%. This compromise is effective, but it limits the improvement that can be realized from feed-forward.

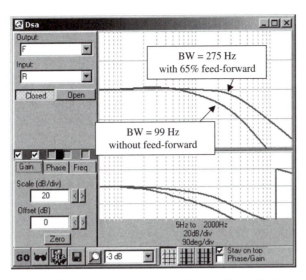

Figure 8-5. Bode plot of command response showing improved bandwidth from feed-forward.

This section will present a means for augmenting the feed-forward signal to account for a nonideal power converter. As will be shown, this will provide further improvement in command response, measured both by trapezoidal response and Bode plots. To understand how to compensate for the effects of the power converter, return to Equation 8.1. Form $T_{IDEAL}(s)$ as in Equation 8.2, but use the general form of $G_{PC}(s)$:

$$T_{IDEAL}(s) \approx \frac{((K_F G_F(s) + G_C(s))G_{PC}(s)G_P(s)}{1 + G_C(s)G_{PC}(s)G_P(s)} = 1 \tag{8.4}$$

$$(K_F G_F(s) + G_C(s))G_{PC}(s)G_P(s) = 1 + G_C(s)G_{PC}(s)G_P(s) \tag{8.5}$$

Equation 8.5 can be satisfied if $K_F = 1$ and $G_F(s) = G_{PC}^{-1}(s)G_P^{-1}(s)$. The result is similar to that gained from Equation 8.3 and, in fact, reduces to that solution for the case where $G_{PC}(s)$ is ideal.

Forming the inverse of the power converter can be challenging. In the model of Experiment 8A, the power converter is a two-pole, low-pass filter:

$$T_{PC}(s) = \frac{\omega^2}{s^2 + 2\zeta\omega s + \omega^2} \tag{8.6}$$

The estimated inverse of the power converter is

$$T_{\text{PC EST}}^{-1}(s) = \frac{s^2 + 2\zeta_{\text{EST}}\omega_{\text{PC EST}}s + \omega_{\text{PC EST}}^2}{\omega_{\text{PC EST}}^2} \tag{8.7}$$

$$= \frac{s^2}{\omega_{\text{PC EST}}^2} + \frac{2\zeta_{\text{EST}}s}{\omega_{\text{PC EST}}} + 1 \tag{8.8}$$

So the feed-forward function, $G_F(s)$, could be set to

$$G_F(s) = \left(\frac{s}{G_{\text{EST}}}\right)\left(\frac{s^2}{\omega_{\text{PC EST}}^2} + \frac{2\zeta_{\text{EST}}s}{\omega_{\text{PC EST}}} + 1\right) \tag{8.9}$$

which requires the evaluation of a third derivative of the command. Such a high-order derivative is usually too noisy to be practical. However, a compromise can be made. A low-pass filter can be cascaded with $G_{\text{PC}}^{-1}(s)$ to reduce noise at high frequencies. The form of the power converter compensator becomes

$$T_{\text{PC EST}}^{-1}(s) = \frac{s^2 + 2\zeta_{\text{PC EST}}\omega_{\text{PC EST}}s + \omega_{\text{PC EST}}^2}{s^2 + 2\zeta_{\text{LP}}\omega_{\text{LP}}s + \omega_{\text{LP}}^2} \cdot \frac{\omega_{\text{LP}}^2}{\omega_{\text{PC EST}}^2} \tag{8.10}$$

where the numerator is the estimated inverse power converter denominator and the denominator is the cascaded low-pass filter. This form, a quadratic polynomial in the numerator and another in the denominator, is called a *bi-quadratic*, or *bi-quad*, filter.

In Equation 8.10, $\omega_{\text{PC EST}}$ is the estimated frequency of the power converter (in rad/sec) and ω_{LP} is the frequency of the noise-reducing low-pass filter. The value for ω_{LP} is application dependent; higher values of ω_{LP} make Equation 8.10 closer to the ideal inverse power converter but also make the function more sensitive to noise. The effect of the bi-quad filter here is for the numerator of Equation 8.10 to cancel the power converter in the command path, leaving the denominator. The denominator will degrade the performance of the feed-forward path but less than the power converter would have because the frequency of the denominator should be significantly higher than that of the power converter.

8.2.1 Experiment 8B: Power Converter Compensation

The *Visual ModelQ* model Experiment 8B (Figure 8-6) will be used to show the benefits of power converter compensation in the feed-forward signal. This system is similar to Experiment 8A except that an alternative feed-forward path has been added in the upper left. The new path follows from *GEst* through "Digital BiQuad" and the feed-forward gain, *KF2*. The bi-quad is set to have a numerator with a break frequency of 350 Hz and a damping ratio of 0.82, which is approximately the same as a power converter denominator. The denominator has a break frequency of 2000 Hz. As discussed earlier, this shifts the feed-forward signal to behave as if the power

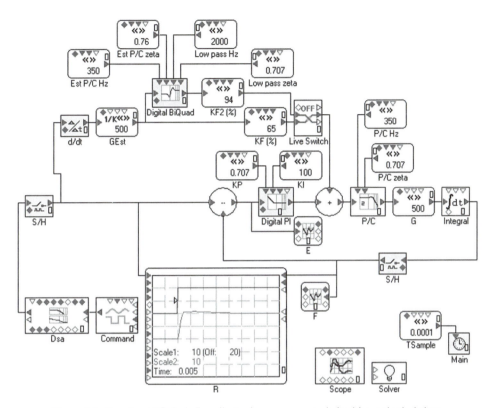

Figure 8-6. Experiment 8B: Including effects of power converter in feed-forward calculations.

converter had a bandwidth of 2000 Hz, allowing higher feed-forward gains for a given level of overshoot.

Experiment 8B introduces a new *Visual ModelQ* block, the *Live Switch*. The *Live Switch* allows you to reconfigure the model during execution. Double-clicking on a *Live Switch* when the model is running toggles the switch state from OFF to ON or ON to OFF. As shown in Figure 8-7, when the switch is OFF, the switch crosses the conductors as they pass through the switch. When the switch is ON, the switch passes the conductors directly from left to right. As shown in Experiment 8B, the default switch position is OFF; this connects the original feed-forward of Experiment 8A.

The improvement offered by power converter compensation is apparent in Figure 8-8. Figure 8-8a shows the noncompensated feed-forward signal where the feed-forward gain was limited to 65%. Figure 8-8b shows the augmented feed-forward signal according to Equations 8.5 and 8.10. Here, the feed-forward gain can be raised to 94% with minimal overshoot. As a point of comparison, the interested reader can load Experiment 8B; run the model, ensure the *Live Switch* is off (to enable

Figure 8-7. Two Live Switches. The left one is "OFF" and the right one "ON."

non-power-converter compensated feed-forward), and set $K_F = 94\%$; the overshoot is about 30%, which is much higher than Figure 8-8b.

The trapezoidal response of the system with uncompensated and compensated feed-forward is shown in Figure 8-9. In order to show more detail, the scale of the *Live-Scope* plots has been expanded to 10 units per division (from 50 in Figure 8-4). The command, which ranges from 0 to 100, goes well off scale. However, the return to zero is shown in greater detail than it was in Figure 8-4. The plant-based feed-forward system overshoots much more than the power converter/plant compensated feed-forward system.

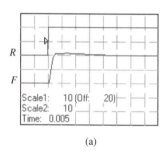
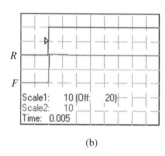

(a)   (b)

Figure 8-8. Step response of system with (a) 65% plant-only feed-forward and (b) 94% feed-forward with power converter compensation.

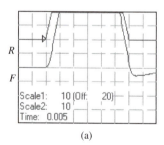
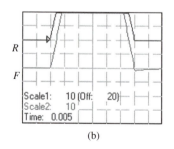

(a)   (b)

Figure 8-9. Trapezoidal response of system with (a) 65% plant-only feed-forward and (b) 94% feed-forward with power converter compensation.

The command response Bode plot of Figure 8-10 shows the improvement offered by compensating for the power converter. The plant-only feed-forward with a gain of 65% provides a 275-Hz bandwidth, which is a significant improvement over the system with 0% feed-forward. However, the power-converter-compensated system with 94% feed-forward provides a 1700-Hz bandwidth, a considerable improvement.

8.2.2 Increasing the Bandwidth vs. Feed-Forward Compensation

The effects on command response of raising the power converter bandwidth are similar to using the bi-quad filter for cancellation. Given a choice between raising the power converter bandwidth or compensating for it in the feed-forward path, designs can benefit more by raising the bandwidth. First, a higher bandwidth will support higher loop gains and thus superior disturbance response. The feed-forward path is outside the loop and thus provides no response to disturbances. Also, the actual power converter is known only to a limited accuracy. For example, the dynamic performance of a power converter may vary over its operating range. Variation between the power converter and the compensation function in the feed-forward path will degrade performance, especially when high feed-forward gains are used. This topic will be covered in a later section.

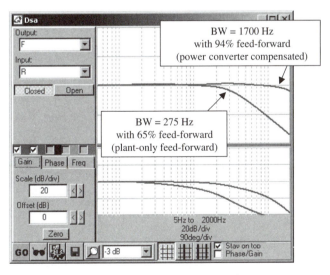

Figure 8-10. Bode plot of command response showing improvement from power converter compensation.

8.3 Delaying the Command Signal

The technique of the previous section took into account the behavior of the power converter by placing its approximate inverse transfer function in the feed-forward path. Another common problem in calculating feed-forward is the presence of a time delay in the control system. There is no practical means of calculating a time advance, which would be the inverse transfer function. As a result, an alternate technique, that of delaying the command signal, will be presented.

Consider the system of Figure 8-11. It is the system of Figure 8-1 augmented with a block to modify the command path ($G_R(s)$). In addition, the feedback function, $H(s)$, is assumed to be a delay, $H_D(s)$. Using Mason's signal flow graphs, the transfer function is developed and shown in Equation 8.11.

$$T_{IDEAL}(s) \approx \frac{(K_F G_F(s) + G_R(s)G_C(s))G_{PC}(s)G_P(s)}{1 + G_C(s)G_{PC}(s)G_P(s)H_D(s)} = 1 \tag{8.11}$$

$$(K_F G_F(s) + G_R(s)G_C(s))G_{PC}(s)G_P(s) = 1 + G_C(s)G_{PC}(s)G_P(s)H_D(s) \tag{8.12}$$

which can be satisfied if $K_F = 1$, $G_F(s) = G_{PC}^{-1}(s)G_P^{-1}(s)$ and $G_R(s) = H_D(s)$.

The delay represented in $H_D(s)$ comes from at least two sources. First, as discussed in Chapter 4, digital systems have an effective delay of half the sample interval, due to the sample-and-hold. A second source is the implicit delay caused by cascading digital differentiation in the feed-forward calculations and analog integration in the plant. This produces another half sample of delay, similar to velocity estimation (Section 4.2.3). Together, the two sources add to a delay of one sample; that total is assumed to be in the feedback path, for simplicity. Accordingly, setting $G_R(s) = 1/z$ should approximately compensate for the effect.

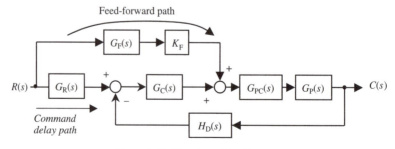

Figure 8-11. Plant-based feed-forward.

8.3.1 Experiment 8C: Command-Path Delay

Experiment 8C, shown in Figure 8-12, will be used to experiment with command signal delay. This is similar to Experiment 8A, with a few exceptions. The power converter bandwidth was raised from 350 Hz to 2 kHz while the sample rate was reduced from 10 kHz to 1 kHz. This was done to make loop delay the dominant problem where, in Experiment 8B, the power converter was dominant. (To demonstrate this with a rough comparison, a 2000-Hz filter has a 90° phase lag, equivalent to 1/4 cycle at 2000 Hz; the delay at that frequency is then $1/4 \times 0.5$ msec, or 0.125 msec. This is considerably smaller than 1 msec, a one-interval delay with 1-kHz sampling.) So the sample delay is dominant, shifting the focus to the delay problem.

The feed-forward path in Experiment 8C is based wholly on the plant; power converter compensation is not necessary here since delay is the dominant problem. The gain K_F defaults to 94%. The *Live Scope* shows the step response, which has substantial overshoot ($\sim 30\%$). This overshoot results almost entirely from the sample delay, which the interested reader can confirm by running Experiment 8C, setting *TSample* (which is shown in the lower right of Figure 8-12) to 0.0001 sec, and noticing the near-total elimination of the overshoot. A *Live Switch* provides the means to reconfigure the system to place a delay of one sample in the command path. When this delay is switched in, the feed-forward gain can be set to 94% without inducing significant overshoot.

Figure 8-13 shows the step response of the system with $K_F = 94\%$. Figure 8-13a is the system without a delay in the command path; Figure 8-13b has the delay. The improvement

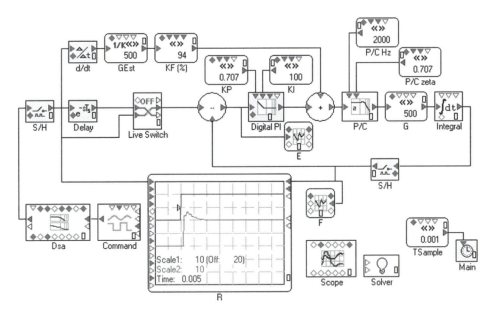

Figure 8-12. Experiment 8C: Delaying the command signal to improve system response.

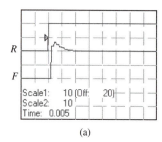
(a)

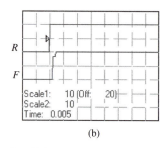
(b)

Figure 8-13. Step response of system with 94% feed-forward and (a) plant-only feed-forward and with (b) feed forward and a delay in the command path.

in following the square wave is dramatic. Figure 8-14 shows a similar comparison for the trapezoidal command response, again amplified so return to zero is shown in greater detail. Overshoot is removed and the system follows the command almost flawlessly.

8.3.2 Experiment 8D: Power Converter Compensation and Command Path Delay

Experiment 8D, shown in Figure 8-15, shows a system with delays from the power converter and sampling at about the same order of magnitude. This allows experimentation with both effects present in significant amounts. The system is similar to other experiments in this chapter. The power converter bandwidth is 350 Hz (a delay of approximately 1/4 cycle at 350 Hz, or about 0.7 msec); the sample time is set to 1.0 msec, so a single-interval delay is 1 msec. Two *Live Switches* allow for the inclusion of power conversion compensation and command path delay independently. Figure 8-15 shows the step response, with the feed-forward calculations augmented for both effects (i.e., with both *Live Switches* ON).

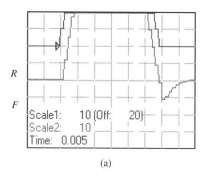
(a)

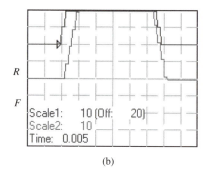
(b)

Figure 8-14. Trapezoidal response of system with 94% feed-forward and (a) plant-only feed-forward and (b) a delay in the command path.

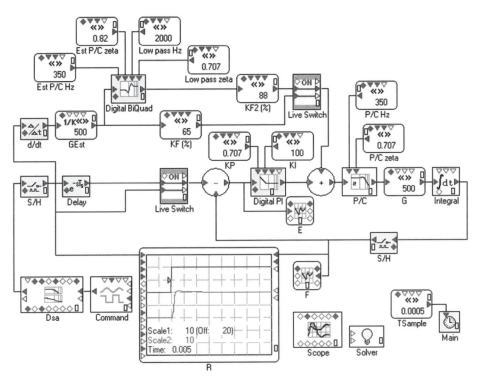

Figure 8-15. Experiment 8D, feed-forward in a system with significant delays from both power-conversion and sampling.

The step response of Figure 8-15 shows some overshoot. However, the trapezoidal response is probably more valuable in determining the ability of the system to follow an aggressive command. The trapezoidal response of the system with plant-only feed-forward gain (both *Live Switches* off) and a feed-forward gain of 65% is shown in Figure 8-16a. This produces an overshoot of almost 10 units to a command of 100, almost 10%. Raising K_F above or below 65% increases the overshoot. Figure 8-16b shows the system with both a command path delay of one cycle and power converter compensation for a 350-Hz bi-quad filter. Here, overshoot is nearly eliminated.

The Bode plot of command response for Experiment 8D is shown in Figure 8-17. Here, the system with plant-only feed-forward and $K_F = 65\%$ is shown with a bandwidth of about 300 Hz and 1.3-dB peaking at about 65 Hz. The system with both command path delay and power converter compensation has a bandwidth of 435 Hz and no peaking below 200 Hz. This accounts for the superior performance of the trapezoidal response shown in Figure 8-16b.

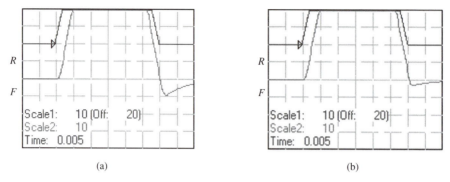

Figure 8-16. Trapezoidal response of system with (a) plant-only feed-forward and 65% K_F vs. (b) command-delayed and power-converter-compensated feed-forward and 88% K_F.

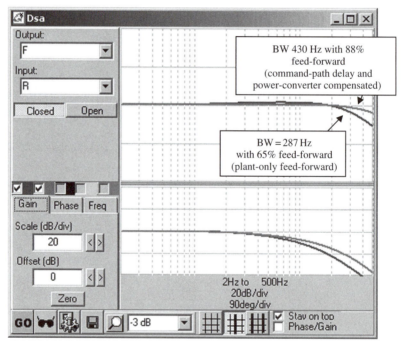

Figure 8-17. Bode plot system with (a) plant-only feed-forward (K_F = 65%) vs. (b) command-delayed and power-converter-compensated feed-forward (K_F = 88%).

8.3.3 Tuning and Clamping with Feed-Forward

Tuning refers to setting loop gains. Feed-forward gains are outside the loop and do not directly affect margins of stability. However, feed-forward gains can mask both

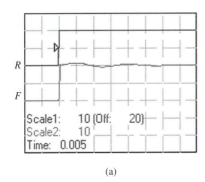
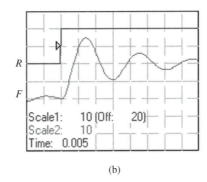

(a)   (b)

Figure 8-18. When $K_P = 0.42$ and $K_I = 707$, (a) feed-forward hides tuning problems, a condition rectified by (b) turning feed-forward off.

low margins of stability and slow loop response. For example, Figure 8-18 shows Experiment 8B where the control law gains are set poorly: $K_P = 0.42$ and $K_I = 707$. In Figure 8-18a, the step response appears to be acceptable. However, when all feed-forward is removed (turn both *Live Switches* off and set $K_F = 0\%$), signs of marginal stability are obvious.

Feed-forward in Figure 8-18a masked the tuning problem because the feed-forward path provided nearly all the response to the command. Low margins of stability are evident only in proportion to the signal exiting the control law. By temporarily eliminating feed-forward, as in Figure 8-18b, the control law provided the entire plant excitation, making tuning problems evident. This teaches the principle that feed-forward gains should be zeroed or, at least, greatly reduced when tuning. After tuning is complete, feed-forward gains can be reinstated.

The issue of clamping in Experiments 8B–8D should also be mentioned. Since feed-forward is added outside the control law, neither clamping nor windup control is ensured (clamping and windup control are an implicit part of all *Visual ModelQ* control laws). If these functions increase the power converter command above saturation levels, the model power converter would follow its command where a practical system would be limited. This will have a significant effect, especially on large-amplitude commands; should this be a concern, augment the model by adding a clamp function just left of the power converter.

8.4 Variation in Plant and Power Converter Operation

Thus far in this chapter, the feed-forward path has been calculated based on perfect knowledge of the plant and power converter. The scaling gain G_{PEst} can be set precisely to G in a model; in practical systems, G is not always known with great accuracy. Similarly, the power converter operation is not always predictable. This section will investigate some of the effects that proceed from imperfect knowledge.

8.4.1 Variation of the Plant Gain

What if the plant gain varied during normal operation, such as inductor saturation, loss of thermal mass in a temperature controller, or inertia decline in a motion controller? In all cases, the actual plant gain ($G_P(s)$) would increase. However, assuming this change is unknown to the controller, the feed-forward gain, $G_F(s) = G_{P\,Est}^{-1}(s)$, would remain unchanged. An uncompensated increase in the actual plant gain of just 20% can cause significant overshoot. This is demonstrated in Figure 8-19, which is from Experiment 8D; Figure 8-19a is Experiment 8D fully compensated but with the plant gain 20% low ($G = 400$), and Figure 8-19b has the gain set 20% high ($G = 600$). Neither system follows the trapezoid as well as the system with accurate gain estimation (Figure 8-16b), although both still follow better than the system with simple plant-based feed-forward, as the interested reader can confirm with Experiment 8D.

Increasing the gain of the actual plant relative to the modeled plant is equivalent to raising the feed-forward gain, K_F. This can be seen in Equations 8.13 and 8.14 by observing that raising $G_P(s)/G_{PEST}(s)$ is equivalent to raising K_F. These equations assume the feed-forward gain ($G_F(s)$) is set to the estimated inverse plant ($G_{P\,EST}^{-1}(s)$). From Equation 8.3,

$$T(s) \approx \frac{\left(K_F G_{P\,EST}^{-1}(s) + G_C(s)\right) G_P(s)}{1 + G_C(s) G_P(s)} \tag{8.13}$$

$$\approx \frac{K_F \frac{G_P(s)}{G_{P\,EST}(s)} + G_C(s) G_P(s)}{1 + G_C(s) G_P(s)} \tag{8.14}$$

The simple correction to dealing with changing plant dynamics that are unknown to the controller is to reduce the feed-forward scaling (K_F). Setting K_F in the operating condition with the highest plant gain will ensure reasonable feed-forward gains across the operating range. Of course, if the variation of the plant is known, it can be used to rescale K_F as the system runs.

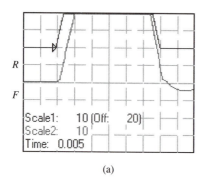

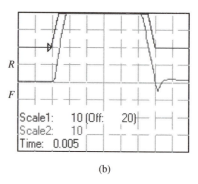

(a)   (b)

Figure 8-19. Experiment 8D, fully compensated, with plant gain (a) 20% low and (b) 20% high (compare to Figure 8-16b).

8.4.2 Variation of the Power Converter Operation

As with the plant gain, the behavior of the power converter is often known with limited accuracy. Inaccuracy can result from lack of measurements, unit-to-unit variation, and variation during normal operation. In each case, the power converter compensation function is only an approximation to the actual power converter. Fortunately, the effect of variation in the power converter is often modest.

Figure 8-20 shows the fully compensated feed-forward function of Experiment 8D with the plant gain set at its nominal value (500) and the power converter bandwidth varied from its nominal value. Figure 8-20a shows the response to a trapezoid with the power converter bandwidth about 25% lower than nominal (250 Hz); Figure 8-20b shows the response with the bandwidth about 40% high (500 Hz). Both overshoot more than the system with nominal power converter gains (Figure 8-16a), but both still perform better than the system with plant-only feed-forward.

As has been demonstrated in this section, command response will degrade when feed-forward parameters vary from actual system values. Designers wishing to maximize command response should make every effort to measure these values where practical and to adjust gains to system operation. However, the performance of systems can be significantly improved, even when there is significant variation between feed-forward gains and system parameters.

8.5 Feed-Forward for the Double-Integrating Plant

The previous discussion shows feed-forward for the single-integrating plant. When feed-forward is applied to systems with a double-integrating plant, such as a motor with position control, the feed-forward function ($G_F(s)$) becomes double differentiation. For a motor, the plant ($G_P(s)$) would be K_T/Js^2, so $G_F(s)$ would be Js^2/K_T. The appearance of a

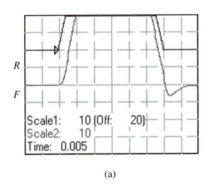

(a)

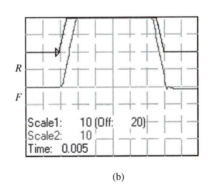

(b)

Figure 8-20. Experiment 8D, fully compensated, with power converter bandwidth set (a) low (250 Hz) and (b) high (500 Hz).

second derivative causes some concern; differentiation is a process so noisy that it is often impractical, and differentiating twice seems unrealistic. However, this is not the case here.

In most motion-control applications, the command is generated by the controller; it is completely known and, for digital controllers, can be produced almost without noise. In fact, the command and its derivatives are often produced in reverse order: the highest-order derivative can be produced first and integrated, sometimes multiple times, to produce the command. For example, a position profile calculated for a motor is often calculated by first determining acceleration, integrating this to create the velocity command, and then integrating velocity to create the position command. So the second derivative of the position command, the acceleration command, can be provided without explicit differentiation. Bear in mind that the second derivative in a double-integrating plant may be comparable to the first derivative of a single-integrating plant. This stands opposed to Equation 8.9, where the notion of taking the three derivatives was rejected; that would compare to two derivatives beyond acceleration.

Most of the discussion concerning feed-forward in single-integrating plants can be applied to double-integrating plants. When full feed-forward is used in double-integrating plants, the system will overshoot. The overshoot is caused by imperfect knowledge of the plant, feedback, and power converter. Again, the normal solution is to reduce the feed-forward gains to eliminate overshoot. As before, other solutions are to reduce the imperfections of the power converter and feedback as much as possible, and to compensate for those imperfections that remain. Chapter 17 will discuss feed-forward in motion-control systems in detail.

8.6 Questions

1. a. If the power converter and feedback signals are ideal, what is the command response of a control system using full feed-forward? *Hint: Use Equation 8.1.*
   b. Why is full feed-forward impractical in most systems?
2. For the given simplified control loop:

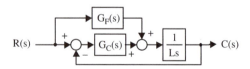

   a. What is the ideal feed-forward function if $G_C(s)$ is a PI control law?
   b. What if $G_C(s)$ is a PID control law?
3. What is the effect of full feed-forward gains on disturbance response? on phase margin and gain margin?

4. For the system of Experiment 8A:
   a. Set the sample rate at 1 kHz (TSample = 0.001) and zero feed-forward. Tune the PI controller so KV is at the maximum value without overshoot and KI causes about 10% overshoot to a square wave. What are the gains?
   b. What is the bandwidth? *Note: Since the controller samples at 1 kHz, aliasing begins at 500 Hz. Since the FFT shows frequencies up to 2 kH, you will see the effects of aliasing; concentrate on the frequencies below 500 Hz.*
   c. Repeat parts a and b for a sample time of 0.0001 seconds.
   d. What is the relationship between TSample and bandwidth?
5. For the system of Experiment 8A, install the values of gains from 4a.
   a. What is the maximum feed-forward gain that causes no more than 20% overshoot.
   b. What is the bandwidth?
   c. Compare the results of adding feed-forward to raising the sample rate.

Chapter 9

Filters in Control Systems

Filters are used extensively in control systems. They reduce noise, eliminate aliasing, and attenuate resonances. The most common filters in control systems are low-pass filters. They are used to remove noise from a variety of origins: electrical interconnections, resolution limitations, EMI, and intrinsic noise sources in feedback devices. As discussed in Chapter 3, the chief shortcoming of low-pass filters is the instability they induce by causing phase lag at the gain crossover frequency. Notch filters are also used in control systems but less regularly; their effects, both positive and negative, are more limited than those of low-pass filters.

Most discussions of filters in the technical literature relate to their application in communication systems. The traditional measures of filter performance, such as how much distortion is introduced and how quickly the filter makes the transition from passing a signal to blocking it, are oriented toward communication applications. Control systems engineers are concerned with different issues. Usually, controls engineers want to attenuate high-frequency signals while generating minimal phase lag at the gain crossover frequency.

This chapter will consider filters from the point of view of the control system. The chapter begins with a discussion of the many uses for filters throughout the system. This is followed by a presentation of several common filters and how to implement these filters in the analog and digital domains.

9.1 Filters in Control Systems

Filters are used throughout a control system. As shown in Figure 9-1, they are present in the controller, the feedback devices, and the power converter. Even the plant will sometimes include a filter.

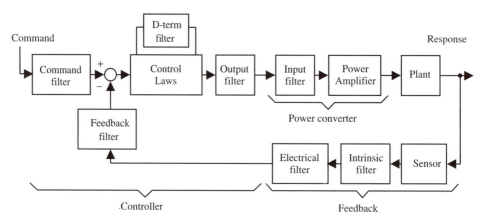

Figure 9-1. Seven common filters in a control system.

9.1.1 Filters in the Controller

The controller makes the most extensive use of filters. They are used to remove noise, reduce aliasing, and calm resonances. Sometimes the filter characteristics are fixed and other times they can be adjusted as part of a tuning procedure. The more flexible the controller filters, the more options you will have when commissioning a control system.

9.1.1.1 Using Low-Pass Filters to Reduce Noise and Resonance

Low-pass filters attenuate all signal components above a specified frequency; the Bode plot for such a filter is shown in Figure 9-2, which is taken from Experiment 9A (not shown). Most control system filters are low-pass filters that are designed to reduce high-frequency noise. Such filters can be applied to the command or feedback signals as they enter the control system or to elements of the control law that are particularly susceptible to noise, especially derivative ("D") terms.

Low-pass filters can also be used to remove resonance, a phenomenon in which a plant has high gain around one frequency called the *resonant frequency*. If the resonance occurs where the loop phase is near 180°, the increased gain at resonance reduces gain margin; in fact, sometimes all the gain margin can be removed, causing instability to manifest itself as sustained oscillations at or near the resonant frequency. Electrical resonance commonly occurs in current and voltage controllers; inductances and capacitances, either from components or from parasitic effects, combine to form L-C circuits that have little resistive damping. Also, mechanical resonance occurs in motion control systems; this is discussed in Chapter 16.

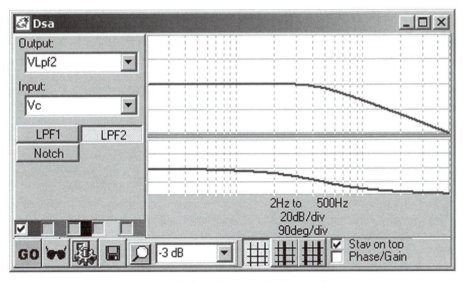

Figure 9-2. Bode plot for a two-pole low-pass filter.

9.1.1.2 Using Low-Pass Filters to Reduce Aliasing

Low-pass filters are also used in the controller to prevent aliasing in an analog command or feedback signal. As discussed in Section 5.3, aliasing is a phenomenon in which frequencies higher than half the sample frequency alias down to lower frequencies. The goal of antialiasing filters is to eliminate nearly all frequency content above $0.5 \times F_{SAMPLE}$. Often, the main concern for aliasing is in the command. This is because high-frequency noise can alias down to low frequencies where the control system will respond. For example, for a system sampled at 1 kHz, a 100,001-Hz noise signal would alias down to 1 Hz, where the effects of even small-amplitude noise will be noticeable.

Aliasing in feedback signals may be of less concern than in the command. One reason is that the control loop output frequency is limited to half the sample frequency; the control system cannot normally produce output that causes sustained aliasing. Also, if the feedback device and controller are from a single vendor, that vendor can thoroughly characterize the feedback noise. If a feedback sensor is known to have a content in a certain frequency range, filters can be designed to attenuate just that range before sampling occurs. Because the command signal often comes from outside a vendor's product, noise content of the command signal may not be as well characterized; as a result, antialiasing filters for the command will usually be more severe than those for the feedback. Finally, the phase lag induced by antialias filters in the command path causes fewer problems; command filtering slightly reduces command response, whereas the phase lag induced by filters in the feedback path is within the control loop and thus contributes to instability.

9.1.1.3 Using Notch Filters for Noise and Resonance

Notch filters are also used in controllers, albeit less regularly. Where low-pass filters attenuate all signals above a specified frequency, notch filters remove only a narrow band of frequencies; as seen in the Bode plot of Figure 9-3, notch filters pass the frequency components below and above the notch frequency. The fact that notch filters pass high frequencies leads to their strongest attribute: they usually cause little phase lag at the gain crossover frequency, assuming the notch frequency is well above that. Notch filters can be useful on the command for a fixed-frequency noise source such as that from line frequency (50 or 60 Hz) noise. Notch filters are also used to remove resonances from the system. Both notch and low-pass filters can cure resonance; notch filters do so while creating less phase lag in the control loop.

Although notch filters improve the primary shortcoming of low-pass filters (the reduction of phase margin), they are still used less regularly. Notch filters work on only a narrow band of frequencies. To be useful, the notch filter must be tuned to the frequency of resonance or of noise generation. If the offending frequency is known, digital notch filters can be set to filter it with great accuracy. However, there are many cases where the noise or resonant frequency will vary. For example, resonant frequencies often vary slightly from one system to another. In such a case, the notch filter may have to be manually tuned for every control system. In the case of analog systems, notch filters become more complicated to configure. Because the value of analog passive components varies from one unit to another and, to a lesser extent, over time and temperature, analog notch filters often must be "tweaked in" on every application. Worse yet, time and temperature variation may force adjustment of the filter frequency after the controller is placed in operation.

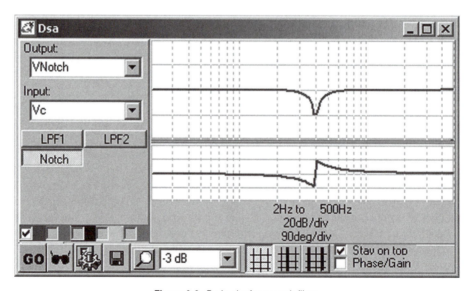

Figure 9-3. Bode plot for a notch filter.

9.1.2 Filters in the Power Converter

Power converters often have low-pass filters at the input of the power command signal. This may be a simple R-C circuit added to reduce the effect of noise on the power converter command signal. Various filters can also be applied within the power amplifier to reduce circuit noise.

Another role of filters in regard to power conversion is found in modeling. Power converters are often closed-loop controllers themselves, such as when a current controller is used to create torque in a motion control system. Here, the power converter can be modeled as a filter, assuming a filter can be designed with similar phase and gain characteristics over the frequency range of interest. This is the basis for the power converter model used in many of the *Visual ModelQ* experiments in this book.

9.1.3 Filters in the Feedback

Feedback devices can include both explicit and implicit filters. Some physical processes act like filters. For example, many temperature sensors, such as thermistors, have thermal inertia. When the object being measured changes temperature, it takes time for the sensor's thermal mass to change. That delay causes the sensor to behave as if it were in line with a low-pass filter.

Many electrical and electronic sensors use active amplification to convert low-voltage signals, often at levels of millivolts or even microvolts, to levels appropriate for transmission to the controller. Amplification is subject to noise. Such sensors commonly include filters to reduce noise.

Also, like power converters, sensors can incorporate embedded closed-loop controllers. For example LEM (www.lem.com) manufactures a family of such current sensors. Resolvers, which are position-sensing devices used in motion systems (see Section 14.3), also use closed-loop techniques to convert the sensor's raw output into a form directly usable by the controller. These closed-loop sensors behave as if they were in line with a low-pass filter that had a bandwidth equivalent to that of the sensor conversion loop.

If you are modeling your control system, you need to model accurately the effects of filters in the feedback sensors. Usually the sensor vendor can provide information on the speed of the feedback sensor; at the very least, you should expect the bandwidth of the sensor to be specified. The overall sensor should be modeled as the relationship from the physical effect to the sensor output.

9.2 Filter Passband

The four common filter passbands are low-pass, notch, high-pass, and bandpass. This chapter will discuss low-pass and notch filters, since they are more common in control systems.

9.2.1 Low-Pass Filters

Low-pass filters attenuate all frequencies above a specific frequency. Low-pass filters are often characterized by their bandwidth, the frequency at which the signal is attenuated by 3 dB. The transfer functions of most low-pass filters have poles but no zeros. The order of the filter is equivalent to the number of poles.

9.2.1.1 First-Order Low-Pass Filters

The simplest filters are single-pole low-pass filters. The transfer function is shown in Equation 9.1.

$$T_{1-\text{POLE}}(s) = \frac{K}{s + K} \tag{9.1}$$

Note that the gain of Equation 9.1 at DC ($s = 0$) is 1. All filters discussed in this chapter will have unity DC gain. The filter time constant, τ, is equal to $1/K$. Single-pole filters are sometimes written in terms of the time constant:

$$T_{1-\text{POLE}}(s) = \frac{1}{\tau s + 1} \tag{9.2}$$

The value of the pole, K, is equal to the filter bandwidth in rad/sec; in other words, the gain at $f = K/(2\pi)$ Hz is -3 dB.

9.2.1.2 Second-Order Low-Pass Filters

Two-pole filters have an s^2 term in the denominator. The poles are usually complex (as opposed to real or imaginary). The two-pole filter is commonly given in the form of Equation 9.3:

$$T_{2-\text{POLE}}(s) = \frac{\omega_N{}^2}{s^2 + 2\zeta\omega_N s + \omega_N{}^2} \tag{9.3}$$

The poles of this equation are $s = -\zeta\omega \pm j\omega_N\sqrt{1 - \zeta^2}$, where ω_N is the natural frequency of the filter and ζ is the damping ratio. Below $\zeta = 1$, the filter poles are complex. In this case, the filter will overshoot in response to a step command. Lower ζ causes more overshoot; when ζ is low enough, the step response will have pronounced ringing. Also, below $\zeta = 0.707$, the filter will have peaking in the gain; the lower ζ is below 0.707, the more peaking it will demonstrate. Figure 9-4 shows the step response of a two-pole filter with a damping ratio of 1.0 and of 0.4. Figure 9-5 shows the Bode plot of the same filters.

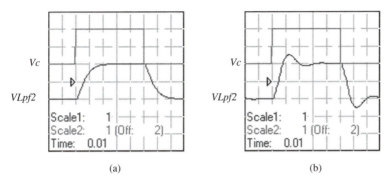

Figure 9-4. Step response of a two-pole low-pass filter with (a) $\zeta = 1$ and (b) $\zeta = 0.4$.

Two-pole low-pass filters are commonly used in control systems. The damping ratio ζ is often set to 0.707 because it is the lowest damping ratio that will not generate peaking in a two-pole low-pass filter. Two-pole filters with lower damping ratios ($\zeta < 0.707$) are sometimes preferred because these filters can provide the same attenuation at a given high frequency while inducing less phase lag at a lower frequency. This measure of filters, balancing the attenuation at high frequency against the phase lag at low frequency, is appropriate for control systems. According to this standard, filters with lower damping ratios perform better in control systems.

For example, consider using a two-pole filter to achieve at least 20-dB attenuation for all frequencies above 200 Hz while inducing the minimum phase lag at 20 Hz, the gain cross-over frequency. If the damping ratio is set to 0.7, the minimum filter bandwidth (ω_N) to meet this criterion is 63 Hz (this was determined experimentally using Experiment 9A).

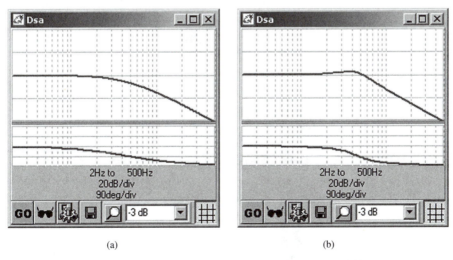

Figure 9-5. Bode plot of a two-pole low-pass filter with (a) $\zeta = 1$ and (b) $\zeta = 0.4$.

The phase lag at 20 Hz is $-26°$. A one-pole filter or a two-pole filter with a larger damping ratio produces poorer results. For a single-pole filter, the settings to achieve at least 20-dB attenuation for all frequencies above 200 Hz imply that the bandwidth is just 21 Hz (again, this was determined experimentally). The resulting phase lag at 20 Hz is 43°. For two identical one-pole filters in cascade, 20 dB of attenuation at 200 Hz implies a bandwidth of 66 Hz. This filter produces 33° phase lag at 20 Hz. However, if the damping ratio is set to 0.4, the filter bandwidth moves to 60 Hz, and the phase lag is just 16°. These results are shown in Table 9-1. (Note that having two single-pole filters with the same bandwidth is equivalent to a two-pole filter at the same bandwidth with a damping ratio of 1.0.)

When the damping ratio falls below 0.707, the two-pole filter will have peaking. If the peaking is large enough, it can reduce the gain margin. When two-pole filters are used, a damping ratio between 0.4 and 0.7 often works well. However, single-pole filters are easier to implement; in analog form they can be created with two passive components, and in digital form they require fewer calculations. As a result, both one-pole and two-pole filters are commonly used in controllers.

9.2.1.3 A Simple Model for a Closed Loop System

The two-pole low-pass filter with a damping ratio less than 1 is the simplest filter that models effects of marginal stability: overshoot and ringing. As a result, two-pole filters are frequently used as a model of a closed-loop control system. For example, engineers will often characterize a control system as having a *damping ratio*. Broadly speaking, this implies that the controller has a command response similar to that of a two-pole filter with a specified bandwidth and damping ratio. Such a comparison is limited; closed-loop control systems are more complex than a two-pole filter. However, the comparison provides a useful shorthand.

9.2.1.4 Higher-Order Low-Pass Filters

High-order filters are used because they have the ability to roll off gain after the bandwidth at a sharper rate than low-order filters. The attenuation of a filter above the bandwidth grows proportionally to the number of poles. When rapid attenuation is required, higher-order filters are often employed.

TABLE 9-1 PHASE LAG FOR DIFFERENT FILTERS WITH 20-dB ATTENUATION AT 200 Hz

Filter order	Damping ratio	Filter bandwidth	Attenuation at 200 Hz	Phase lag at 20 Hz
1	N/A	21 Hz	20 dB	43°
2	1.0	66 Hz	20 dB	33°
2	0.7	63 Hz	20 dB	26°
2	0.4	60 Hz	20 dB	16°

The *s*-domain form of higher order filters is

$$T(s) = \frac{A_0}{s^M + A_{M-1}s^{M-1} + \cdots + A_1 s + A_0} \tag{9.4}$$

where M is the number of poles (the order) and A_0 through A_{M-1} are the coefficients of the filter.

High-order filters are often shown in the cascade form (that is, as a series of two-pole filters). Both analog and digital filters can be built as a series of two-pole filters (for a complex or real pole pair) and an optional single-pole filter. This requires an alternative form of the transfer function as shown in Equation 9.4. Equation 9.5 shows the form for an even number of poles; for an odd number of poles, a single pole can be added in cascade.

$$T(s) = \left(\frac{C_1}{s^2 + B_1 s + C_1}\right)\left(\frac{C_2}{s^2 + B_2 s + C_2}\right) \cdots \left(\frac{C_{M/2}}{s^2 + B_{M/2}s + C_{M/2}}\right) \tag{9.5}$$

The form of Equation 9.5 is preferred to the form of Equation 9.4 because small variations in the coefficients of Equation 9.4 can move the poles enough to affect the performance of the filter substantially [72]. In analog filters, variation comes from value tolerance of passive components; in digital filters, variation comes from resolution limitation of microprocessor words. Variations as small as 1 part in 65,000 can affect the performance of large-order filters.

An alternative to the cascaded form is the *parallel* form, where Equation 9.5 is divided into a sum of second-order filters, as shown in Equation 9.6. Again, this is for an even number of poles; a real pole can be added to the sum to create an odd-order filter.

$$T(s) = \frac{D_1}{(s^2 + B_1 s + C_1)} + \cdots + \frac{D_{M/2}}{(s^2 + B_{M/2}s + C_{M/2})} \tag{9.6}$$

Both the cascaded form and the parallel form have numerical properties superior to the direct form of Equation 9.4 for higher-order filters. For more on the subject of sensitivity of higher-order filters, see Refs. 1, 16, 45, and 72.

9.2.1.5 Butterworth Low-Pass Filters

Butterworth filters are called *maximally flat* filters because, for a given order, they have the sharpest roll-off possible without inducing peaking in the Bode plot. The two-pole filter with a damping ratio of 0.707 is the second-order Butterworth filter. Butterworth filters are used in control systems because they do not induce any peaking. The requirement to eliminate all peaking from a filter is conservative. Allowing some peaking can be beneficial because it allows equivalent attenuation with less phase lag in the lower frequencies; this was demonstrated in Table 9-1. Still, the Butterworth filter is a natural selection for organizing the many poles of higher order filters used in control systems.

The general formula for Butterworth filters depends on whether the order is odd or even. For odd orders, the formula is

$$T(s) = \left(\frac{\omega_N}{s+\omega_N}\right)\prod_1^{(M-1)/2}\left(\frac{\omega_N^2}{s^2+2\cos(\theta_i)\omega_N s+\omega_N^2}\right), \quad \theta_i = i \times 180/N \quad (9.7)$$

(The symbol $\prod$ indicates a series of products, just as Σ indicates a sum.) M indicates filter order. For example, a fifth-order Butterworth filter is

$$T(s) = \left(\frac{\omega_N}{s+\omega_N}\right)\left(\frac{\omega_N^2}{s^2+2\cos(36°)\omega_N s+\omega_N^2}\right)\left(\frac{\omega_N^2}{s^2+2\cos(72°)\omega_N s+\omega_N^2}\right)$$

For even orders, the formula is

$$T(s) = \prod_1^{M/2}\left(\frac{\omega_N^2}{s^2+2\cos(\theta_i)\omega_N s+\omega_N^2}\right), \quad \theta_i = (i-0.5) \times 180/N \quad (9.8)$$

For example, a fourth-order Butterworth filter is

$$T(s) = \left(\frac{\omega_N^2}{s^2+2\cos(22.5°)\omega_N s+\omega_N^2}\right)\left(\frac{\omega_N^2}{s^2+2\cos(67.5°)\omega_N s+\omega_N^2}\right)$$

In addition to Butterworth filters, there are various high-order filters, such as Bessel (*linear phase or constant time delay*) filters, Chebyshev filters, and elliptical (Cauer) filters. These filters are used only occasionally in control systems and so are beyond the scope of this book. References 1, 16, 45, and 72 present detailed discussions of filters.

9.2.2 Notch

The traditional notch or *bandstop* filter is second order, with two poles and two zeros. The form is

$$T(s) = \left(\frac{s^2+\omega_N^2}{s^2+2\zeta\omega_N s+\omega_N^2}\right) \quad (9.9)$$

The attenuation of the ideal notch filter at the notch frequency, ω_N, is complete ($-\infty$ dB). This can be seen by observing that the gain of Equation 9.9 will be zero when the input frequency is ω_N rad/sec. This is because $s = j\omega_N$ ($j = \sqrt{-1}$), so $s^2 = -\omega_N^2$ forces the numerator of Equation 9.9 to 0.

Normally, the damping ratio (ζ) in a notch filter is set below 1.0 and often is much lower. The lower the damping ratio, the sharper the notch. This is shown in Figure 9-6a, which has a lower damping ratio and a sharper notch than Figure 9-6b (0.25 and 1.0, respectively.) The lower damping ratio not only causes a sharper notch but also reduces the phase lag caused below the notch center frequency. For example, the phase lag at 30 Hz (the first division to the left of the 40-Hz notch center) is smaller for the lower damping ratio (Figure 9-6a vs. b).

Note that in Figure 9-6, the high-damping-ratio notch appears to attenuate more at the notch center than does the low-damping-ratio filter. Actually, attenuation at the notch center is about 130 dB for both filters. The resolution of the DSA data causes the illusion that the sharper notch has more attenuation. For the model used to generate Figure 9-6, the attenuation was measured by applying a 40-Hz sine wave and using the scope to measure the attenuation. Although the ideal notch filter would produce $-\infty$ dB, practical notch filters pass some of the input. One reason notch filters do not perfectly attenuate the signal at the notch frequency is numerical imprecision in the filter coefficients.

Another observation that can be made from Equation 9.9 is that when the applied frequency is significantly above or below ω_N, the numerator and denominator of Equation 9.9 are nearly equal, so the gain is approximately unity. When the input frequency is much larger than ω_N, the s^2 term will dominate both the denominator and numerator, so the transfer function is approximately s^2/s^2, or unity. Similarly, when the input frequency is very small, the s terms in Equation 9.9 will vanish, leaving only ω_N^2/ω_N^2, or, again, unity. Only when the frequency is near ω_N does the notch filter have significant effect. This behavior can be seen in the near-unity gain of the notch in Figure 9-6 at frequencies away from the notch frequency.

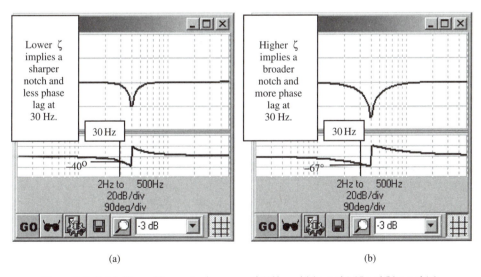

(a)   (b)

Figure 9-6. Notch filters with a center frequency of 40 Hz and (a) a ζ of 0.25 and (b) a ζ of 1.0.

Notch filters rarely have more than two poles. This is because the numerator zeros attenuate frequencies near ω_N, so a higher-order denominator is unnecessary. Although there are limits to the attenuation a practical two-pole notch filter can provide, these limits are not normally improved by increasing the filter order. In practice, a notch filter cannot cancel the notch frequency perfectly. Digital filters have round-off errors in the filter coefficients and in the mathematics. Analog filters have passive components with values that are specified only to a certain tolerance, and the active components that are common in analog filters generate low-amplitude noise at all frequencies, including the notch center frequency. In practice, two-pole notch filters attenuate the notch frequency well enough for most control system applications. However, multiple notch filters can be cascaded when there is more than one frequency that needs strong attenuation. Still, this is better thought of as multiple two-pole notch filters than as a single notch with more than two poles.

9.2.3 Experiment 9A: Analog Filters

Experiment 9A provides three filters: a single-pole low-pass, a two-pole low-pass, and a two-pole notch filter. The parameters for each of these filters can be adjusted. You can view the time-domain response on three *Live Scopes* and the frequency-domain response through the DSA. Use Experiment 9A to repeat the results of Table 9-1 using the following procedure:

1. Determine the frequency above which all components must be attenuated (e.g., 200 Hz).
2. Determine the level of attenuation (e.g., at least 20 dB above 200 Hz).
3. Specify the crossover frequency where phase lag should be minimized; it should be much lower than the frequency from step 1 (e.g., 20 Hz).
4. Using the DSA, adjust the filter bandwidth until the specified attenuation and frequency from steps 1 and 2 are met.
5. Measure the phase lag at the specified crossover frequency.
6. For two-pole filters, adjust the damping ratio (e.g., 1.0, 0.7, 0.4) and repeat.

This process will allow you to compare these filters for use in control systems. In most cases, the two-pole low-pass filter with a damping ratio of about 0.4 will be superior. However, a very low damping ratio can generate excessive peaking near the filter bandwidth. When the gain of the system is such that the filter bandwidth is near the phase crossover frequency, this peaking can reduce the gain margin.

9.2.4 Bi-Quad Filters

The most general two-pole filter is the bi-quadratic, or *bi-quad*, filter, so named because there are two quadratic terms, one in the numerator and another in the denominator. The unity-DC-gain form of the bi-quad is

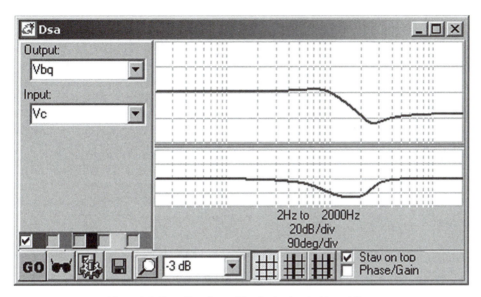

Figure 9-7. From Experiment 9B, a Bode plot of a bi-quad filter.

$$T(s) = \left(\frac{s^2 + 2\zeta_N\omega_N s + \omega_N^2}{s^2 + 2\zeta_D\omega_D s + \omega_D^2}\right)\left(\frac{\omega_D^2}{\omega_N^2}\right) \tag{9.10}$$

The filter constants, ζ and ω, are subscripted with N or D according to their position; N-terms are in the numerator and D-terms are in the denominator. The constant term at the right of the equation adjusts the DC gain ($s = 0$) for unity.

The bi-quad filter has two degrees of freedom beyond the traditional notch filter of Equation 9.9. First, the zero and pole natural frequencies can be different. Second, the damping term in the numerator (ζ_N) can be nonzero, so ω_N is not completely attenuated. The bi-quad filter can be used to form filters that have zeros, such as elliptical filters. The bi-quad filter is simulated in Experiment 9B. This was used to produce the Bode plot of Figure 9-7 ($\omega_N = 2\pi 250$, $\delta_N = 0.2$, $\omega_D = 2\pi 90$, $\delta_D = 0.4$).

9.3 Implementation of Filters

The implementation of filters falls into at least five categories: passive analog filters, active analog filters, switched capacitor filters, infinite impulse response (IIR) digital filters, and finite impulse response (FIR) filters.

9.3.1 Passive Analog Filters

Passive filters are the simplest filters to implement. The most common example is the resistor-capacitor (RC), shown in Figure 9-8. RC filters cost less than $0.02 on a surface-mounted board, take little space, and work at much higher frequency than active analog and digital filters. This explains their ubiquitous use to eliminate noise within a circuit and on incoming signals. The inductor-resistor-capacitor, or L-R-C, filter in Figure 9-8 is less commonly employed, in part because the inductor is relatively expensive and large. The key limitation of passive filters is the restricted order. An RC circuit produces one pole; the inductor adds a second. For filters with more poles, active analog or digital filters are required.

9.3.2 Active Analog Filters

The area of active filters is well studied and many topologies have been developed. References 16 and 72 provide numerous circuits for various bi-quad filters. The most common two-pole filter is probably that from Sallen and Key, which is shown in Appendix A. This appendix also includes an example of an active analog notch filter.

9.3.3 Switched Capacitor Filters

Switched capacitor technology allows high-order filters to be implemented in monolithic integrated circuits (ICs). The two-pole cascaded filters of the previous section depend on stable, accurate resistor-capacitor (RC) time constants. Unfortunately, this structure is difficult to implement on an IC. Although resistors and capacitors can be fabricated on monolithic ICs, tight tolerances are difficult to hold and the component values are nonlinear and are sensitive to temperature. In addition, passive components require a lot of silicon. The switched capacitor filter (SCF) avoids these disadvantages by using switches and op-amps, both of which

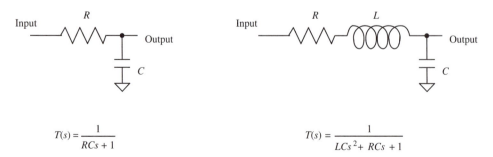

$$T(s) = \frac{1}{RCs + 1}$$

$$T(s) = \frac{1}{LCs^2 + RCs + 1}$$

Figure 9-8. RC and L-R-C filters.

can be realized in monolithic form efficiently; these active components can be combined with a small number of external passive components to create low-cost, precise, high-order filters.

SCFs are analog filters that have many of the advantages of digital filters. Filter characteristics such as the center frequency can be programmed so that changes can be made through software. These filters take less printed circuit board area. They also have less unit-to-unit variation. However, these filters are sometimes avoided in control systems because they have high DC offsets, and the DC offset is sensitive to temperature. For more information on SCFs, see Refs. 34 and 72.

9.3.4 IIR Digital Filters

Infinite impulse response, or *IIR*, filters are recursive; that is, the new output value is based in part on old values of the output. The name is based on the characterization that their response to an impulse function is nonzero for all time. For digital filters, an impulse function is a pulse that has a width of one time step but an area of 1. When such a pulse is applied to an IIR filter, the response steps up and then declines to zero. As the term *infinite* indicates, the decline may approach zero but, ignoring resolution limitations, will never reach it. The output never reaches zero but instead declines a fixed percentage for each time step.

9.3.4.1 First-Order Low-Pass IIR Filter

The single-pole low-pass filter from Table 5-1 is an IIR filter:

$$T(z) = z(1 - e^{-\omega T})/(z - e^{-\omega T}) \tag{9.11}$$

For example, if a filter is designed for a bandwidth of 200 Hz ($\omega = 2\pi \times 200 = 1256$ rad/sec) and the sample frequency is 4 kHz ($T = 0.00025$ sec), the transfer function will be

$$T(z) = z(0.2696)/(z - 0.7304)$$

Using the techniques from Section 5.5, the time-based algorithm for this filter is

$$C_N = 0.2696 \times R_N + 0.7304 \times C_{N-1}$$

After the first time step, R_N (the impulse function) falls to zero and the filter becomes a never-ending exponential decline: $C_N = 0.7304 \times C_{N-1}$.

9.3.4.2 Second-Order IIR Filter

The two-pole low-pass filter from Table 5-1 is also an IIR filter:

$$T(z) = (1 + B_1 + B_2)z^2/(z^2 + B_1 z + B_2)$$

$$B_1 = -2e^{-\zeta \omega T} \cos(\omega_N T \sqrt{1 - \zeta^2}), B_2 = e^{-2\zeta \omega_N T}$$

For example, for a 50-Hz filter with a damping ratio of 0.5 and sampling at 1 kHz, the constants would be $B_1 = -1.6464$ and $B_2 = 0.7304$. The filter could be implemented as

$$C_N = 0.0840 \ R_N + 1.6464 \ C_{N-1} - 0.7304 \ C_{N-2}$$

The second-order notch filter from Table 5-1 is

$$K(z^2 + A_1 z + A_2)/(z^2 + B_1 z + B_2)$$
$$B_1 = -2e^{-\zeta \omega_N T} \cos(\omega_N T \sqrt{1 - \zeta^2})$$
$$B_2 = e^{-2\zeta \omega_N T}$$
$$A_1 = -2\cos(\omega_N T)$$
$$A_2 = 1$$
$$K = (1 + B_1 + B_2)/(1 + A_1 + A_2)$$

The bi-quad filter is the most general second-order filter. The IIR form of a bi-quad filter is

$$K(z^2 + A_1 z + A_2)/(z^2 + B_1 z + B_2)$$
$$B_1 = -2e^{-\zeta_D \omega_D T} \cos\left(\omega_D T \sqrt{1 - \zeta_D^2}\right)$$
$$B_2 = e^{-2\zeta_D \omega_D T}$$
$$A_1 = -2e^{-\zeta_N \omega_N T} \cos\left(\omega_N T \sqrt{1 - \zeta_N^2}\right)$$
$$A_2 = e^{-2\zeta_N \omega_N T}$$
$$K = (1 + B_1 + B_2)/(1 + A_1 + A_2)$$

Note that the notch filter is a special case of the bi-quad filter where $\omega_N = \omega_D$ and $\zeta_N = 0$.

9.3.4.3 Experiment 9C: Digital Filters

Experiment 9C provides the digital equivalents of the filters of Experiment 9A: a single-pole low-pass, a two-pole low-pass, and a two-pole notch filter. The same parameters for each of these filters can be adjusted. Compare the step response and

Bode plots from these filters with those of the analog filters in Experiment 9A. In general, the results will be similar as long as the excitation frequency is below about half the sampling frequency. After that, the effects of aliasing become apparent in the digital filter. When the model is launched, the DSA analyzes only up to 500 Hz, or half the sample frequency, so aliasing is not seen. To see the effects of aliasing, increase the sample time (T_{SAMPLE}) to 0.002 and rerun the Bode plot.

9.3.4.4 Higher-Order Digital Filters

The general form for a z-domain IIR filter is

$$T(z) = K\frac{z^M + B_{M-1}z^{M-1} + \cdots + B_1z + B_0}{z^M + A_{M-1}z^{M-1} + \cdots + A_1z + A_0} \tag{9.12}$$

As was discussed with Equations 9.4 and 9.5 for the s-domain, limitations on the accuracy of the filter constant often cause designers to implement high-order filters as cascaded two-pole filters. The same effects apply to z-domain filters, causing digital filters to be implemented according to Equation 9.13. Equation 9.13 is for even-order filters; for odd-order filters, cascade a single-pole/single-zero term:

$$T(z) = K\left(\frac{z^2 + E_1z + F_1}{z^2 + C_1z + D_1}\right) \cdots \left(\frac{z^2 + E_{M/2}z + F_{M/2}}{z^2 + C_{M/2}z + D_{M/2}}\right) \tag{9.13}$$

Using Equation 9.13, higher-order filters, such as Butterworth filters, can be implemented with cascaded bi-quad filters. Filter constants are commonly available in the s-domain, such as Equations 9.7 and 9.8. The s-domain filters can be converted to the z-domain by:

1. Converting the filter to s-domain cascade form (Equation 9.5).
2. Converting the series of second-order terms to the bi-quad form (Equation 9.10) by calculating ζ_D, ω_D, ζ_N, and ω_N. Add a single-pole filter if the order is odd.
3. Implementing those bi-quad filters in the z-domain (Equation 9.13). Add the single-pole z-domain filter if the order is odd.

9.3.5 FIR Digital Filters

Finite impulse response (FIR) filters are nonrecursive filters: The output depends only on a history of input values. A simple example of an FIR filter is the moving-average filter of order M:

$$C_N = (R_N + R_{N-1} + R_{N-2} + R_{N-3} + \cdots + R_{N-M})/M \tag{9.14}$$

Moving-average filters are among the most intuitive filters; in fact, they are commonly used to smooth data by people who otherwise do not employ filters. The z-transform of the moving-average filter is

$$T(z) = \left(1 + \frac{1}{z} + \frac{1}{z^2} + \frac{1}{z^3} + \cdots + \frac{1}{z^M}\right)\bigg/ M \tag{9.15}$$

Some controls engineers have used moving-average filters but are unfamiliar with their characterization in the z-domain. In fact, these filters can be analyzed in the frequency domain just like an IIR or s-domain filter.

The general form of an FIR is similar to that of the moving-average filter except that the coefficients of the delayed input terms can vary. These terms were fixed at $1/M$ in Equation 9.15. Equation 9.16 shows the general FIR. Any IIR filter can be approximated by an FIR filter, assuming enough delayed terms of the input are available.

$$T(z) = A_0 + \frac{A_1}{z} + \frac{A_2}{z^2} + \frac{A_3}{z^3} + \cdots + \frac{A_M}{z^M} \tag{9.16}$$

The key advantage of FIRs is their ability to represent high-order filters with only modest sensitivity to coefficient quantization and without creating limit cycles [1]. This makes the FIR filters ideal for communications applications, such as telephony applications, where the sharp roll-off characteristics of high-order filters are required.

The chief disadvantage of FIRs is the larger number of terms required to implement filters. Whereas a typical IIR filter has between 4 and 10 coefficients, many FIR filters have over 100 coefficients. Each coefficient of the FIR implies memory for storing a delayed input (called a *tap*) and the need for a multiplication and an addition (*multiply-and-accumulate*). An FIR with 100 taps requires 100 multiply-and-accumulate operations. FIRs are often executed on digital signal processors (DSPs) because they require more computational power than is available from general microprocessors. In fact, the ability to provide high-speed FIRs was a key factor in the early success of the DSP.

Aside from moving-average filters, FIRs are not commonly used in control applications. The need for high-order filters is minimal in control systems, and lower-order filters can be implemented with fewer computational resources with IIR filters. If you do require an FIR, most digital filter texts provide techniques for the design of FIRs, for example, Refs. 1 and 45.

9.4 Questions

1. An input that is sampled at 5 kHz is corrupted with a noise input with a frequency of 194 kHz. What frequency will the noise alias to?

2. Design the following analog filters:
    a. Two-pole low-pass with a break frequency of 500 Hz and $\zeta = 1.0$.
    b. Two-pole notch with a break frequency of 500 Hz and $\zeta = 1.0$.
    c. Three-pole Butterworth low-pass with a break frequency of 200 Hz.
3. What steps are necessary to configure a bi-quad filter as a notch filter?
4. Which of the following are IIR filters?
    a. $C_N = 0.1R_N + 0.4R_{N-1} + 0.1R_{N-2}$
    b. $C_N = C_{N-1} + 0.1R_N + 0.4R_{N-1} + 0.1R_{N-2}$
    c. $C_N = -0.2C_{N-1} + 0.2C_{N-2} + R_N$

Chapter 10

Introduction to Observers in Control Systems

10.1 Overview of Observers

Most concepts in control theory are based on having sensors to measure the quantity under control. In fact, control theory is often taught assuming the availability of near-perfect feedback signals. Unfortunately, such an assumption is often invalid. Physical sensors have shortcomings that can degrade a control system. For example, sensors are expensive. In many cases, the sensors and their associated cabling are among the most expensive components in the system. Also, sensors reduce the reliability of control systems. In addition, some signals are impractical to measure because the objects being measured may be inaccessible for such reasons as harsh environments. Finally, sensors usually induce significant errors, such as noise, deterministic errors, and limited responsiveness.

Observers can be used to augment or replace sensors in a control system. Observers are algorithms that combine sensed signals with other knowledge of the control system to produce *observed* signals. These observed signals can be more accurate, less expensive to produce, and more reliable than sensed signals. Observers offer designers an inviting alternative to adding new sensors or upgrading existing ones.

The principle of an observer is that by combining a measured feedback signal with knowledge of the control-system components (primarily the plant and feedback system), the behavior of the plant can be known with greater precision than by using the feedback signal alone. As shown in Figure 10-1, the observer augments the sensor output and provides a feedback signal to the control laws.

In some cases, the observer can be used to enhance system performance. It can be more accurate than sensors or can reduce the phase lag inherent in the sensor. Observers

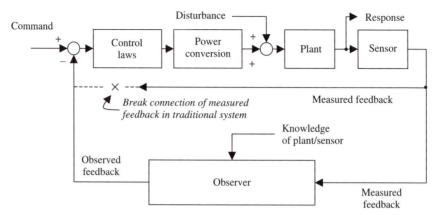

Figure 10-1. Role of an observer in a control system.

can also provide observed disturbance signals, which can be used to improve disturbance response. In other cases, observers can reduce system cost by augmenting the performance of a low-cost sensor so that the two together can provide performance equivalent to a higher-cost sensor. In the extreme case, observers can eliminate a sensor altogether, reducing sensor cost and the associated wiring. For example, in a method called *acceleration feedback*, which will be discussed in Chapter 18, acceleration is observed using a position sensor and thus eliminating the need for a separate acceleration sensor.

Observer technology is not a panacea. Observers add complexity to the system and require computational resources. They may be less robust than physical sensors, especially when plant parameters change substantially during operation. Still, an observer applied with skill can bring substantial performance benefits, and they do so, in many cases, while reducing cost or increasing reliability.

The Luenberger observer is perhaps the most practical observer form. It combines five elements:

- A sensor output, $Y(s)$
- A power converter output (plant excitation), $P_C(s)$
- A model (estimation) of the plant, $G_{PEst}(s)$
- A model of the sensor, $G_{SEst}(s)$
- A PI or PID observer compensator, $G_{CO}(s)$

The general form of the Luenberger observer is shown in Figure 10-2.

10.1.1 Observer Terminology

The following naming conventions will be used. *Estimated* will describe components of the system model. For example, the estimated plant is a model of the plant that is run

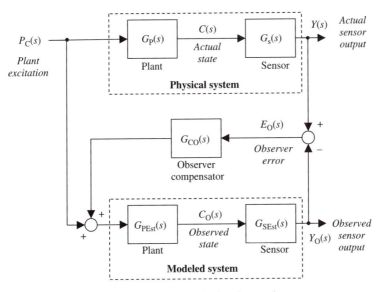

Figure 10-2. General form of the Luenberger observer.

by the observer. *Observed* will apply to signals derived from an observer; thus, the state (C_o) and the sensor (Y_o) signals are *observed* in Figure 10-2. Observer models and their parameters will be referred to as *estimated*. Transfer functions will normally be named $G(s)$, with identifying subscripts: $G_p(s)$ is the plant transfer function and $G_{PEst}(s)$ is the estimated or modeled plant.

10.1.2 Building the Luenberger Observer

This section describes the construction of a Luenberger observer from a traditional control system, adding components step by step. Start with the traditional control system shown in Figure 10-3. Ideally, the control loop would use the actual state, $C(s)$, as feedback. However, access to the state comes through the sensor, which produces $Y(s)$, the feedback variable. The sensor transfer function, $G_s(s)$, often ignored in the presentation of control systems, is the focus here. Typical problems caused by sensors are phase lag, attenuation, and noise.

Phase lag and attenuation can be caused by the physical construction of the sensor or by sensor filters, which are often introduced to attenuate noise. The key detriment of phase lag is the reduction of loop stability. Noise can be generated by several forms of electromagnetic interference (EMI). Noise causes random behavior in the control system, corrupting the output and wasting power. All of these undesirable characteristics are represented by the term $G_s(s)$ in Figure 10-3. The ideal sensor can be defined as $G_{S–IDEAL}(s) = 1$.

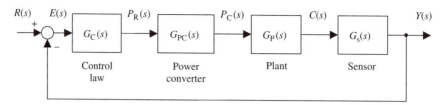

Figure 10-3. Traditional control system.

The first step in dealing with sensor problems is to select the best sensor for the application. Compared to using an observer, selecting a faster or more accurate sensor will provide benefits that are more predictable and more easily realized. However, limitations such as cost, size, and reliability will usually force the designer to accept sensors with undesirable characteristics, no matter how careful the selection process. The assumption from here forward will be that the sensor in use is appropriate for a given machine or process; the goal of the observer is to make the best use of that sensor. In other words, the first goal of the Luenberger observer will be to minimize the effects of $G_s(s) \neq 1$.

For the purposes of this development, only the plant and sensor, as shown in Figure 10-4, need to be considered. Note that the traditional control system ignores the effect of $G_s(s) \neq 1$; $Y(s)$, the sensor output, is used in place of the actual state under control, $C(s)$. But $Y(s)$ is not $C(s)$; the temperature of a component is not the temperature indicated by the sensor. Phase lag from sensors often is a primary contributor to loop instability; noise from sensors often demands correction by the addition of filters in the control loop, again contributing phase lag and ultimately reducing margins of stability.

10.1.2.1 Two Ways to Avoid $G_s(S) \neq 1$

So, how can the effects of $G_s(s) \neq 1$ be removed? One alternative is to follow the sensed signal with the inverse of the sensor transfer function: $G_{SEst}^{-1}(s)$. This is shown in Figure 10-5. On paper, such a solution appears workable. Unfortunately, the nature of $G_s(s)$ makes taking its inverse impractical. For example, if $G_s(s)$ were a low-pass filter, as is common, its inverse would require a derivative, as shown in Equation 10.1. Derivatives are well known for being too noisy to be practical in most cases; high-frequency noise, such as that from quantization and EMI, processed by a derivative generates excessive high-frequency output noise.

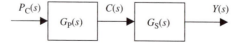

Figure 10-4. Plant and sensor.

Figure 10-5. An impractical way to estimate $C(s)$: adding the inverse sensor transfer function.

$$\text{If } G_{\text{SEst}}(s) = \frac{K_{\text{EST}}}{s + K_{\text{EST}}}, \quad \text{then } G_{\text{SEst}}^{-1}(s) = 1 + \frac{s}{K_{\text{EST}}} \tag{10.1}$$

Another alternative to avoid the effects of $G_s(s) \neq 1$ is to simulate a model of the plant in software as the control loop is being executed. The signal from the power converter output is applied to a plant model, $G_{\text{PEst}}(s)$, in parallel with the actual plant. This is shown in Figure 10-6. Such a solution is subject to drift because most control system plants contain at least one integrator; even small differences between the physical plant and the model plant will cause the estimated state, $C_{\text{Est}}(s)$, to drift. As a result, this solution is also impractical.

The solution of Figure 10-5, which depends wholly on the sensor, works well at low frequency but produces excessive noise at high frequency. The solution of Figure 10-6, which depends wholly on the model and the power converter output signal, works well at high frequency but drifts in the lower frequencies. The Luenberger observer, as will be shown in the next section, can be viewed as combining the best parts of these two solutions.

10.1.2.2 Simulating the Plant and Sensor in Real Time

Continuing the construction of the Luenberger observer, augment the structure of Figure 10-6 to run a model of the plant and sensor in parallel with the physical plant and sensor. This configuration, shown in Figure 10-7, drives a signal representing the power conversion output through the plant model and through the sensor model to generate the observed sensor output, $Y_O(s)$. Assume for the moment that the models are exact replicas of the physical components. In this case, $Y_O(s) = Y(s)$ or, equivalently, $E_O(s) = 0$. In such a case, the observed state, $C_O(s)$, is an accurate representation of the actual state. So $C_O(s)$ could be used to close the control loop; the phase lag of $G_S(s)$ would have no effect on the system. This achieves the first goal of observers, the

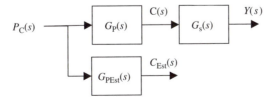

Figure 10-6. Another impractical solution: deriving the controlled state from a model of the plant.

elimination of the effects of $G_s(s) \neq 1$, but only for the unrealistic case where the model is a perfect representation of the actual plant.

10.1.2.3 Adding the Observer Compensator

In any realistic system $E_O(s)$ will not be zero because the models will not be perfect representations of their physical counterparts and because of disturbances. The final step in building the Luenberger observer is to route the error signal back to the model to drive the error toward zero. This is shown in Figure 10-8. The observer compensator, $G_{CO}(s)$, is usually a high-gain PI or PID control law.

The gains of $G_{CO}(s)$ are often set as high as possible so that even small errors drive through the observer compensator to minimize the difference between $Y(s)$ and $Y_O(s)$. If this error is small, the observed state, $C_O(s)$, becomes a reasonable representation of the actual state, $C(s)$; certainly, it can be much more accurate than the sensor output, $Y(s)$.

One application of the Luenberger observer is to use the observed state to close the control loop; this is shown in Figure 10-9, which compares to the traditional control system of Figure 10-3. The sensor output is no longer used to close the loop; its sole function is to drive the observer to form an observed state. Typically, most of the phase lag and attenuation of the sensor can be removed, at least in the frequency range of interest for the control loop.

10.2   Experiments 10A–10C: Enhancing Stability with an Observer

Experiments 10A–10C are *Visual ModelQ* models that demonstrate one of the primary benefits of Luenberger observers: the elimination of phase lag from the control loop and the resulting increase in margins of stability. Experiment 10A represents the traditional system. Experiment 10B restructures the loop so the actual state is used as the feedback variable. Of course, this is not practical in a working system (the actual state is not accessible) and is only used here to demonstrate the

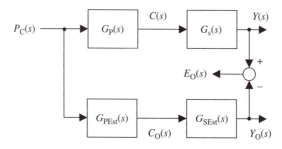

Figure 10-7. Running models in parallel with the actual components.

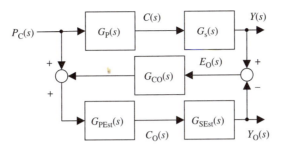

Figure 10-8. The Luenberger observer.

negative effects of the sensor. Experiment 10C adds a Luenberger observer to the control system. The result will be that the system performance will be equal to that with the actual state; the destabilizing effects of the phase lag from the sensor will be eliminated.

The system in Experiment 10A, shown in Figure 10-10, is a typical PI control loop. The power converter, G_{PC}, is modeled as a 50-Hz-bandwidth two-pole low-pass filter with a damping ratio of 0.707. The plant is a single integrator scaled by 50. The sensor, G_S, is modeled by a single-pole low-pass filter with a bandwidth of 20 Hz. The PI control law is tuned aggressively so that the margins of stability are low. The sensor is the primary contributor to phase lag in the loop and thus the primary cause of low margins of stability.

Experiment 10A includes a *Live Scope* display of C, the actual state. This signal is the system response to R, the square wave command. The signal C shows considerable overshoot and ringing, which indicates marginal stability. The gains of G_C, the PI controller ($K_P = 1.5$ and $K_I = 30$), are set too aggressively for the phase lag of this loop.

The marginal stability of Figure 10-10 is caused in large part by the phase lag created by G_S. Were G_S an ideal sensor ($G_S = 1.0$), the system would behave well with the gains

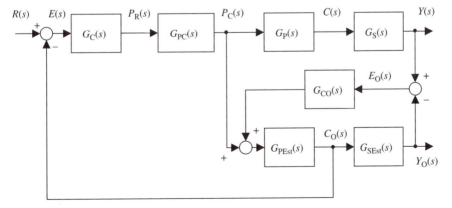

Figure 10-9. A Luenberger observer-based control loop.

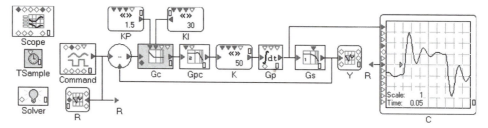

Figure 10-10. Experiment 10A: a traditional control system.

used in that model. This is demonstrated in Experiment 10B (see Figure 10-11), which is identical in all respects to Figure 10-9 except that the loop feedback is the actual state (C), not the measured output (Y). Of course, this structure is impractical; loops must be closed on measured signals. In traditional control systems, the use of sensors with significant phase lag[1] usually requires reducing control law gains, which implies a reduction in system performance. In other words, the most common solution for the stability problems of Figure 10-10 is to reduce the PI gains and accept the lower level of performance. An observer can offer a better alternative.

Experiment 10C (see Figure 10-12) is Experiment 10A with an observer added. Several new blocks are used to construct the observer of Figure 10-8:

- *Subtraction* A subtraction block is added on the right to take the difference of Y, the actual feedback signal, and Y_O, the observed feedback signal. This forms the observer error signal.
- G_{CO} A PID control law is used as the observer compensator. G_{CO} was tuned experimentally using a process that will be described later in this chapter. G_{CO} samples at 1 kHz, as do the other digital blocks (Delay, G_{PEst}, and G_{SEst}).
- *Addition* An addition block is added to combine the output of the observer compensator and the power converter output.
- K_{Est} An estimate of the gain portion of the plant. Here, the gain is set to 50, the gain of the actual plant $(K_{Est} = K)$.
- G_{PEst} A digital integrator, which is a summation scaled by $1/T$. It is used to estimate the second part of the plant, G_P.
- G_{SEst} A digital filter used to model the sensor transfer function. This is a single-pole, low-pass filter with a bandwidth of 20 Hz. This is an accurate representation of the sensor.
- C_O Variable block C_O, the observed state.
- *Delay* A delay of one sample time. This delay recognizes that there must be a sample-hold delay at some position in the digital-observer loop. In this case, it is assumed that during each cycle of the observer, Y_O is calculated to be used in the succeeding cycle.

---

[1]Significant in comparison to the other sources of phase lag in the loop, such as internal filters and the power converter.

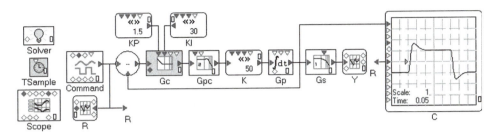

Figure 10-11. From Experiment 10B: An idealized system, which uses the actual state for feedback, has conservative margins of stability.

The other significant change in Experiment 10C is that the feedback for the control loop is taken from the observed state, not the sensor output, as was the case in Experiment 10A. The results of these changes are dramatic. The actual state portrayed in Figure 10-12 shows none of the signs of marginal stability that were evident in Figure 10-10. This improvement is wholly due to reduction of phase lag from the sensor. In fact, the response is so good that it is indistinguishable from the response of the system closed using the actual state as shown in Figure 10-11 (the desirable, albeit impractical, alternative).

Here, the Luenberger observer provided a practical way to eliminate the effects of sensor phase lag in the loop. Further, these benefits could be realized with a modest amount of computational resources: a handful of simple functions running at 1 kHz. Altogether, these calculations represent a few microseconds of computation on a modern DSP and not much more on a traditional microprocessor. In many cases, the calculations are fast enough that they can be run on existing hardware using uncommitted computational resources.

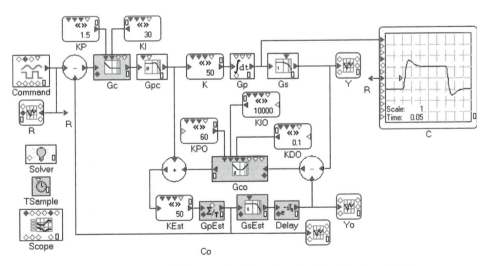

Figure 10-12. Model of observer-based control system, from Experiment 10C.

10.2.1 Experiment 10D: Elimination of Phase Lag

A brief investigation can verify the elimination of phase lag demonstrated with Experiments 10A–10C. Experiment 10D (see Figure 10-13) displays three signals on *Live Scopes*: C, the actual state, Y, the measured state, and C_O, the observed state. At first glance, all three signals appear similar. However, upon closer inspection, notice that Y (rightmost) lags C (top) by about a division. For example, C crosses two vertical divisions at two time divisions after the trigger time, which is indicated by a small triangle; Y requires three time divisions to cross the same level, about 10 ms longer. Note that the time scale has been reduced to 0.01 sec in Experiment 10D to show this detail. Note also that all three *Live Scopes* are triggered from R, the command, and so are synchronized.

Now compare the observed state, C_O, to the actual state. These signals are virtually identical. For example, C_O crosses through two vertical divisions at the second time division, just like the actual state. The phase lag from the sensor has been eliminated by the observer. This explains why the margins of stability were the same whether the system used the actual state (Experiment 10B) or the observed state (Experiment 10C) for feedback and why the margins were lower in the system using the measured state (Experiment 10A).

It should be pointed out that this observer has several characteristics that will not be wholly realized in a practical system: The plant and sensor models are near-perfect representations of the actual plant and system, there are no disturbances, and the sensor

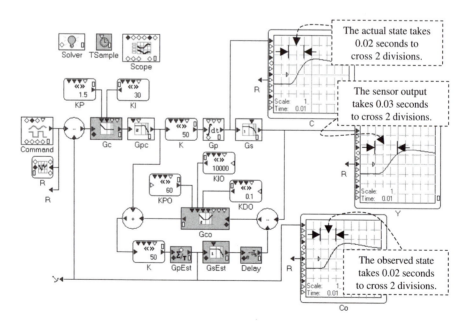

Figure 10-13. Experiment 10D: three signals in an observer-based system.

signal is noiseless. Certainly, these imperfections will limit the benefits of observers in real-world control systems. Even so, the principle shown here is reliable in working machines and processes: The Luenberger observer can produce substantial benefits in the presence of sensor phase lag, and it can do so with modest computational resources.

10.3 Filter Form of the Luenberger Observer

The Luenberger observer can be analyzed by representing the structure as a transfer function[5]. This approach can be used to investigate system response to nonideal conditions: disturbances, noise, and model inaccuracy. The form is not normally used for implementation because of practical limitations, which will be discussed. However, the filter form is useful because it provides insight into the operation of observers.

The observer transfer function has two inputs, $P_C(s)$ and $Y(s)$, and one output, $C_O(s)$. In this analysis, the actual model and sensor are considered a black box. This is shown in Figure 10-14. The focus is on understanding the relationship between the inputs to the observer and its output. Signals internal to the observer, such as $E_O(s)$ and $Y_O(s)$, are ignored. In fact, they will become inaccessible as the block diagram is reduced to a single function.

Using Mason's signal flow graphs to build a transfer function from Figure 10-14 produces Equation 10.2. There is a single path, P_1, from $Y(s)$ to the observed state, $C_O(s)$. Also, there is one path, P_2, from $P_C(s)$ to the $C_O(s)$. Finally, there is a single loop, L_1. The equations for P_1, P_2, and L_1 can be read directly from Figure 10-14:

$$P_1 = G_{CO}(s) \times G_{PEst}(s)$$
$$P_2 = G_{PEst}(s)$$
$$L_1 = -G_{CO}(s) \times G_{PEst}(s) \times G_{SEst}(s)$$

Using Mason's signal flow graphs, Equation 10.2 can be written by inspection:

$$C_O(s) = \frac{Y(s) \times P_1 + P_C(s) \times P_2}{1 - L_1} \tag{10.2}$$

$$C_O(s) = \frac{Y(s) \times G_{CO}(s) \times G_{PEst}(s) + P_C(s) \times G_{PEst}(s)}{1 + G_{PEst}(s) \times G_{CO}(s) \times G_{SEst}(s)}$$

$$= Y(s) \frac{G_{PEst}(s) \times G_{CO}(s)}{1 + G_{PEst}(s) \times G_{CO}(s) \times G_{SEst}(s)} \tag{10.3}$$

$$+ P_C(s) \frac{G_{PEst}(s)}{1 + G_{PEst}(s) \times G_{CO}(s) \times G_{SEst}(s)}$$

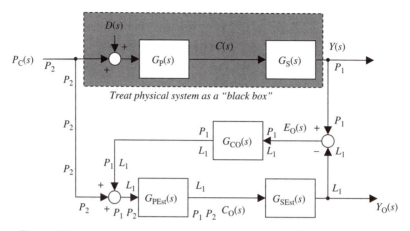

Figure 10-14. Luenberger observer as transfer function between $Y(s)$, $P_C(s)$, and $C_O(s)$.

Equation 10.2 is the sum of two factors. These factors are separated in Equation 10.3. The first factor, dependent on the sensor output, $Y(s)$, is shown in Equation 10.4:

$$Y(s) \times G_{SEst}^{-1}(s) \frac{G_{PEst}(s) \times G_{CO}(s) \times G_{SEst}(s)}{1 + G_{PEst}(s) \times G_{CO}(s) \times G_{SEst}(s)} \tag{10.4}$$

Note that the term $G_{SEst}(s)$ has been multiplied through the fractional term and divided out of the scaling term, $Y(s)$. Equation 10.4 can be viewed as the sensor output, multiplied by the inverse estimated sensor transfer function, and then filtered by the term on the far right; as will be shown, the term on the right is a low-pass filter. So Equation 10.4 is the form shown in Figure 10-5 followed by a low-pass filter.

The second factor of Equation 10.3, dependent on the power converter output, is shown in Equation 10.5:

$$P_C(s) \times G_{PEst}(s) \frac{1}{1 + G_{PEst}(s) \times G_{CO}(s) \times G_{SEst}(s)} \tag{10.5}$$

Here, the estimated plant transfer function, $G_{PEst}(s)$, has been pulled out to scale the filter term. The scaling term is equivalent to the form shown in Figure 10-6 used to calculate $C_{Est}(s)$. As will also be shown, the term on the right is a high-pass filter. So Equation 10.5 is the form shown in Figure 10-6 followed by a high-pass filter.

10.3.1 Low-Pass and High-Pass Filtering

The rightmost term in Equation 10.4 can be shown to be a low-pass filter. That term is shown in Equation 10.6:

$$\frac{G_{PEst}(s) \times G_{CO}(s) \times G_{SEst}(s)}{1 + G_{PEst}(s) \times G_{CO}(s) \times G_{SEst}(s)} \tag{10.6}$$

To see this, first consider the individual terms of Equation 10.6. $G_{PEst}(s)$ is a model of the plant. Plants for control systems usually include one or more integrals. At high frequencies, the magnitude of this term declines to near zero. $G_{SEst}(s)$ is a model of the sensor. Sensor output nearly always declines at high frequencies because most sensors include low-pass filters, either implicitly or explicitly. The term $G_{CO}(s)$ is a little more difficult to predict. It is constructed so the open-loop gain of the observer $(G_{SEst}(s) \times G_{PEst}(s) \times G_{CO}(s))$ will have sufficient phase margin at the observer crossover frequencies. Like a physical control loop, the compensator has a high enough order of differentiation to avoid 180° of phase shift while the open-loop gain is high. However, the maximum order of the derivative must be low enough so that the gain at high frequency declines to zero. So evaluating the product of $G_{PEst}(s)$, $G_{SEst}(s)$, and $G_{CO}(s)$ at high frequency yields a small magnitude; by inspection, this will force the magnitude of Equation 9.6 to a low value at high frequency. This is seen because the "1" will dominate the denominator, reducing Equation 10.6 to its numerator.

Using similar reasoning, it can be seen that Equation 10.6 will converge to "1" at low frequencies. As has been established, $G_{PEst}(s)$ will usually have one order of integral; at low frequencies, this term will have a large magnitude. $G_{CO}(s)$ will add one order of integration or will at least have a proportional term. Typically, $G_{SEst}(s)$ will be a low-pass filter with a value of unity at low frequencies. Evaluating the product of $G_{PEst}(s)$, $G_{SEst}(s)$, and $G_{CO}(s)$ at low frequency yields a large magnitude; by inspection, this will force the magnitude of Equation 10.6 to 1. (This can be seen because the "1" in the denominator will be insignificant, forcing Equation 10.6 to 1.) These two characteristics, unity gain at low frequency and near-zero gain at high frequency, are indicative of a low-pass filter.

The right-hand term of Equation 10.5 can be investigated in a similar manner. This term, shown in Equation 10.7, has the same denominator as Equation 10.6, but with a unity numerator. At high frequency, the denominator reduces to approximately 1, forcing Equation 10.7 to 1. At low frequencies, the denominator becomes large, forcing the term low. This behavior is indicative of a high-pass filter.

$$\frac{1}{1 + G_{PEst}(s) \times G_{CO}(s) \times G_{SEst}(s)} \tag{10.7}$$

10.3.2 Block Diagram of the Filter Form

The filter form of the Luenberger observer is shown as a block diagram in Figure 10-15. This demonstrates how the observer combines the input from $Y(s)$ and $P_C(s)$. Both of these inputs are used to produce the observed state. The term from $Y(s)$ provides good low-frequency information but is sensitive to noise; thus, it is intuitive that this term should be followed by a low-pass filter. The term from $P_C(s)$ provides poor low-frequency information because it is subject to drift when integral gains are even slightly inaccurate. On the other hand, this term is not as prone to generate high-frequency noise as was the $Y(s)$ term. This is because the plant, which normally contains at least one integral term, acts as a filter, reducing the noise content commonly present on the power converter output. It is intuitive to follow such a term with a high-pass filter. The observed state is formed two different ways from the two different sources and uses filtering to combine the best frequency ranges of each into a single output.

10.3.3 Comparing the Loop and Filter Forms

The filter form (Figure 10-15) of the Luenberger observer is presented to aid under-standing. The loop form (Figure 10-8) is more practical for several reasons. First, in most cases, it is computationally more efficient. Also, the loop form observes not only the plant state, but also the disturbance [28]. The filter form does offer at least one advantage over the traditional form: The process of tuning the observer is exchanged for a configuration of a filter; for some designers, this process might be more familiar.

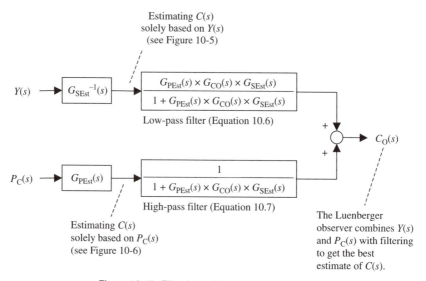

Figure 10-15. Filter form of the Luenberger observer.

The standard form of the Luenberger observer requires the tuning of the $G_{CO}(s) - G_{PEst}(s) - G_{SEst}(s)$ loop. This process is like tuning a physical control loop. Like a physical control loop, the loop form of observers can become unstable. Assuming the two filters of Figure 10-15 are implemented in the z-domain as a ratio of two polynomials, this type of instability can be avoided entirely. However, this advantage is small because the instability in the loop form is easily avoided. This book will focus on the loop form but use the filter form for analysis.

10.4 Designing a Luenberger Observer

Section 10.2 provided experiments that demonstrated performance of the Luenberger observer. In that section, a fully configured observer was provided as part of a control system. Before the model for that system could be used, several steps were required to design that observer. Those steps are the focus of this section. There are four major components of observer design: modeling the sensor, modeling the plant, selecting the observer compensator, and tuning that compensator. This section will provide details on each of those components and then conclude with a step-by-step procedure.

This presentation of observers is based on classical controls: block diagrams and transfer functions in the s-domain. Readers may have considered that observers are commonly presented in the matrix-based *state-space* form. State space is a useful means of representing observer-based systems, especially when the system is complex, for example, when high-order plants are used. The weakness of the state-space representation is that it hinders intuition. The abstract form simplifies mathematical manipulation but leaves most designers puzzled: How can I implement this? Under what conditions will I see benefit and in what quantity? How do I debug it?

This book presents observers in the classical form, even though the approach is limited to lower-order systems. This approach is well used in application-oriented writing concerning observers [30, 73, 78, 97, 98], even though it is generally avoided by authors of academic texts [33, 35, 36, 77] in favor of the more general state-space form. The classical approach is used here in the belief that a great deal of benefit can be gained from the application of observers to common control systems, but the dominance of the state-space representation has limited its application by working engineers.

10.4.1 Designing the Sensor Estimator

The task of designing the sensor estimator is to derive the transfer function of the sensor. The key parameters are filtering and scaling. For example, the model in Section 10.2 used a low-pass filter with unity gain to model the effects of the sensor. One benefit of modeling the entire control system (as was done in Section 10.2) is that the transfer function of the sensor is known with complete certainty. In the system of Section 10.2, the sensor was selected as a low-pass filter, so designing the sensor estimator was trivial ($G_{\text{SEst}}(s) = G_{\text{S}}(s)$). Of course, in a practical system the sensor transfer function is not known with such precision. The designer must determine the sensor transfer function as part of designing the observer.

10.4.1.1 Sensor Scaling Gain

The scaling gain of a sensor is of primary importance. Errors in sensor scaling will be reflected directly in the observed state. Most sensor manufacturers will provide nominal scaling gains. However, there may be variation in the scaling, either from one unit to another or in a particular unit during operation. (If unit-to-unit variation is excessive, the scaling of each unit can be individually measured.) Fortunately, sensors are usually manufactured with minimum variation in scaling gains, so with modest effort these devices can often be modeled with accuracy.

Offset, the addition of an erroneous DC value to the sensor output, is another problem in sensors. If the offset is known with precision, it can be added to the sensor model; in that case, the offset will not be reflected in the observed state. Unfortunately, offset commonly varies with operating conditions as well as from one unit to another. Offset typically will not affect the dynamic operation of the observer; however, an uncompensated offset in a sensor output will be reflected as the equivalent DC offset in the observed state. The offset then will normally have the same effect, whether the loop is closed on the sensor output or the observed state: The offset will be transferred to the actual state. In that sense, the offset response of the observer-based system will be the same as that of the traditional system.

10.4.2 Sensor Filtering

The filtering effects in a sensor model may include explicit filters, such as when electrical components are used to attenuate noise. The filtering effects can also be implicit in the sensor structure, such as when the thermal inertia of a temperature sensor produces phase lag in sensor output. The source of these effects is normally of limited concern at this stage; here, attention is focused on modeling the effects as accurately as possible, whatever the source. The filtering effects can be thought of more broadly as the dynamic performance: the transfer function of the sensor less the scaling gain. As was the case with scaling gains, most manufacturers will provide nominal dynamic performance in data sheets, perhaps as a Bode plot or as an s-domain transfer function. Again, there can be variation between the manufacturer's data and the parts. If the variation is excessive, the designer must evaluate the effects of these variations on system performance.

In some cases, varying parameters can be measured. Since gain and offset terms can be measured at DC, the process to measure these parameters is usually straightforward. Measuring the dynamic performance of a sensor can be challenging. It requires that the parameter under measurement be driven into the sensor input; the sensor is then calibrated by comparing the sensor output to the output of a calibrating sensing device, which must be faster and more accurate than the sensor. Fortunately, such a procedure is rarely needed. Small deviations between the filtering parameters of the actual and model sensors have minimal effect on the operation of the observer.

Since most observers are implemented digitally, filtering effects usually need to be converted from the s-domain to the z-domain. This was the case in Experiment 10C; note that $G_{PEst}(s)$ and $G_{SEst}(s)$ are digital equivalents to their analog counterparts. The conversion can be accomplished using Table 5-1. This table gives the conversion for one- and two-pole filters; higher-order filters can be converted to a product of single- and double-pole filters. The conversion to the z-domain is not exact; fortunately, the z-domain filters in Table 5-1 provide a slight phase lead compared to their s-domain equivalents, so the digital form can be slightly phase advanced from the analog form. However, when adding the delay introduced by sampling, which affects only the digital form, the result is that both the digital and the analog forms have about the same phase lag below half the sample frequency (the Nyquist frequency).

10.4.3 Designing the Plant Estimator

The task of designing the plant estimator is similar to that of designing the sensor estimator: Determine the transfer function of the plant. Also like the sensor, the plant can be thought of as having DC gain and dynamic performance. The plant often provides more challenges than the sensor. Plants are usually not manufactured with the precision of a sensor. The variation of dynamic performance and scaling is often substantial. At the same time, variation in the plant is less of a concern because an appropriately designed observer compensator will remove most of the effects of such variations from the observed state. The following discussion will address the process of estimating the plant in three steps: estimating the scaling gain, the order of integration, and the remaining filtering effects. In other words, the plant will be assumed to have the form

$$G_P(s) = K \times \frac{1}{s^N} \times G_{LPF}(s) \qquad (10.8)$$

where $G_P(s)$ is the total plant transfer function, K is the scaling gain, N is the order of integration, and $G_{PF}(s)$ is the plant filtering. The task of estimating the plant is to determine these terms.

10.4.3.1 Plant Scaling Gain (K)

The scaling gain, K, is the overall plant gain. As with sensors, the nominal gain of a plant is usually provided by the component manufacturer. For example, a current controller using an inductor as a plant has a gain $1/Ls$, where L is the inductance, so the plant gain, K, is $1/L$; since inductor manufacturers provide inductance on the data

sheet, determining the gain here seems simple. However, there is usually more variation in the gain of a plant than in that of a sensor. For example, inductance often is
specified only to ±20% between one unit and another. In addition, the gain of the
plant during operation often varies considerably over operating conditions. Saturation
can cause the incremental inductance of an iron-core inductor to change by a factor of
five times or more when current is increased from zero to full current. This magnitude
of variation is not common for sensors.

Another factor that makes determination of K difficult is that it may depend on
multiple components of the machine or process; the contributions from some of
those components may be difficult to measure. For example, in a servo system, the
plant gain K is K_T/J, where K_T is the motor torque constant (the torque per amp) and
J is total inertia of the motor and the load. K_T is usually specified to an accuracy of
about 10% and the motor inertia is typically known to within a few percentage
points; this accuracy is sufficient for most observers. However, the load inertia is
sometimes difficult to calculate and even harder to measure. Since it may be many
times greater than the motor inertia, the total plant gain may be virtually unknown
when the observer is designed. Similarly, in a temperature-controlled liquid bath, the
gain K includes the thermal mass of the bath, which may be difficult to calculate and
inconvenient to measure; it might also vary considerably during the course of
operation.

The problems of determining K, then, fall into two categories: determining nominal
gain and accounting for variation. For the cases when nominal gain is difficult to
calculate, it can be measured using the observer with modest effort. A process for this
will be the subject of Section 10.4.3.4. The problems of in-operation variation are
another matter. Normally, if the variation of the plant is great, say, more than
20–50%, the benefits of the observer may be more difficult to realize. Of course, if
the variation of the gain is well known and the conditions that cause variation are
measured, then the estimated scaling gain can be adjusted to follow the plant. For
example, the variation of inductance is repeatable and relatively easy to measure. In
addition, the primary cause of the variation, the current in the inductor, is often
measured in the control system; in such a case, the variation of the gain can be coded
in the observer's estimated plant. However, in other cases, accounting for such variation may be impractical.

10.4.3.2 Order of Integration

Most control system plants have one order of integration, as is the case with all
the plants of Table 2-2. In other cases, the plant may be of higher order because
multiple single-order plants are used, such as when a capacitor and inductor are
combined into a second-order "tank" circuit. Higher-order plants are common in
complex systems. In any event, the order of the plant should be known before
observer design begins. This is a modest requirement for observers since even a

traditional control system cannot be designed without knowledge of the order of the plant. The assumption here is that the order of the plant (N in Equation 10.8) is known.

10.4.3.3 Filtering Effects

After the scaling gain and integration have been removed, what remains of the plant is referred to here as *filtering effects*. Like sensors, the filtering effects of plants can generally be determined through manufacturer's data. In most cases, these effects are small or occur at high-enough frequencies that they have little influence on the operation of the control system. For example, servomotors have two filtering effects that are generally ignored: viscous damping and interwinding capacitance. Viscous damping contributes a slight stabilizing effect below a few hertz. It has so little effect that it is normally ignored. Parasitic capacitance that connects windings is an important effect but usually has little to do directly with the control system, because the frequency range of the effect is so far above the servo controller bandwidth.

10.4.3.4 Experiment 10E: Determining the Gain Experimentally

As discussed in Section 10.4.3.1, the plant scaling gain often must be determined experimentally. The process to do this is simple and can be executed with an ordinary Luenberger observer, assuming the observer error signal (the input to the observer compensator) is available and that there are no significant disturbances to the control loop during this process. The system should be configured as follows:

1. Configure the observer estimated sensor and plant. Use the best guess available for the estimated plant scaling gain.
2. Configure the system to close the control loop on the sensor feedback.
3. Set the observer gains to very low values.
4. Minimize disturbances or operate the product-process in regions where disturbances are not significant. (Disturbances reduce the accuracy of this procedure.)
5. Excite the system with fast-changing commands.
6. Monitor $E_O(s)$, the observer error.
7. Adjust the estimated plant scaling gain until the observer error is minimized.

Experiment 10E modifies the model of Experiment 10C for this procedure. The estimated sensor is accurate; the estimated plant less the scaling gain is also configured accurately. Only the scaling gain is in error (20 here, instead of 50). The observer error, $E_O(s)$, has large excursions owing to the erroneous value of K_{Est}. The result is shown in

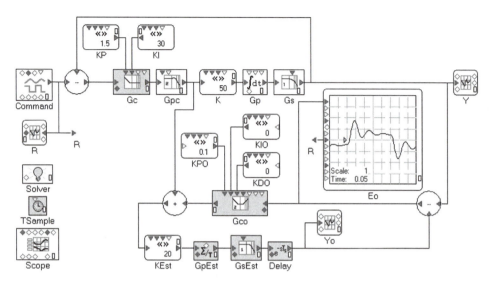

Figure 10-16. Output of Experiment 10E with K_{Est} inaccurate ($K_{Est} = 20$ and $K = 50$).

Figure 10-16. The effect of changing the estimated scaling gain is shown in Figure 10-17. If K_{Est} is either too large (a) or too small (c), the error signal grows. The center (b) shows K_{Est} adjusted correctly ($K_{Est} = 50$). Only when K_{Est} is adjusted correctly is E_O minimized.

The process to find K_{Est} is demonstrated here with a near-perfect sensor estimator ($G_s(s) = G_{SEst}(s)$). However, it is still effective with reasonable errors in the estimated sensor filtering effects. The reader is invited to run Experiment 10E and to introduce error in the estimated sensor and to repeat the process. Set the bandwidth of the estimated sensor to 30 Hz, a value 50% high (double-click on the node at the top center of G_{SEst} and set the value to 30). Notice that the correct value of K_{Est} still minimizes error, though not to the near-zero value that was attained with an accurate estimated sensor in Figure 10-17.

The procedure that started this section stated that low gains should be placed in the compensator. Experiment 10E used $K_{IO} = 0$, $K_{DO} = 0$, and $K_{PO} = 0.1$, low values indeed. The only purpose here is to drive the DC error to zero. Otherwise, the fully integrating plant would allow a large offset in E_O even if K_{Est} were correct.

Another requirement of compensator gains is that they provide a stable loop. A single integrator in the combined estimated plant and sensor can be stabilized with a proportional gain. This explains why Experiment 10E has K_{PO} as its only nonzero gain. A double integration in that path requires some derivative term. In all cases, the gains should be low; the low gains allow the designer to see the observer error more easily. High gains in a well-tuned compensator drive the observer error to zero too rapidly for this process.

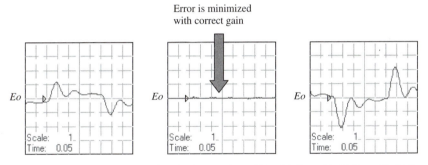

Figure 10-17. Effect of changing K_{Est} on $E_O(s)$ for three settings of K_{Est}: (a) $K_{Est} = 30$; (b) $K_{Est} = 50$; (c) $K_{Est} = 85$.

10.4.4 Designing the Observer Compensator

Observer-compensator design is the process of selecting which gains will be used in the compensator; here, this amounts to selecting which combination of P, I, and D gains will be required. The derivative gain is used for stabilization. The order of integration in the $G_{PEst} - G_{SEst}$ path determines the need for a derivative gain in G_{CO}. If the order is two, a derivative gain in G_{CO} will normally be necessary; without it, the fixed 180° phase lag of double integration makes the loop difficult to stabilize. In addition, the phase lag of the $G_{PEst} - G_{SEst}$ path at and around the desired bandwidth of the observer must be considered. If the phase lag is about 180° in that frequency range, a derivative term will normally be required to stabilize the loop.

The need for a derivative term and its dependence on the behavior of $G_{PEst} - G_{SEst}$ is demonstrated by comparing the compensators of Experiments 10C and 10E. In Experiment 10E, the derivative gain is zeroed because the single-integrating plant did not require a derivative term. In Experiment 10C, a derivative term was required. The reason for the difference is the change in observer bandwidth. In Experiment 10E, the observer bandwidth is purposely set very low, well under the sensor bandwidth. In Experiment 10C, the observer is configured to be operational and its bandwidth is well above the 20-Hz bandwidth of the sensor; well above the sensor bandwidth, the single-pole low-pass filter of the sensor has a 90° phase lag like an integrator. That, combined with the 90° phase lag of a single-integrating plant, generates 180° in the $G_{PEst} - G_{SEst}$ path and so must be compensated; the derivative gain of $G_{CO}(s)$ is a simple way to do so.

Beyond the cases where a derivative gain is required in the compensator, derivative gains can be used to enhance stability of the loop (this is generally the case in control systems). However, derivative gains amplify noise. Observers are particularly sensitive to noise and the derivative gains can needlessly increase that sensitivity.

The goal of the observer compensator is to drive observer error to zero. A fully integrating plant will normally require an integral gain for this purpose. Without the integral gain, disturbances will cause DC errors in the observed state. Because disturbances are present in most control systems and because most applications will not tolerate an unnecessary DC error in the observed state, an integral gain is required in most observer-based systems. If an integral gain is used, a proportional gain is virtually required to stabilize the loop. Thus, most systems will require a PID or, at least, a PI compensator.

The PI-PID compensator is perhaps the simplest compensator available. However, other compensators can be used. The goal is to drive the observer error to zero and to do so with adequate margins of stability. Any method used to stabilize traditional control loops is appropriate for consideration in an observer loop. This book focuses on relatively simple plants and sensors, so the PID compensator will be adequate for the task.

One final subject to consider in the design of G_{CO} is saturation. Traditional control loop compensators must be designed to control the magnitude of the integral when the power converter is heavily saturated. This is because the command commonly changes faster than the plant can respond and large-scale differences in command and response can cause the integrator to grow to very large values, a phenomenon commonly called *windup*. If windup is not controlled, it can lead to large overshoot when the saturation of the power converter ends. This is not normally a concern with observers. Since they follow the actual plant, they are usually not subjected to the conditions that require windup control. For these reasons, PID observer compensators normally do not need windup control. When working in *Visual ModelQ*, be aware that all of the standard *Visual ModelQ* compensators have windup control. Be sure to set the integral saturation levels in the compensator very high so that they do not inadvertently interfere with the operation of the observer.

10.5 Introduction to Tuning an Observer Compensator

Tuning an observer compensator is much like tuning a traditional control system. Higher gains give faster response but greater noise susceptibility and, often, lower margins of stability. A major difference is that the observer plant and sensor parameters are set in software; thus, they are known accurately and they do not vary during operation. This allows the designer to tune observers aggressively, setting margins of stability lower than might be acceptable in traditional control loops.

In many cases the gains of the observer compensator are set as high as margins of stability will allow. Because observers are almost universally implemented digitally, the main cause of instability in the observer is delay from the sample time of the observer loop. Typically, the observer should operate at a bandwidth much higher than the control loop for which it provides feedback. For this reason, the observer will sometimes need to sample faster than the control loop it supports.

Other issues in observer tuning that must be taken into consideration include noise, disturbances, and model inaccuracy. These are discussed briefly here but are discussed in detail in [28]. Noise must be considered when tuning an observer. Like any control system, higher gains cause increased noise susceptibility. Further, because of the structure of Luenberger observers, they are often more susceptible to sensor noise than are most control systems. In many cases, sensor noise, not stability concerns, will be the primary upper limit to observer gains. However, for the purposes of this discussion, the assumed goal is to tune the observer for maximum gains attainable with adequate margins of stability.

In order to provide intuition to the process of tuning, the Luenberger observer of Figure 10-8 can be drawn as a traditional control loop, as shown in Figure 10-18. Here, the sensor output, $Y(s)$, is seen as the command to the observer loop. The loop is closed on the observed sensor output, $Y_O(s)$; the power converter output, $P_C(s)$, appears in the position of feed-forward term. (Just to be clear, this diagram shows the observer and no other components of the control system.) Accordingly, the process of tuning the observer loop is similar to tuning a traditional control system with feed-forward.

The procedure here is:

1. Temporarily configure the observer for tuning.
2. Adjust the observer compensator for stability.
3. Restore the observer to the normal Luenberger configuration.

10.5.1 Step 1: Temporarily Configure the Observer for Tuning

The observer should be temporarily configured with a square wave command in place of Y and without the feed-forward (P_C) path. This is done in Experiment 10F and is shown in Figure 10-19. The command square wave generator, R, has been connected in the normal position for Y. The estimated sensor output, Y_O, is the primary output for this procedure.

The square wave generator is used because it excites the observer with a broader range of frequencies than the physical plant and sensor normally can. The rate of change of a sensor is limited by physics: Inertia makes instantaneous changes unattainable. There is no such limitation on command generators. The use of the square wave command to excite the observer allows the designer to more clearly see its response to high frequencies. The sharper edges of a square wave reveal more about the observer's margins of stability than do the gentler output signals of a physical sensor.

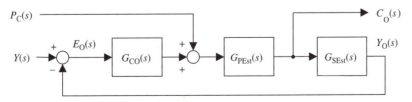

Figure 10-18. Luenberger observer drawn as a traditional control loop.

10.5.2 Step 2: Adjust the Observer Compensator for Stability

As with physical loops, the observer compensator can be tuned experimentally with a zone-based tuning procedure (see Section 3.5). For a PID observer compensator, this implies that the proportional and integral terms are zeroed while the derivative gain is tuned. The proportional gain is tuned and then, finally, the integral term is tuned. For a PI observer compensator, zero the I-term and tune the P-term; then tune the I-term. The results of applying this method to the PID compensator in Experiment 10F are shown in Figure 10-20. Note that in many cases, a DC offset will appear in the feedback signal when K_{PO} is zero; if the offset is large, it can move the error signal off screen. If this occurs, temporarily set K_{PO} to a low value until the offset is eliminated; in Experiment 10F, use $K_{PO} = 0.1$ for this effect.

During the process of tuning, the term driven by P_C should be zeroed. This is because the P_C term acts like a feed-forward to the observer loop, as shown in Figure 10-18. Zone-based tuning recommends that all feed-forward paths be temporarily eliminated because they mask the performance of the loop gains. After the loop is tuned, the path from P_C can be restored.

10.5.2.1 Modifying the Tuning Process for Nonconfigurable Observers

If it is not possible to reconfigure the observer to the form shown in Figure 10-19, the observer can still be tuned by selecting the gain margin. First the system must be configured so that the control loop is closed on the sensed signal, since this method will temporarily generate instability in the observer. The low-frequency gains are zeroed and the high-frequency gain (typically K_{DO}) is raised until the observer becomes unstable, indicating 0-dB gain margin. The gain is then lowered to attain the desired gain margin. For example, raise K_{DO} in steps of 20% until the observer oscillates, and

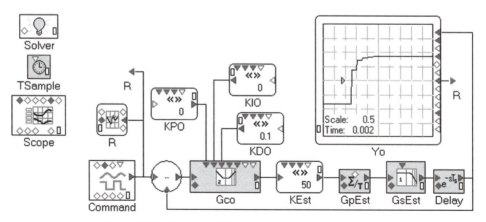

Figure 10-19. Experiment 10F: tuning an observer compensator.

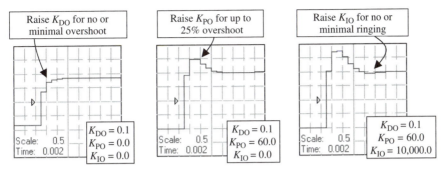

Figure 10-20. Tuning the observer of Experiment 10F in zones. Start with zero gains. (a) Raise K_{DO}, but avoid overshoot. (b) Raise K_{PO} for about 25% overshoot. (c) Raise K_{IO} until ringing just starts.

then reduce it by, say, 12 dB (a factor of 4) to yield 12 dB of gain margin for the observer.

You may wish to verify this with Experiment 10F. Configure the waveform generator *Command* (press F2 and click on the waveform generator to adjust its properties) for gentler excitation, similar to what might come from a physical plant and sensor: Set Waveform to "S-Curve" and set Frequency to 20 Hz. Zero K_{PO} and K_{IO}. (Temporarily set K_{PO} to 0.1 to eliminate DC offset if necessary.) Raise K_{DO} in small steps until the observer becomes unstable — this occurs at $K_{DO} \sim 0.32$. Notice that there are no signs of instability with the s-curve command until the system becomes self-oscillatory. Reduce K_{DO} by 12 dB, or a factor of 4, to 0.08 to get 12 dB of gain margin; this is essentially the same value that was found earlier with the tuning procedure ($K_{DO} = 0.1$). Repeat for the remaining gains.

10.5.2.2 Tuning the Observer Compensator Analytically

Observers can be tuned analytically; this is generally easier than tuning a physical system analytically because the transfer function of the observer is known precisely. After the transfer function is found, the gains can be set to modify the poles of the transfer function. Pole placement methods are presented in Ref 33.

10.5.2.3 Frequency Response of Experiment 10G

This section will investigate the frequency response of the observer. Experiment 10G is reconfigured to include a dynamic signal analyzer, or DSA (see Figure 10-21). The DSA is placed in the command path; this means the DSA will inject random excitation into the command and measure the response of certain signals. The closed-loop response is shown by the DSA as the relationship of Y_O to R. A variable block has been added for E_O, the observer error, so the open-loop response of the observer can be displayed. The relationship of Y_O to E_O in the DSA is the open-loop gain of the observer.

The closed-loop Bode plot taken from Experiment 10G is shown in Figure 10-22. The plot shows the closed-loop response as having high bandwidth compared to the sample rate. The observer is sampled at 1000 Hz as defined by the digital controller (not shown in Figure 10-22). However, the response is still 0 dB at 250 Hz, one-fourth of the sample rate. (The performance above 250 Hz is difficult to determine here since the FFT also samples at 1000 Hz and does not provide reliable data above one-fourth of its sample frequency.) There is about 4 dB of peaking at 100 Hz, a substantial amount for a physical control system but reasonable for an observer. Recall that observers do not need robustness for dealing with parameter changes as do ordinary control systems, and so stability margins can be lower than physical control systems.

Experiment 10G confirms that the observer gains are set quite high for the sample rate. In fact, the only limitation is the sample rate itself. Were the observer sampled faster, the gains could be raised. The reader may wish to experiment here. The sample time (*TSample*) can be reduced to 0.00025. Using the same procedure as before, the gains can be raised to approximately $K_{DO} = 0.6$, $K_{PO} = 500$, and $K_{IO} = 500,000$. In this case the observer bandwidth is about 1000 Hz, still one-fourth of the sample frequency.

One point that needs to be reinforced is that it is often not appropriate to maximize the bandwidth of the observer. High observer bandwidth maximizes the response to sensor noise. (None of the experiments in this chapter has noise sources.) One strength of observers is that they can be used to filter sensor noise while using the power converter signal to make up for any phase lag. So sensor noise will often be the dominant limitation on observer bandwidth, in which case the observer bandwidth may be intentionally set lower than is possible to achieve

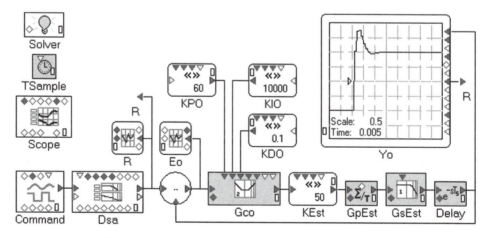

Figure 10-21. Experiment 10G, adding a DSA to Experiment 10F.

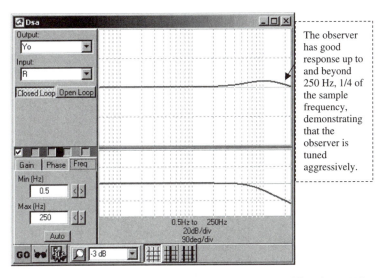

Figure 10-22. Frequency response of the Luenberger observer of Experiment 10G.

based on stability margins. So this procedure reaches the upper limit of observer loop response; be prepared to return and lower the response, for example, by lowering K_{DO} and reducing the other gains according to Figure 10-20.

10.5.3 Step 3: Restore the Observer to the Normal Luenberger Configuration

Restore the observer to the normal Luenberger configuration as shown in Figure 10-8. Remove the connection to the waveform generator and reconnect $Y_O(s)$. Reconnect the path from $P_C(s)$. The observer should be ready to operate.

10.6 Questions

1. Compare the rise time of observer- and nonobserver-based systems to a step command.
   a. Open Experiment 10A and retune the system for reasonable margins of stability (e.g., find maximum K_P without overshoot and maximum K_I for 30% overshoot to step).
   b. What is the settling time?
   c. Repeat parts a and b using the observer-based system of Experiment 10C.
   d. Why is there a large difference in settling times? How does that difference relate to the observer?
2. Retune an observer for a lower bandwidth.

    a. Open Experiment 10G and retune the system starting with $K_{DO} = 0.05$, $K_{PO} = 0$, and $K_{IO} = 0$. For limits allow 15% overshoot with K_{PO} and slight ringing with K_{IO}.

    b. What is the bandwidth of the observer?

    c. Using Experiment 10C, place the values of part a in the observer and see the effect on command response. Evaluate settling time and compare to results with original observer gains (see Question 1c).

3. Using hot connection on the *Live Scope* in Experiment 10C, compare the plant output (C) to the sensor output (Y).

    a. Are they similar?

    b. If so, does this imply sensor phase lag is not a significant factor in closed-loop performance?

4. Show that having the exact value for estimated sensor bandwidth is not required for the experimental process to find K_{Est} discussed in Section 10.4.3.4. In Experiment 10E, corrupt the value of the sensor bandwidth by changing it from 20 to 30 Hz (top node of G_{SEst} block); be careful to restore the value to 20 before saving the file, because changing the value of a node permanently modifies the *mqd* file.

    a. Adjust K_{Est} to minimize error signal, E_O.

    b. What conclusion can you draw?

Section II
Modeling

Chapter 11

Introduction to Modeling

Modeling is the foundation of practical control techniques. Applying the many methods of controls to improve command response, stability, and disturbance rejection requires a thorough understanding of the objects under control. How does a disturbance couple into the plant? What delay does the feedback device inject? How will the power converter limit the responsiveness of the system? To answer these questions, you need an accurate mathematical description of the system. In other words, you need a good model.

An important distinction for a model is whether or not it is linear, time invariant (LTI) as discussed in Section 2.2.2. LTI models are comparatively simple to analyze because they can be represented in the frequency domain. However, no realistic system is completely LTI. Non-LTI behavior, such as saturation, deadband, and parameter variation, usually requires time-domain models for analysis. Fortunately, as has been demonstrated by the examples throughout this book, given the right modeling environment, a single model can be used for both time-domain and frequency-domain analysis.

This chapter will discuss modeling techniques for the frequency and time domains. The frequency domain will be used to model an LTI element — a digital filter — including an example using a Microsoft Excel spreadsheet. The time domain will be discussed in more detail; it deserves more effort because it is more complex.

11.1 What Is a Model?

A model is a mathematical expression that describes the important relationships between the input and output of a system or component. The simplest models are proportional. For example, the model of an ideal electrical resistor is $V = IR$. Models of other electrical elements require derivatives or integrals with respect to time. As elements become more complex, the number of integrals and derivatives (the "order" of the system) increases. A simple model for a permanent-magnet brush motor requires two integrals and five parameters (inductance, resistance, back-EMF, inertia, and viscous damping) to describe just the linear relationship between the applied voltage and motor velocity.

11.2 Frequency-Domain Modeling

The frequency domain allows simple models to provide comprehensive information about linear control systems. Frequency-domain models give insight into control systems not available from time-domain models. Frequency-domain analysis is based on evaluating transfer functions of s or z. Bode plots are one alternative to displaying the results of frequency-domain modeling. In Bode plots, the system is assumed to be excited with sinusoids across a range of frequencies; in this case, s is limited to values of $j\omega$ (see Section 2.2.1), and the transfer function can affect only the phase and gain of the excitation frequencies.

Another class of frequency-domain techniques is *root locus*. Root locus solves for the roots of the closed-loop transfer function for a continuum of loop gain values. Those roots follow a trajectory, called a *locus*, as the gain increases. Root locus is commonly taught as part of engineering control system classes. Discussion of the root locus method can be found in Refs. 32, 33, 69, and 85 and almost all other controls books written for college programs.

11.2.1 How the Frequency Domain Works

The frequency domain provides transfer functions in either s, the Laplace operator, or z. Table 2-1 provides the s-domain equivalents for common time-domain functions; Table 5-1 provides z-domain equivalents. Table 2-2 provides linear transfer functions for many common plants. Using these tables and knowledge of the plant, controller, and feedback device, s-domain transfer functions can be derived that adequately represent system behavior for a wide range of operating conditions.

Once the function of s or z is derived, Bode plots make the evaluation of that function at any frequency straightforward. Replace each instance of s with $j\omega$. For example, consider a low-pass filter, $K/(s + K)$. This can be evaluated for all values of s from 10 Hz to 1000 Hz at the rate of 20 steps per decade. In this case, $f_0 = 10$ and $f_{N+1} = f_N \times 10^{1/20} = f_N \times 10^{0.05}$. Such an evaluation is so straightforward that it can be done in a Microsoft Excel spreadsheet. Such a spreadsheet is shown in Figure 11-1.

The example of Figure 11-1 assumes that complex (as opposed to real and imaginary) functions are not available in Excel. Here, the numerator and denominator must be evaluated separately and then combined. The numerator in this case is simply K. The denominator is $s + K$, where s is $j2\pi f$ and $j = \sqrt{-1}$. So the A column is f, starting at f_0 (10 Hz) and stepping in geometric increments of $10^{0.05}$. The formulas for columns B–G are shown for reference in row 5 of Figure 11-1. The B column converts hertz to radians/second and so is 2π times the A column. The C column is the numerator, which is fixed at the value K. The D column is the magnitude of the denominator, which is the sqrt([B column]^2 + [C column]^2). The E column is the angle of the denominator, atan2([C column], [B column]).

	A	B	C	D	E	F	G
1	K	628.00					
2							
3	F	ω	Num	Den-magnitude	Den-angle	T-magnitude	T-angle
4	Hz	Rad/s			Radians		Rad
5		A5*2*PI()	B$1	SQRT(C5^2+B5^2)	ATAN2(C5,B5)	C5/D5	-E5
6	10.00	62.83	628.00	631.14	0.10	1.00	-0.10
7	11.22	70.50	628.00	631.94	0.11	0.99	-0.11
8	12.59	79.10	628.00	632.96	0.13	0.99	-0.13
9	14.13	88.75	628.00	634.24	0.14	0.99	-0.14
10	15.85	99.58	628.00	635.85	0.16	0.99	-0.16
11	17.78	111.73	628.00	637.86	0.18	0.98	-0.18
12	19.95	125.37	628.00	640.39	0.20	0.98	-0.20
13	22.39	140.66	628.00	643.56	0.22	0.98	-0.22
14	25.12	157.83	628.00	647.53	0.25	0.97	-0.25
15	28.18	177.08	628.00	652.49	0.27	0.96	-0.27
16	31.62	198.69	628.00	658.68	0.31	0.95	-0.31
17	35.48	222.94	628.00	666.40	0.34	0.94	-0.34
18	39.81	250.14	628.00	675.98	0.38	0.93	-0.38
19	44.67	280.66	628.00	687.86	0.42	0.91	-0.42
20	50.12	314.91	628.00	702.53	0.46	0.89	-0.46
21	56.23	353.33	628.00	720.57	0.51	0.87	-0.51
22	63.10	396.44	628.00	742.66	0.56	0.85	-0.56
23	70.79	444.82	628.00	769.57	0.62	0.82	-0.62
24	79.43	499.09	628.00	802.17	0.67	0.78	-0.67
25	89.13	559.99	628.00	841.41	0.73	0.75	-0.73
26	100.00	628.32	628.00	888.35	0.79	0.71	-0.79
27	112.20	704.98	628.00	944.13	0.84	0.67	-0.84
28	125.89	791.01	628.00	1009.99	0.90	0.62	-0.90
29	141.25	887.52	628.00	1087.24	0.95	0.58	-0.95
30	158.49	995.82	628.00	1177.30	1.01	0.53	-1.01
31	177.83	1117.33	628.00	1281.72	1.06	0.49	-1.06
32	199.53	1253.66	628.00	1402.16	1.11	0.45	-1.11
33	223.87	1406.63	628.00	1540.45	1.15	0.41	-1.15
34	251.19	1578.26	628.00	1698.62	1.19	0.37	-1.19
35	281.84	1770.84	628.00	1878.90	1.23	0.33	-1.23
36	316.23	1986.92	628.00	2083.80	1.26	0.30	-1.26
37	354.81	2229.36	628.00	2316.12	1.30	0.27	-1.30
38	398.11	2501.38	628.00	2579.01	1.32	0.24	-1.32
39	446.68	2806.60	628.00	2876.00	1.35	0.22	-1.35
40	501.19	3149.05	628.00	3211.06	1.37	0.20	-1.37
41	562.34	3533.29	628.00	3588.67	1.39	0.17	-1.39
42	630.96	3964.42	628.00	4013.85	1.41	0.16	-1.41
43	707.95	4448.15	628.00	4492.27	1.43	0.14	-1.43
44	794.33	4990.91	628.00	5030.27	1.45	0.12	-1.45
45	891.25	5599.89	628.00	5635.00	1.46	0.11	-1.46
46	1000.00	6283.19	628.00	6314.49	1.47	0.10	-1.47

Figure 11-1. Spreadsheet evaluating $T(s) = K/(s + K)$.

The F column is the transfer function magnitude, which is the magnitude of the numerator, C column, divided by the magnitude of the denominator, D column. The G column, the transfer function angle, is the angle of the numerator (which is zero) less the angle of the denominator as calculated in the E column. Using an environment that supports complex numbers (as do some versions of Excel) makes this process less tedious.

Frequency-domain models provide the simplest models for analyzing control systems. As just demonstrated, even a spreadsheet can provide some level of analytical support. However, frequency-domain models are limited to linear, time-invariant systems. This limitation is severe enough that for many control systems, frequency-domain models are not sufficient for detailed analysis. In those cases, the time domain can be used.

11.3 Time-Domain Modeling

Time-domain models use time-based, or *temporal*, differential equations to define system operation. Simulation is the process of solving these differential equations numerically in discrete, closely spaced time steps. The many solutions, one separated from the next by a small increment of time, combine to provide a history of solutions to the differential equations. These solutions can be displayed in the time domain, such as by plotting them on a software oscilloscope or dynamic signal analyzer. Time-domain models are capable of simulating nonlinear and time-varying behavior. This is a major advantage of time-based models when compared to frequency-based models.

11.3.1 State Variables

In time-domain modeling, the differential equations are usually arranged as a set of first-order equations according to the form shown in Equation 11.1. The variables x_0, $x_1, \ldots$ in Equation 11.1 are called *state variables*, and the differential equations are called *state equations*. The number of state variables in a model is called the *order*. The inputs to the system are $u_0, u_1, \ldots$ In the general case, each state derivative can be any function of all state variables, of all inputs, and of time.

$$\frac{d}{dt}x_0 = f_0(x_0, x_1, x_2, \ldots, u_0, u_1, u_2, \ldots, t)$$
$$\frac{d}{dt}x_1 = f_1(x_0, x_1, x_2, \ldots, u_0, u_1, u_2, \ldots, t) \tag{11.1}$$
$$\cdots$$

11.3.1.1 Reducing Multiple-Order Equations

Some functions, especially filters, are provided as a single multiple-order differential equation. Fortunately, any such equation can be written in state-equation form. For example, the third-order equation

$$\frac{d^3}{dt^3}y(t) + 4\frac{d^2}{dt^2}y(t) + 7\frac{d}{dt}y(t) + 8y(t) = 10f(t) \tag{11.2}$$

can be written as a set of three first-order differential equations. One process to make this conversion is described in the following paragraphs.

First, determine the order, N, which is equal to the number of derivatives in the original equation; for the example of Equation 11.2, the order is 3. Then define state variables as the original variable and $N-1$ of its derivatives. For the example of Equation 11.2, the state variables are

$$x_0(t) = y(t); \qquad x_1(t) = \frac{d}{dt}y(t); \quad x_2(t) = \frac{d^2}{dt^2}y(t)$$

Now write the original Nth-order equation as N equations in terms of the derivatives of the state variables, $x_0(t) - x_{N-1}(t)$. Continuing the example,

$$\frac{d}{dt}x_0(t) = x_1(t)$$

$$\frac{d}{dt}x_1(t) = x_2(t)$$

$$\frac{d}{dt}x_2(t) = 10f(t) - 4[x_2(t)] - 7[x_1(t)] - 8[x_0(t)]$$

Note that this agrees with the form of Equation 11.1.

In general, for N equations there will be $N-1$ equations of the form

$$\frac{d}{dt}x_j(t) = x_{j+1}(t)$$

and one equation that has the coefficients of the original equation but with the state variables substituted for the original variable and its derivatives.

11.3.1.2 Matrix Equations

Equation 11.1 is often written in matrix form, as shown in Equation 11.3; the state variables are arranged in a vector called $\mathbf{x}$ and the inputs are arranged in the vector $\mathbf{u}$. The entire set of equations is written as a single, first-order matrix differential equation:

$$\frac{d}{dt}\mathbf{x} = \mathbf{f}(\mathbf{x}, \mathbf{u}, t) \tag{11.3}$$

When modeling in the time domain, Equation 11.3 is usually solved numerically. A differential equation solver uses the state variable initial values and evaluates the derivatives to approximate the values for the next increment of time. Those solutions, in turn, become the initial values for solving the equations in the next increment of time. The time increment must be kept small in order to keep the accuracy high. Simulation is a process of solving Equation 11.3 many times, with each successive solution inching time forward. The individual solutions act like the still frames of a movie reel to give the appearance of a continuous solution.

It should be noted that sometimes it is possible to provide a closed-form solution to Equation 11.3. However, for most practical cases this limits the system to linear, time-invariant behavior. In such cases, the frequency domain is usually preferred because it provides more useful information for most control system designers and is easier to work with. The assumption here is that the time-domain simulations will be solved numerically in an iterative process.

11.3.1.3 Time-Based Simulation

The many solutions of Equation 11.3 are the values of $\mathbf{x}(t)$ at many points in time, $\mathbf{x}(t_1)$, $\mathbf{x}(t_2)$, $\mathbf{x}(t_3)$, and so on; as a shorthand, $\mathbf{x}(t_M)$ is often written as $\mathbf{x}_M$. So a simulation is the sequence of $\mathbf{x}_M$, where M traverses 0 to the end of simulation time.

Figure 11-2 provides a detailed flowchart of a typical simulation. First, initial conditions must be set by the particular model being simulated and simulation time must be initialized to $M = 1$. (The first solution is calculated at time t_1, because the solution at t_0 must be given as the initial condition.) Then a model must provide the values of the derivatives (i.e., the right side of Equation 11.3). Next, a differential equation solver provides the values of the state variables for the next increment of time; those values become the initial conditions for the next increment in time. The process repeats either indefinitely or until t_M passes the maximum simulation time.

11.3.2 The Modeling Environment

As shown in Figure 11-2, the tasks required for time-domain simulation fall into two categories: the modeling environment and the model itself. The modeling environment provides the control of simulation time and the differential equation solver. The control of simulation time is straightforward; simulation time must be initialized and incremented throughout the solving process.

11.3.2.1 The Differential Equation Solver

The differential equation solver is the backbone of time-based simulation. The process of solving the differential equations at any one time step combines the previous solution with the slope of the derivatives as given by Equation 11.3. In the simplest case, Euler's

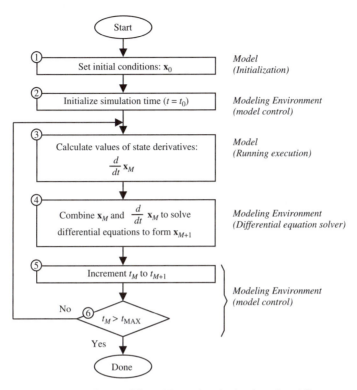

Figure 11-2. Solving differential equations for time-based modeling.

integration, the derivative is multiplied by the time step to estimate the difference from the new value. This is shown in Figure 11-3 for a single-order (nonmatrix) differential equation:

$$\frac{d}{dt}x_1 = f(x_1, u, t)$$

The operation of Euler's integration is shown in Figure 11-3. At the start of simulation time, t_0, the value of x_0 must be provided to the solver. The derivative of $x(t)$ evaluated at $t_0(\frac{d}{dt}x_0)$ is provided to the differential equation solver by the modeling equations, about which more will be said in due time. The derivative (i.e., slope) is shown in Figure 11-3 as a short, bold segment evaluated at times t_0, t_1, and t_2. That bold segment is extended to the width of the time step to estimate the next value of $x(t)$; this is shown by the arrowheads extending out of the bold segment, proceeding the width of the time step. Euler's integration is

$$x_M = x_{M-1} + \frac{d}{dt}x_{M-1} \times \Delta t \tag{11.4}$$

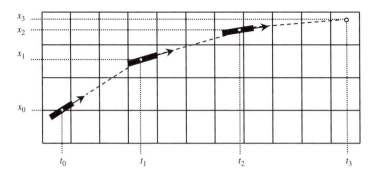

Figure 11-3. Euler's integration used to solve a single-order differential equation.

This process is easily extended to higher-order differential equations, which are simply multiple single-order equations. Although the model equations (f_0, f_1, f_2, and so on) may define interdependencies of the state variables, the differential equation solves the many single-order equations separately.

11.3.2.2 Advanced Differential Equation Solvers

Euler's integration demonstrates the principles of numerically solving differential equations. However, because Euler's integration is prone to error, it is not commonly used. It relies solely on the derivative values at the start of the time increment. Of course, the values of derivatives change as time increases, so the derivatives may vary significantly, even over a short span of time. Many differential equation solvers have been developed to correct this shortcoming. For example, *Visual ModelQ* relies on a method known as the fourth-order Runge–Kutta method (see Appendix C), which uses a weighted average of the derivative evaluated under four conditions to estimate the mean value of the derivative over the entire time interval.

The fourth-order Runge–Kutta method requires four evaluations of the state variable derivatives for each time step, whereas Euler's method requires only one. For a given model, fourth-order Runge–Kutta would require about the same computational effort as Euler's if its time step were four times longer. Actually, the Runge–Kutta method is usually more efficient than Euler's because, even with a time step four times longer than Euler's method, Runge–Kutta still has less error; in other words, Runge–Kutta is more efficient because it will provide equivalent results to Euler's with a time step longer than four times that of Euler's. In addition to Runge–Kutta, Ref. 82 lists numerous differential equation solvers.

11.3.2.3 Selecting ΔT

Selecting the increment of time (ΔT) depends on many factors, including the dynamics of the control system and the desire to display data. Differential equation solvers give good estimates only when the time increment is short enough; make it too long and the solver can actually diverge, a phenomenon in which the state variables rise to unbounded values. Make the time increment too short, and computational resources are wasted; simulations can take much longer than is necessary.

The simplest method is to fix the time increment based on the controller. For digital control systems, the time increment can often be selected as equal to the sample rate. Assuming the sample rate is appropriate for the control system, this method will produce good simulation of effects that occur slower than the sample rate. It will not, of course, provide information on phenomena that occur within a cycle. For analog controllers, the time increment can often be based on the bandwidth of the fastest loop being simulated; it should be no larger than $0.1/f_{BW}$. For example, a 100-Hz loop often requires a time increment no longer than 1 msec.

A class of differential equation solvers called *variable time step* methods can dynamically select the maximum sample time to achieve a specified accuracy in the simulation. These methods solve the differential equations and estimate the error at each time step. If the error is under an error level specified by the user, the time increment can be lengthened; if the error is too large, the time increment can be shortened. Variable time step methods can provide great benefit for systems that have sharp periodic changes, such as pulse-width-modulated (PWM) stages that turn on and off at discrete points in time; with such systems, the time increment may need to be very short near the sharp changes but can be relaxed away from the edges. The variation of time increment can be substantial; in some systems it can decrease by a factor of more than 1000 near sharp edges. Variable time step methods are discussed in Ref. 82.

11.3.3 The Model

The simulation consists of two sections: the modeling environment and the model. In most cases, modelers purchase a modeling environment such as *Matlab* or *Visual ModelQ*. The modeling environment provides model control, display and analysis tools, and differential equation solvers that are general. Usually, the modeler is responsible for building the model. This requires two steps: setting initial conditions and providing the model equations (i.e., the functions **f** in Equation 11.3).

11.3.3.1 Initial Conditions

In many cases, the initial conditions are of limited importance and often just default to zero. However, wise use of initial conditions can speed the simulation process in some models. For example, if a modeler wanted to examine the behavior of a motor rotating at 5000 RPM, but it required several minutes of simulation time for the motor to accelerate from standstill to 5000 RPM, the speed of the motor could be set to 5000 RPM as an initial condition.

11.3.3.2 Writing the Modeling Equations

Writing the modeling equations requires skill. The modeler must be familiar with both the s and the z domains. The modeler must also be cognizant of the functions of the plant, the feedback, the power converter, and the control laws. The simulation will predict only those behaviors that are described accurately in the model equations.

In this section, several specific models will be reviewed to develop example model equations. If you are using a modeling environment that accepts graphical input, be aware that the environment goes through a similar process to convert your graphical entries into equivalent model equations. Even if you use or plan to use such an environment, you should be aware of the techniques to write model equations so that you can better understand the models that your environment generates.

11.3.3.3 Modeling an RC Circuit

For most modelers, models are best derived from block diagrams. Integrals in the block diagram should be segregated from other blocks to facilitate the formation of state equations. This is true whether or not the model is for LTI systems. For example, consider the simple circuit in Figure 11-4.

The circuit in Figure 11-4 can be drawn as a block diagram, as shown in Figure 11-5. The two operations you need to understand are that the capacitor current, I_C, is equal to $(V_{IN} - V_{OUT})/R$ and that the capacitor voltage (V_{OUT}) is the integral of the capacitor current divided by the capacitance ($I_C \times 1/sC$). The block diagram should be formed so that each integration ($1/s$) is in a block by itself. Then the state variables are defined as the output of the integrators. The number of integrals defines the order of the system. Figure 11-5 represents a first-order system, which can be described with a single state equation:

$$\frac{d}{dt}x_0 = \frac{I_C}{C} = \frac{V_{IN} - x_0}{RC}$$

The algorithm to run such a model is shown in Program 11-1. This program assumes that a differential equation solver has been configured to call a routine named *Model*. *Model* fills the value of the state derivative vector, **DX**, which is assumed here to be of length 1. *Model* is assumed to have access to the model constants, R and C, as well as to the input, V_IN. *Model* also uses the state variable value, which is stored in the vector **X**, which is also assumed to be of length 1. Note that *Model* does not set the values of the state variable (**X**) directly; this task is left to the differential equation solver that calls *Model*. *Model* implements block 3 of Figure 11-2, setting the value of the state variable derivative.

```
START MODEL (X, DX):
  DX (0)= (V_IN-X(0))/(R*C)
END MODEL
```

Program 11-1. A model for an R-C filter.

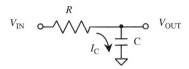

Figure 11-4. Schematic of an R-C filter.

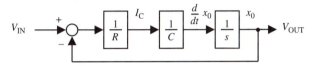

Figure 11-5. A block diagram of an R-C filter.

11.3.3.4 Modeling a Two-Pole Low-Pass Filter

Sometimes the system to be modeled is described as an *s*-domain transfer function. To be modeled in the time domain, the transfer function must be converted to time-based differential equations. A two-pole low-pass filter provides an example. The function, taken from Table 2-1, is

$$\frac{C(s)}{R(s)} = \frac{\omega_N^2}{s^2 + 2\zeta\omega_N s + \omega_N^2} \tag{11.5}$$

Equation 11.5 is a second-order equation, as indicated by the s^2 in the denominator. The order of a transfer function is equal to the highest order of s in the equation, assuming that there is at least one constant (s^0) term and no terms with a negative power of s (e.g., s^{-1}). The number of rows in the matrix of Equation 11.3 is equal to the order of the system, so there will be two rows for Equation 11.5.

The first step in converting an equation of order N into N first-order equations is to rearrange the Nth-order equation to the form

$$C(s)s^N = f(s) \tag{11.6}$$

This form can be created for Equation 11.5 with algebra:

$$C(s)s^2 = R(s)\omega_N^2 - 2\zeta\omega_N C(s)s - \omega_N^2 C(s) \tag{11.7}$$

Next, form the differential equation, recognizing that each power of s is equivalent to differentiation:

$$\frac{d^2}{dt^2}C(t) = \omega_N^2 R(t) - 2\zeta\omega_N \frac{d}{dt}C(t) - \omega_N^2 C(t) \tag{11.8}$$

At this point, the second-order equation is converted to two first-order time-domain equations, as was done for Equation 11.2:

$$x_0 = C(t) \tag{11.9}$$

$$x_1 = \frac{d}{dt}C(t) = \frac{d}{dt}x_0 \tag{11.10}$$

$$\frac{d}{dt}x_1 = \omega_N^2 R(t) - 2\zeta\omega_N x_1 - \omega_N^2 x_0 \tag{11.11}$$

The block diagram for the low-pass filter can be drawn from Equations 11.9–11.11, as shown in Figure 11-6. To verify this process, the transfer function of Equation 11.5 can be derived directly from Figure 11-6 using Mason's signal flow graph (section 2.4.2) or by twice applying the $G/(1 + GH)$ rule, first to the middle loop and then to the outer loop.

The algorithm to run a model of a low-pass filter is shown in Program 11-2. This program assumes that a differential equation solver has been configured to call a routine named *Model*. *Model* fills the values of the vector **DX**, which is assumed to be of length 2. *Model* is assumed to have access to the model constants, OMEGA and ZETA, as well as the input, R. *Model* also uses the state variable values, which are stored in the vector **X**, which is also assumed to be of length 2. Again, the differential equation solver estimates the values of **X** based on the values of **DX** set by *Model*, and *Model* implements only block 3 of Figure 11-2.

```
START MODEL (X, DX):
  DX(0) = X(1)
  DX(1) = OMEGA ^2 * R - 2 * ZETA * OMEGA * X(1) - OMEGA ^2 * X(0)
END MODEL
```

Program 11-2. A model for a low-pass filter.

11.3.3.5 Modeling an Analog PI Controller

Modeling a control system requires combining models of the controller, the plant, the feedback, and the power converter. The order of the system model is the sum of the

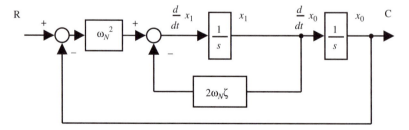

Figure 11-6. Block diagram for Equations 11.9–11.11.

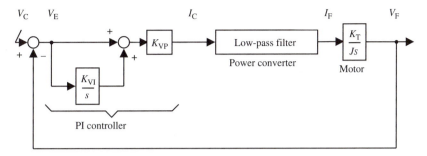

Figure 11-7. Block diagram of motion control system.

orders of the separate components of the system. For example, consider the velocity controller of Figure 11-7. It is a PI controller with a power converter (modeled itself by a low-pass filter) and a motor.

The velocity controller of Figure 11-7 is drawn in state-space form in Figure 11-8. The conversion from standard control system block diagrams to state space requires only a few steps. The low-pass filter for the power converter is exchanged for the state-space form developed in Figure 11-6. The integral terms ($1/s$) are separated, and each is labeled with a state-space variable, x_0 through x_3.

The state-space equations for the system of Figure 11-8 can be written out. One technique to simplify this process for numerical solutions is to write the equations in terms of physically meaningful intermediate results. For example, current command (I_C) and velocity error (V_E) can be used to collect the results of the PI controller. The state-space equations for Figure 11-8 are:

$$V_E = V_C - x_0 \tag{11.12}$$

$$\frac{d}{dt} x_3 = K_{VI} \times V_E \tag{11.13}$$

$$I_C = K_{VP} \times (V_E + x_3) \tag{11.14}$$

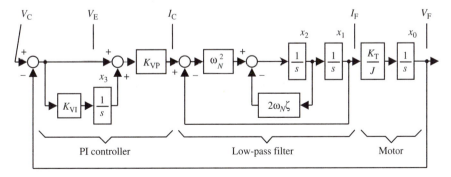

Figure 11-8. Block diagram of motion control system in state-space form.

$$\frac{d}{dt}x_2 = \omega_N^2 I_C - 2\zeta\omega_N x_2 - \omega_N^2 x_1 \tag{11.15}$$

$$\frac{d}{dt}x_1 = x_2 \tag{11.16}$$

$$\frac{d}{dt}x_0 = \frac{K_T}{J}x_1 \tag{11.17}$$

The algorithm to run such a model is shown in Program 11-3. *Model* fills the values of the state derivative vector (**DX**), which is assumed here to be of length 4. *Model* is assumed to have access to the model constants, KVI, KVP, KT, J, ZETA, and OMEGA. *Model* also uses the state variable values, which are stored in the vector **X**, which is assumed to be of length 4.

```
START MODEL (X, DX):
  VE = VC − X (0)
  DX(3) = KVI * VE
  IC = KVP * (VE + X(3))
  DX(2) = OMEGA ^2 * IC-2 * ZETA * OMEGA * X(2) − OMEGA ^2 * X(1)
  DX(1) = X(2)
  DX(0) = KT * X(1)/J
END MODEL
```

Program 11-3. A model for an analog PI control system.

11.3.3.6 Modeling a Digital PI Controller

The analog PI controller can be converted to a digital controller by replacing the integral of the analog controller with a summation and by adding a sample-and-hold. This assumes the current loop is still analog; this can be implemented by configuring the PI controller to calculate a current command, which is delivered to an analog current loop through a D/A converter. The digital controller is shown in Figure 11-9.

When modeling digital control systems, only the analog components contribute states; the summations and differences of digital controllers are calculated at each sample by the *Model*. The system of Figure 11-9 would require the differential equation solver to solve just three states, the states from Figure 11-8, excluding the control law integrator (x_3). This is shown in Figure 11-10.

The state equations for the digital PI controller are the same as the equations for states x_0–x_2 of Figure 11-7 (Equations 11.15–11.17.) However, Equations 11.12–11.14 are replaced by the digital PI algorithms:

$$V_E = V_C - x_0 \tag{11.18}$$

$$V_{E\_INT} = V_{E\_INT} + T \times V_E \tag{11.19}$$

$$I_C = K_{VP} \times (V_E + K_{VI} \times V_{E\_INT}) \tag{11.20}$$

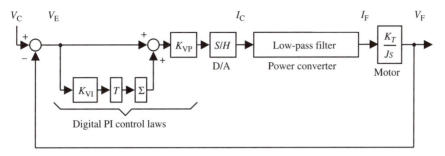

Figure 11-9. Block diagram of a digital motion control system.

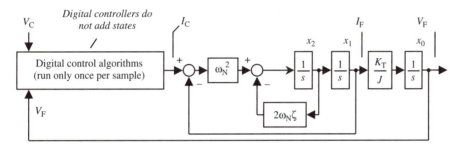

Figure 11-10. State block diagram of digital PI controller.

These algorithms must be executed once and only once for each cycle of the digital controller. Because of this, the routine *Model* must be able to determine the times at which the controller will sample data. This requires *Model* to have another calling argument, TIME, in addition to the **X** and **DX** vectors. A typical algorithm to simulate digital sampling is shown as the first six lines of Program 11-4.

```
START MODEL (X, DX, TIME):
  START IF (TIME > T_NEXT_SAMPLE)
    VE = VC - X(0)
    VE_INT = VE_INT + T * VE
    IC = KVP * (VE + KVI * VE_INT)
    T_NEXT_SAMPLE = T_NEXT_SAMPLE + T_SAMPLE
  END IF
  DX(2) = OMEGA ^2 * IC-2 * ZETA * OMEGA * X(2) - OMEGA ^2 * X(1)
  DX(1) = X(2)
  DX(0) = KT * X(1)/J
END MODEL
```

Program 11-4. A model for a digital PI control system.

11.3.3.7 Adding Calculation Delay

Program 11-4 does not include delay for the time it takes the digital controller to calculate the control laws. This can be done by adding a few new variables. The variable t_{OUTPUT} holds the estimated output time based on sample time and calculation time, $t_{CALCULATE}$. The variable I_{C_STORE} holds the new output of the control law until simulation time passes t_{OUTPUT}. (Before t_{OUTPUT}, it is assumed I_C will retain the output of the control law from the previous sample.) This is shown in Program 11-5.

```
START MODEL (X, DX, TIME):
  START IF (TIME > T_NEXT_SAMPLE)

    REMARK: THIS IS THE DIGITAL CONTROLLER
    VE = VC - X(0)
    VE_INT = VE_INT + T * VE

    REMARK: STORE IC IN IC_STORE FOR T_CALCULATE SECONDS
    IC_STORE = KVP * (VE + KVI * VE_INT)
    T_NEXT_SAMPLE = T_NEXT_SAMPLE + T_SAMPLE
    T_OUTPUT = TIME + T_CALCULATE
  END IF
  START IF (TIME > T_OUTPUT)
    IC = IC_STORE
  END IF
  DX(2) = OMEGA ^2 * IC-2 * ZETA * OMEGA * X(2) - OMEGA ^2 * X(1)
  DX(1) = X(2)
  DX(0) = KT * X(1)/J
END MODEL
```

Program 11-5. A model for a digital PI control system with calculation time.

11.3.3.8 Adding Saturation

Program 11-5 does not include saturation of the current controller. However, all current controllers have a maximum output current. Program 11-6 is the system of Figure 11-9 augmented to include saturation. As discussed in Section 3.8, provisions should be made in the control law to prevent integral windup while the current controller is saturated. Program 11-6 includes provisions to limit windup to 120% of the current controller's maximum output.

```
START MODEL (X, DX, TIME):
  START IF (TIME > T_NEXT_SAMPLE)

    REMARK: THIS IS THE DIGITAL CONTROLLER
    VE = VC - X(0)
    VE_INT = VE_INT + T * VE

    REMARK: STORE IC IN IC_STORE FOR T_CALCULATE SECONDS
```

```
IC_STORE = KVP * (VE + KVI * VE_INT)

REMARK: POSITIVE ANTI-WIND-UP
START IF (IC_STORE > IMAX * 1.2)
  VE_INT = (IMAX * 1.2/KVP-VE)/KVI
END IF

REMARK: NEGATIVE ANTI-WIND-UP
START IF (IC_STORE < - IMAX * 1.2)
  VE_INT = (-IMAX * 1.2/KVP-VE)/KVI
END IF

REMARK: POSITIVE CLAMPING
START IF (IC_STORE > IMAX)
  IC_STORE = IMAX
END IF

REMARK: NEGATIVE CLAMPING
START IF (IC_STORE < -IMAX)
  IC_STORE = -IMAX
END IF
T_NEXT_SAMPLE = T_NEXT_SAMPLE + T_SAMPLE
END IF
START IF (TIME > T_OUTPUT)
    IC = IC_STORE
END IF
DX(2) = OMEGA ^2 * IC-2 * ZETA * OMEGA * X(2) - OMEGA ^2 * X(1)
DX(1) = X(2)
DX(0) = KT * X(1)/J
END MODEL
```

Program 11-6. A model for a digital PI control system with saturation.

11.3.4 Frequency Information from Time-Domain Models

As has been discussed throughout this book, controls engineers need both time- and frequency-domain information to understand thoroughly the operation of control systems. For this reason, many modeling environments provide the ability to cull frequency-domain information from time-domain models. For example, *Visual ModelQ* provides the software dynamic signal analyzer (DSA), which excites the time-domain model with a pseudorandom input and analyzes the response to calculate the Bode plot of the system or element.

When time-domain models are used to produce frequency-domain information from non-LTI systems, such information may be valid at only a single operating point. For this reason, this type of model is often called an *operating point* model or, equivalently, a *small-signal* or *linearized* model. If the operating point changes, so may the frequency response. For example, if the inertia of a mechanism varied with position, the Bode plot would be valid for the mechanism in only one position; each new position would generate a unique Bode plot. Note that LTI systems have only one Bode plot for all operating conditions.

A detailed discussion of how frequency-domain information is derived from time-domain models is beyond the scope of this book. However, you should be aware of this feature because it is important for controls engineers to have access to a modeling environment that can perform this analysis. Otherwise, you may find yourself in the unenviable position of needing to develop two models: one for the frequency domain and another for the time domain.

11.4 Questions

1. Provide the modeling equations for the following figure. Assume an analog PI controller ($G_C(s)$), a single-pole low-pass filter with a break frequency of 200 Hz for the feedback filter ($H(s)$), and a model of a single-pole low-pass filter with a break frequency of 100 Hz for the power converter ($G_{PC}(s)$).

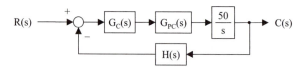

2. Repeat Question 1, but change the control law to PID, with the derivative filtered by a single-pole low-pass filter with a bandwidth of 60 Hz.
3. Repeat Question 1, but change $G_C(s)$ to a digital PI controller sampling at 1 kHz.

Chapter 12

Nonlinear Behavior and Time Variation

Systems with linear, time-invariant (LTI) behavior can be understood using frequency-domain techniques, such as the open-loop method. Unfortunately, all control systems include regions of nonlinear operation, and many have significant parameter variation. This chapter will review non-LTI behavior and provide steps that can be taken to understand it and compensate for its impact.

This chapter will also survey a number of common nonlinear behaviors. Several topics for each behavior will be addressed: what problems it causes, how it is modeled, and how to compensate for it in the controller. There are so many types of non-LTI behavior that it is not practical to cover all of them. Instead, the goal here is to cover several common examples and to provide an approach for understanding and correcting non-LTI behavior in general.

12.1 LTI Versus non-LTI

As discussed in Section 2.2.2, linear, time-invariant behavior requires three characteristics. Consider a system with an input $r(t)$ and an output $c(t)$:

1. *Homogeneity*: If $r(t)$ generates $c(t)$, then $k \times r(t)$ generates $k \times c(t)$.
2. *Superposition*: If $r_1(t)$ generates $c_1(t)$ and $r_2(t)$ generates $c_2(t)$, then $r_1(t) + r_2(t)$ generates $c_1(t) + c_2(t)$.
3. *Time invariance*: If $r(t)$ generates $c(t)$, then $r(t - \tau)$ generates $c(t - \tau)$.

Examples of linear behavior include addition, subtraction, scaling by a fixed constant, integration, differentiation, time delay, and sampling. No practical control system is completely linear, and most vary over time.

When a system is LTI it can be represented completely in the frequency domain. That is, it can be described as a function of s and z. Bode plots for command and disturbance response can be calculated relatively easily. Non-LTI systems can also have Bode plots, but, unlike Bode plots for LTI systems, those plots change depending on the system operating conditions.

12.2 Non-LTI Behavior

Non-LTI behavior is any behavior that violates one or more of the three criteria just listed for LTI systems. The most common nonlinear behaviors are gains that vary as a function of operating conditions. There are many examples:

- The inductance of a steel-core inductor will decline as the current in the inductor increases.
- The gain of most power converters falls sharply when the command exceeds the converter's rating.
- The rotational inertia of a robotic arm varies with the position of some of the axes.
- The capacitance of many electrolytic capacitors declines as the electrolyte ages.
- Many motors produce torque proportional to the product of two variables: field and rotor flux; motors that allow both to vary are nonlinear because the operation is described with the multiplication of two variables.

12.2.1 Slow Variation

Within nonlinear behavior, an important distinction is whether the variation is *slow* or *fast* with respect to the loop dynamics. The simplest case is when the variation is slow. Here, the nonlinear behavior may be viewed as a linear system with parameters that vary during operation. For example, consider Figure 12-1, a thermal controller with ideal power conversion and feedback and with slowly varying thermal capacitance ($C(t)$). (Notice that here $C(t)$ is used for thermal capacitance rather than for *controlled output* as is otherwise the convention in this book.)

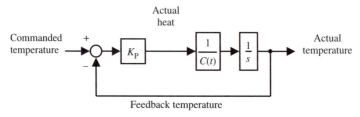

Figure 12-1. A thermal controller with varying thermal capacitance.

If the variation of capacitance is slow with respect to the loop dynamics, the control system can still be described with frequency-domain equations. Assuming slowly varying capacitance, the command response of Figure 12-1 can be approximated with Equation 12.1:

$$T(s) \approx \frac{K_P/C(t)}{s + K_P/C(t)} \tag{12.1}$$

The nonlinear behavior of the capacitance affects the loop dynamics, but because the variation is slow, the dynamics can still be characterized effectively with a transfer function. In Equation 12.1, the sole effect on command response of varying the thermal capacitance is in varying the bandwidth. The value of $C(t)$ at any given time is called the *operating point*. At different operating points, the Bode plot will change according to Equation 12.1.

12.2.2 Fast Variation

If the variation of the loop parameter is *fast* with respect to the loop dynamics, the situation becomes more complicated. Transfer functions cannot be relied upon for analysis. The definition of *fast* depends on the system dynamics. For example, consider the familiar equation for torque and acceleration with a fixed inertia:

$$T(t) = J\frac{d}{dt}\omega(t) \tag{12.2}$$

If the inertia varies, the equation becomes more complex:

$$\begin{aligned} T(t) &= \frac{d}{dt}(J(t)\omega(t)) \\ &= J(t)\frac{d}{dt}\omega(t) + \omega(t)\frac{d}{dt}J(t) \end{aligned} \tag{12.3}$$

The variation of $J(t)$ would be *slow* in Equation 12.3 if the first term dominated and the second could be ignored; it would be *fast* if the second term was significant.

The impact of fast variation on frequency-domain techniques is large. For instance, in Equation 12.3, there is no longer a pure integral relationship between torque and velocity. The block diagrams used in earlier chapters to represent motion systems are not valid. Compare this to the impact of *slow* variation; in that case, Equation 12.2 would still be valid, albeit with variation in J.

For control systems, the line between "fast" and "slow" is determined by comparing the rate at which the gain changes to the settling time of the controller. If parameter variation occurs over a time period no faster than ten times the control loop settling time, the effect can be considered slow for most applications. Fortunately, for most disciplines requiring controls, conditions where fast, substantial nonlinear behavior seriously affects system performance are rare.

12.3 Dealing with Nonlinear Behavior

Nonlinear behavior can usually be ignored if the changes in parameter values affect the loop gain by no more than a couple of decibels, or, equivalently, about 25%. A variation this small will be tolerated by systems with reasonable margins of stability. If a parameter varies more than that, there are at least three courses of action: Modify the plant, tune for the worst-case conditions, or compensate for the nonlinearity in the control loop.

12.3.1 Modify the Plant

Modifying the plant to reduce the variation is the most straightforward solution to nonlinear behavior from the control system designer's perspective. It cures the problem without adding complexity to the controller or compromising system performance. This solution is commonly employed, in the sense that components used in control systems are generally better behaved (that is, closer to LTI) than components used in "open-loop" systems.

Unfortunately, enhancing the LTI behavior of a loop component can increase its cost significantly. This is one reason why components for control systems are often more expensive than open-loop components. For example, "servo" hydraulic valves, valves that can regulate fluid flow almost linearly with the control signal, especially near zero flow, cost substantially more than simple on/off hydraulic valves. As a second example, reducing inductance variation in a steel-core inductor usually requires increasing the physical size and cost of the inductor.

For most mature control systems, the manufacturer has, over several years, chosen the plant, power converter, and feedback devices to be well behaved in the application. An implicit part of this process is usually selecting components that have near-LTI behavior. In many cases, components are selected on an empirical basis and the designers may even be unaware of the concepts of LTI operation; however, they do recognize which components make the control system work well.

After the product has matured, the opportunity to improve the plant declines. If the product has been successful, you can usually assume that the original designers did a reasonable job balancing the need for LTI behavior and low cost. In a few cases, a mature machine may employ an inappropriate component, and the substitution of a counterpart may substantially improve the control system. On the other hand, young engineers arriving in the middle of a project sometimes suggest changes that are based on a better understanding of control systems but without being well grounded in cost-to-value trade-offs implicit in the design of a machine. The engineer who is new to a project would be wise to assume the system components are well chosen absent evidence to the contrary.

The usual reason to reexamine the selection of system components in a mature machine is an advance in technology that makes new components available for the system.

An example in motion control is the advent of highly resolved, highly accurate sine encoders (see Section 14.5.2). These encoders provide many times the accuracy and resolution (both make the encoder more linear) of the traditional, digital encoders, and they do so at a reasonable premium. Several years ago, achieving equivalent accuracy and resolution was not affordable except in exotic applications. It will be several years before all the machines that deserve this improved feedback device will be redesigned to employ it.

12.3.2 Tuning for Worst Case

Assuming that the variation from the non-LTI behavior is slow with respect to the control loop, its effect is to change gains, as shown in Figure 12-1. In that case, the operating conditions can be varied to produce the worst-case gains while tuning the control system. Doing so can ensure stability for all operating conditions.

Tuning the system for worst-case operating conditions generally implies tuning the proportional gain of the inner loop when the plant gain is at its maximum. This ensures that the inner loop will be stable in all conditions; parameter variation will only lower the loop gain, which will reduce responsiveness but will not cause instability. To achieve this worst-case condition, it may be necessary to offset the command by a DC value and then add an AC signal to observe dynamic response. For example, if plant gain is maximized by high DC current, as it is with a saturated inductor, the command might be a small-amplitude square wave with a large DC offset.

The other loop gains (inner loop integral and the outer loops) should be stabilized when the plant gain is minimized. This is because minimizing the plant gain reduces the inner loop response; this will provide the maximum phase lag to the outer loops and again provides the worst case for stability. So tune the proportional gain with a high plant gain and tune the other gains with a low plant gain to ensure stability in all conditions.

The penalty in tuning for worst case is the reduction in responsiveness. Consider first the proportional gain. Because the proportional term is tuned with the highest plant gain, the loop gain will be reduced at operating points where the plant gain is low. As an example, suppose a system under proportional control had a maximum inner loop bandwidth of 10 Hz, and suppose also that the plant varies by a factor of 4 over normal operating conditions. The bandwidth at maximum plant gain might be 10 Hz; but when the plant gain falls by 4, this bandwidth would also fall by about 4 to 2.5 Hz. In general, you should expect to lose responsiveness in proportion to plant variation. To see the difficulties for yourself, try to tune Experiment 6B with one set of gains for values of G from 500 to 2000.

12.3.3 Gain Scheduling

Compensating for nonlinear behavior in the controller requires that a gain equal to the inverse of the non-LTI behavior be placed in the loop. This is called *gain scheduling*.

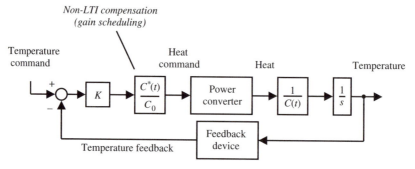

Figure 12-2. Compensating for nonlinear behavior with gain scheduling.

Figure 12-2 shows such compensation for the temperature controller of Figure 12-1. By using gain scheduling, the impact of the non-LTI behavior is eliminated from the control loop.

The gain scheduling shown in Figure 12-2 assumes that the non-LTI behavior can be predicted to reasonable accuracy (generally ±25%) based on information available to the controller. This is often the case. For example, in a motion control system, if the total inertia depends partially on the object being moved and the controller has access to that object's inertia, then the proportional loop gain can be adjusted accordingly. On the other hand, if inertia is changing because material is being wound on a roll and the controller has knowledge of the approximate roll diameter, then the diameter can

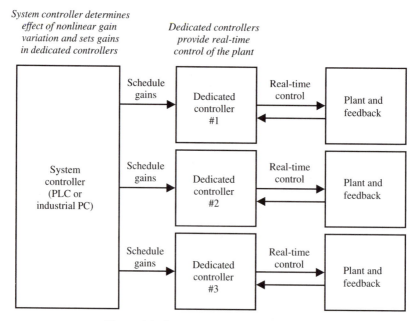

Figure 12-3. System control "scheduling" gains.

be combined with the density of the material to estimate the roll inertia and, again, the proportional gain can be adjusted. Note that in all the controllers used in this book, changing the loop gain (as when compensating for inertia variation) requires only a proportional adjustment to K_P (or K_{VP} for velocity controllers).

Many times, a dedicated control loop will be placed under the direction of a larger system controller, such as an industrial PC or a *programmable logic controller* (PLC). Compensating for non-LTI behavior in dedicated controllers is inconvenient because they are usually difficult to reprogram. In these cases, the more flexible system controller can be used to accumulate information on a changing gain and then to modify gains inside the dedicated controllers to affect the compensation. This is shown in Figure 12-3.

The chief shortcoming of gain scheduling via the system controller is limited speed. The system controller may be unable to keep up with the fastest plant variations. Still, this solution is commonly employed because of the system controller's higher level of flexibility and broader access to information.

12.4 Ten Examples of Nonlinear Behavior

This section will review ten common nonlinear behaviors. For each, the discussion will include a description of the behavior (the model), the problems it creates in control systems, and a method for compensating for the behavior where appropriate.

12.4.1 Plant Saturation

Plant saturation is the decline of a parameter corresponding to an operating condition. Inductance falls as current increases; similarly, torque in motors declines as current increases. In the linear range of operation, the output current of a bipolar transistor is approximately proportional to base current, but the constant of proportionality falls as the output current increases. Figure 12-4 shows an example saturation curve.

Saturation often affects a gain that is in series with the control loop. If the saturating gain is in the denominator of the plant, it will increase the loop gain; if it is in the

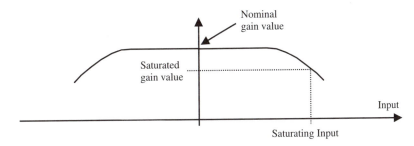

Figure 12-4. Saturating gain vs. input value.

numerator, it will reduce the gain. Raising the plant gain has the same effect as raising the proportional gain (K_P) did in Experiment 3A: the gain crossover will shift to higher frequencies, usually decreasing the phase and gain margins. In the worst case, saturation can raise the loop gain enough to destabilize the loop. For example, consider the proportional current controller in Figure 12-5, which controls an inductor with saturation behavior like that in Figure 12-4.

A model for saturation of an inductor could be mapped to a polynomial such as

$$L(i_{\text{LOAD}}) = L_0 - L_2 \times i_{\text{LOAD}}^2 \tag{12.4}$$

where L_0 and L_2 are constants that are determined empirically. If the loop is tuned with low current where the inductance is maximized, the plant gain and, thus, the loop gain will be minimized. When the inductance saturates (decreases), the loop gain rises; if the inductor saturates sufficiently, an uncompensated loop will become unstable at high currents.

The controller can compensate for this behavior by scheduling the gain factor

$$G_C(i_{\text{LOAD}}) = \left(1 - \frac{L_2}{L_0} \times i_{\text{LOAD}}^2\right)$$

before the power converter; the control laws would see a constant inductance, L_0, independent of the load current. Such a term would compensate the control loop for inductance variation as long as the variation was slow. Experiment 12A models the system of Figure 12-5 for the interested reader, including inductance variation according to Equation 12.4.

12.4.2 Deadband

Deadband is the condition where the gain is zero or very low for small inputs. As shown in Figure 12-6, the average gain of a function with deadband is very low for

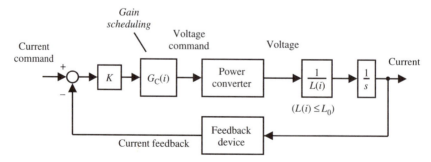

Figure 12-5. Proportional current controller with gain scheduling.

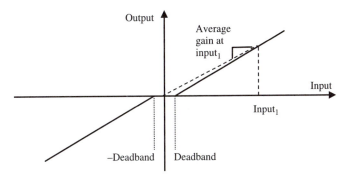

Figure 12-6. Deadband.

small inputs and grows larger as the input increases. Deadband is typical of power output stages from audio speaker drivers (where it is called "crossover distortion") to pulse-width modulation (PWM) inverters.

Deadband can be modeled using a function similar to that shown in Figure 12-6:

```
IF (INPUT > DEADBAND)
  OUTPUT = INPUT − DEADBAND
ELSE IF (INPUT < −DEADBAND)
  OUTPUT = INPUT + DEADBAND
ELSE
  OUTPUT = 0
END IF
```

Note that these equations are scaled for unity gain; that is, they assume that if there were no deadband, *output* would equal *input*.

Deadband can be compensated by gain scheduling with an inverse-deadband function, which has a vertical or nearly vertical gain near zero as shown in Figure 12-7.

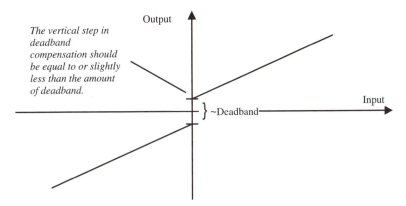

Figure 12-7. Deadband compensation.

This correction can be provided in analog circuits or with algorithms in digital controllers. Deadband is usually not known precisely, so the compensation should use a value that is equal to or less than the actual deadband. If the compensation is less than the actual deadband, the deadband will be reduced but not removed. However, if the compensation is larger than the actual deadband, the deadband will be over-compensated, which can cause sustained oscillations when the input is near the dead-band level.

Figure 12-8 shows a controller with a power converter that has deadband. The deadband compensation feature is added between the control law and the power converter. For the interested reader, Experiment 12B models the system of Figure 12-8.

Integral gain can be used to remove the effect of deadband at low frequencies. It can be used alone or to augment the deadband compensator, which is fast but imperfect at low frequencies because the precise level of deadband is rarely known.

12.4.3 Reversal Shift

Reversal shift is the change in gain when the polarity of an input or output is reversed. The effect is shown in Figure 12-9. The model is

```
IF(INPUT > 0)
    OUTPUT = G_P × INPUT
ELSE
    OUTPUT = G_N × INPUT
END IF
```

For example, heating, ventilation, and air conditioning (HVAC) systems are usually able to source more heat than they can remove. Left uncorrected, this can cause a significant loop-gain variation when the polarity changes. This type of nonlinearity is usually easy to compensate by gain scheduling two gains in the controller, one when the output is positive and the other when it is negative. Experiment 12C models a system with gain reversal and gain scheduling to compensate.

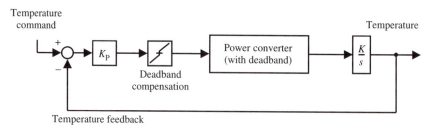

Figure 12-8. Deadband compensation in a control system.

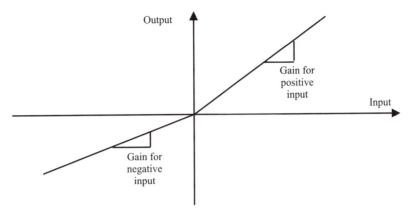

Figure 12-9. Reversal error.

12.4.4 Variation of Apparent Inertia

Apparent inertia is the inertia of the load as sensed by the motor. Apparent inertia in motion control systems frequently varies during machine operation. This variation can result from two sources: Either the mass varies or mechanical linkages change the apparent inertia. The variation of mass is straightforward. In discrete processes (moving one object at a time), the inertia of the object increases proportionally to its mass. In continuous processes, such as where material is rolled and unrolled, the inertia of a material roll will decline as the roll is unwound.

Sometimes mechanical linkages vary rotational inertia without changing mass. For example, on the two-axis robot arm shown in Figure 12-10, the rotational inertia reflected to the shoulder motor varies depending on the angle of the elbow motor (Θ). If the elbow angle is small (Θ_1), the mass is close to the shoulder, so the rotational inertia is small. If the elbow angle is near 180°, the mass is far away from the shoulder, so the inertia increases. The precise description of inertia as a function of angle depends on the dimensions of the mechanism; for this example, it can be referred to as $J(\Theta)$.

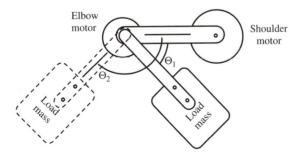

Figure 12-10. The inertia apparent to the shoulder depends on the angle of the elbow.

The impact of inertia variation can be large. The inertia of a two-meter roll of foil is enormous compared with the core upon which it is wound. A small mass on an extended robot arm may appear to the base ten times larger than the same mass pulled in close to the base. The problem is similar to heating liquid baths, where the amount of liquid may vary greatly. In such cases, the variation must be compensated in the control loop with gain scheduling, as shown in Figure 12-11; note that this figure accounts for imperfect knowledge of the rotational inertia, J^*, which will often differ from the actual inertia, J. Variation of inertia is almost always slow when compared to the power converter.

If the inertia grows too large, the system may experience limitations due to mechanical resonance. In this case, it may be impractical to compensate fully for inertia variation with gain scheduling. Resonance is discussed in Chapter 16.

12.4.5 Friction

Friction is force exerted between two objects, one sliding on the other. Friction is dependent on the relative speed of the two objects. A commonly used model of friction shows three components of force: Coulomb (sliding) friction, viscous damping, and stiction. As shown in Figure 12-12, these forces depend on speed. As a first approximation, Coulomb friction is said to have a constant magnitude, changing only in sign with the direction of sliding. Stiction, or *breakaway friction*, is a fixed value when the motor speed is zero and zero when the motor is moving. Viscous damping is proportional to speed. All three forces are opposite to the direction of sliding; that is, all three act to slow motion.

An improved model of friction recognizes that the transition from zero motion (stiction) to sliding (Coulomb friction) is not perfectly sharp [2]. The slope of friction is negative for a small transition of velocity from zero. This is called the *Stribeck effect* and is shown in Figure 12-13. The Stribeck effect occurs at low speeds, generally less than 0.1 RPM.

The Stribeck effect is a destabilizing influence on control systems; it causes a phenomenon known as *stick-slip*. In systems with significant stick-slip, small distances cannot be traversed with accuracy. When the system produces enough force to break free from stiction, sliding begins, but because the frictional force declines, the motion is difficult to stop. The result is often a limit cycle where the motion controller moves

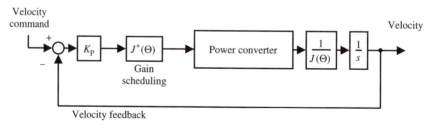

Figure 12-11. Compensation for an inertial variation using gain scheduling.

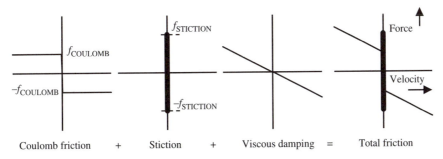

Figure 12-12. Simple model for friction.

past the target point, stops, reverses to correct, passes the target in reverse, and the cycle repeats indefinitely. Typical limit cycles occur at low frequencies (~1 Hz) and small amplitudes (~0.1 mm) [100].

The larger stiction is, compared with Coulomb friction, the more the system will produce stick-slip limit cycles. When stiction is less than 70% of Coulomb friction, stick-slip is eliminated [2]. (Note that although Figure 12-13 shows stiction being larger than Coulomb friction, in a minority of cases Coulomb friction will be the larger of the two.) Stick-slip is exacerbated when the coupling between the motor and the load is compliant. The transmission between a motor and load comprises many components, such as gearboxes, couplings, and belt drives. Each of these components has finite compliance; in other words, each can be thought of as a spring. Each transmission component is connected to the next, so the springs are all in series, as shown in Figure 12-14. The total spring constant of the transmission is the inverse of the sum of the inverted individual spring constants:

$$K_{\text{Total}} = \frac{1}{\frac{1}{K_{\text{Coupling}}} + \frac{1}{K_{\text{Gearbox}}} + \frac{1}{K_{\text{LeadScrew}}}} \tag{12.5}$$

When compliance is relatively high, stick-slip will worsen. A compliant transmission will require the motor to "wind up" the transmission before enough force can be transmitted to break the load free. Moving the load a small distance with a compliant

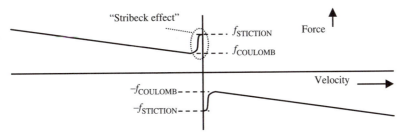

Figure 12-13. Frictional model recognizing the Stribeck effect.

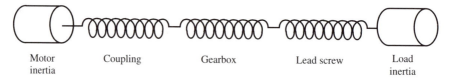

Figure 12-14. Compliance between motor and load comes from transmission components.

coupling is more difficult because, once broken free of the stiction, the load will swing further before coming to rest.

The transition from positive to negative motion is shown as a vertical line in Figures 12-12 and 12-13. During this transition, the frictional force is often modeled according to the Dahl effect. During the Dahl effect there is no continuous motion and the frictional force is proportional to the relative distance between the two objects. When that force exceeds the stiction, the object begins to move and friction is again described by Figure 12-13. The Dahl effect takes place over small distances; the typical distance for an object to break free is 10^{-5} m [2]. However, when the coupling between motor and the load is compliant, the distance is magnified; the motor must wind up the compliant coupling to exert the necessary force on the load, and that can require considerable movement in the motor.

The Dahl effect can be visualized as bristles on a sliding surface. These bristles form a spring when sliding stops: During zero motion, the force of friction is proportional to the relative position of the two objects. When the positional difference is large enough to break the bristle contact (that is, the applied force is larger than the stiction), the bristles break free, sliding begins; the frictional force is then described by Figure 12-13.

The bristle model is depicted in Figure 12-15. In the first frame of that figure (at left) there is no force on the movable object and the bristles are straight. In the second frame, a small force is applied and the bristles begin to bend; here the movable object

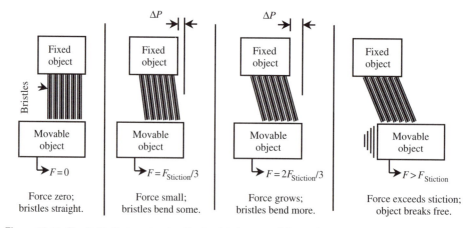

Figure 12-15. The Dahl effect can be visualized as bristles on a sliding surface. In these frames, the force on the movable object increases until it breaks free.

moves a bit and stops. The force is proportional to ΔP, the relative distance between the two objects. In the third frame, the force increases but remains below the stiction; the movable object moves a bit more, the bristles bend more, and again the object comes to rest. In the fourth and final frame, the force applied to the movable object exceeds the stiction and the object breaks free; the bristles slide on the movable object and frictional forces are described no longer by the Dahl effect but by the friction-versus-velocity model of Figure 12-13.

12.4.5.1 Compensating for Friction

Coulomb friction and viscous damping can be compensated for in the controller [54,86]. Uncompensated Coulomb friction prevents the motor from proceeding smoothly through direction changes. Compensation techniques correct this problem. Coulomb friction is often compensated using a feed-forward term based on the velocity command. The Coulomb friction is estimated as a fixed value that is added to or taken away from the motor torque (current) command according to the sign of the velocity command. Using command velocity is often preferred to feedback velocity because Coulomb friction changes instantaneously as the motor goes through zero speed; using the velocity feedback signal implies that the compensating force would jump each time the motor dithers through zero speed, as it will when the velocity command is zero. One alternative is to use the velocity feedback signal with hysteresis larger than the amount of velocity dither; unfortunately, this does not recognize that the Coulomb friction is zero when the motor is at rest.

Viscous damping causes few problems in motion systems and so is compensated less often. When viscous damping is compensated, it is nulled using velocity feedback to estimate a force equal in magnitude but opposite in sign to the viscous damping force. The compensated controller for Coulomb friction and viscous damping is shown in Figure 12-16.

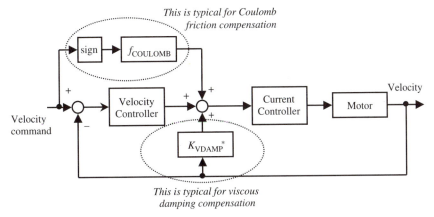

Figure 12-16. Typical compensation techniques for Coulomb friction and viscous damping.

Stiction can also be compensated in the controller, but this is more difficult than compensating for Coulomb friction or viscous damping because the behavior of the Stribeck effect on any given machine is difficult to characterize. The simplest step is to add a small amount of deadband in the position-sensing algorithms [100]; this removes the limit cycle essentially by ignoring the effects of stiction. There are more complex methods of stiction compensation [46], but they are not commonly implemented on general-purpose controllers.

For many machines, the primary means of dealing with stiction is to minimize it through mechanical design. If stiction is reduced to less than 70% of the Coulomb sliding friction, stiction will not induce limit cycles [2]. Stiction in the machine ways is reduced by use of low-friction interfaces such as roller bearings or hydrostatic lubrication [3] or by use of low-friction materials [100]. Also, the use of "way lubricants" greatly reduces stiction; these lubricants form weak bonds to the sliding surfaces and so, unlike conventional lubricants, are not squeezed out from between contacting surfaces when the machine comes to rest [2,37,68]. Reducing compliance in the motor-to-load transmission minimizes the effects of remaining stiction [2]. If you are interested in learning more about friction, Ref. 2 provides a detailed description of friction and its impact on control systems.

12.4.6 Quantization

As discussed in Section 5.9, quantization is a nonlinear effect of digital control systems. It can come from resolution limitations of transducers or of internal calculations. It causes noise in control systems (Section 14.4) and can produce limit cycles (Section 5.9.1).

The straightforward cure for quantization is to increase the resolution of the sensor or mathematical equation that limits resolution. Another technique for improving quantization problems for position-based velocity estimation is the "$1/T$" interpolation method, as discussed in Section 14.5.1. Gain scheduling is not effective for quantization because there usually is no deterministic equation that can be written using available signals.

12.4.7 Deterministic Feedback Error

Errors in feedback sensors can cause nonlinear behavior. Repeatable errors can often be measured when the machine or process is commissioned and then canceled by the controller during normal operation. For example, many machines rely on motor feedback sensors to sense the position of a table that is driven by the motor through a lead screw. The motor feedback, after being scaled by the pitch of the lead screw, provides the position of the table. This is shown in Figure 12-17.

One weakness of this structure is that the lead screw has dimensional errors that degrade the accuracy of the motor feedback signal. However, this can be compensated by measuring the table to great accuracy with an external measuring device such as

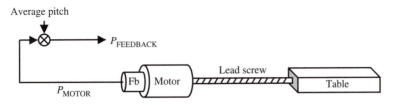

Figure 12-17. A feedback device on a motor is often used to calculate the position of the table.

a laser interferometer. During commissioning, the position of the motor as measured by the motor feedback device is compared with the presumed-accurate signal from the interferometer and the difference is stored in a table. This is shown in Figure 12-18. Many positions (typically hundreds) are measured to fill the table.

During normal operation, the error stored in the table is subtracted from the feedback sensor to compensate for the lead screw inaccuracy. This amounts to using the error table for *offset scheduling*, as shown in Figure 12-19. Of course, the most straightforward cure to deterministic error is to use a more accurate feedback sensor. However, more accurate sensors usually carry penalties, such as being more expensive, larger, or less responsive. In the cases where repeatable errors in feedback devices can be measured, great improvement in accuracy can be gained by using those measurements to offset error during normal operation.

12.4.8 Power Converter Saturation

Power converter saturation is the phenomenon in which the control law may command more output than the power converter can deliver. This happens when the

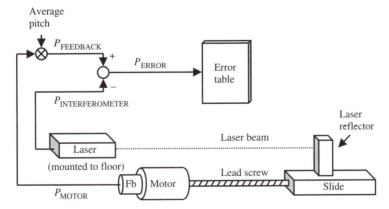

Figure 12-18. System commissioning: Determining lead screw error can be done by relying on the more accurate sensing of a laser interferometer.

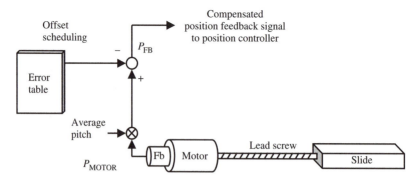

Figure 12-19. Offset scheduling: compensation for lead screw error during normal operation.

command swings large amounts in a small period of time or when disturbance forces exceed the power converter's rating. Modeling power converter saturation is simple because the input is usually clamped, as shown in Figure 12-20. A model is shown here.

```
IF(Input > Output_Max)
  Output = Output_Max
ELSE IF(Input < −Output_Max)
  Output = −Output_Max
ELSE
  Output = Input
END IF
```

Power converter saturation cannot be dealt with in the controller using gain scheduling because the gain of the power converter after saturation falls to zero. If the controller is nonintegrating, the solution is simply to clamp the control output. However, if the controller is integrating, such as PI or PID control, the control law must be protected from integral windup. Windup is the undesirable behavior where the integral state continues to grow when the power converter is saturated. As discussed in Section 3.8,

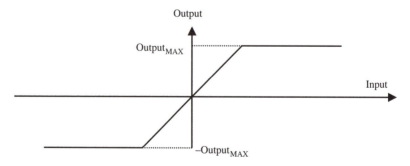

Figure 12-20. Power converter saturation.

if the integral state is permitted to continue growing, or *winding up*, during saturation and the power converter remains in saturation for a long period of time, the integral state can grow to very large values. After the error is finally satisfied, the control loop must unwind the integrator, which can require a long period of reversal. The result is behavior that appears oscillatory even though the unsaturated controller is stable.

Windup should be prevented by ensuring that the integral value is never larger than what is required to command a little more than saturated output [90]. This is done by first checking the control law output against the power converter maximum with some margin (typically 20%–100%). If the control law output is outside this range, the integral output is reset so that the control law output is just at the boundary. This process is called *synchronization* or "antiwindup" and is shown for a PI controller in Figure 12-21. The output can then be clamped to the actual power converter maximum (note that this step is often unnecessary because power converters commonly clamp implicitly as a part of normal operation).

In digital controllers, synchronization must be explicitly calculated after the control laws (see Section 11.3.3.8). In analog circuits, synchronization is often performed by adding a zener diode across the resistor and capacitor that form the PI control law, as shown in Figure 12-22. This circuit clamps the integral as it limits the output so that windup is not a problem. If the output maximum is about 80% of the op-amp power

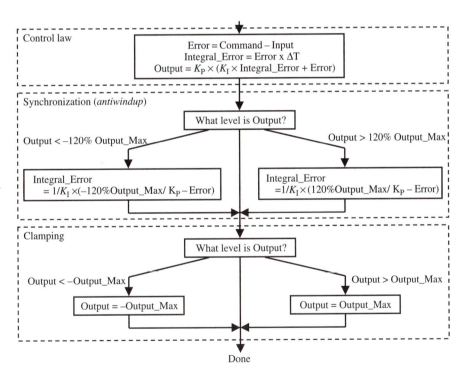

Figure 12-21. Using a PI controller requires that synchronization be performed before saturation clamping.

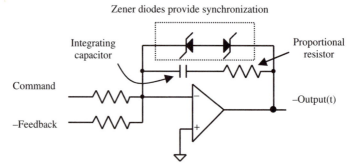

Figure 12-22. Synchronization in an analog PI controller.

supply, zener diodes are often not required. For example, if the power converter saturates when the op-amp output is 12 volts and the op-amp is supplied with ±15 volts, then the clamping diodes may not be required; the circuit will remain in synchronization (or close enough), relying only on the op-amp saturation characteristics.

The choice of 120% of saturation for synchronization is based on experience. The number must be significantly more than 100% because otherwise the integral will discharge when the control output barely saturates. Leaving a little margin (in this case, 20%) allows the integral to remain accurate when the system is on the border where noise pushes the system in and out of saturation. Values as large as 200% are commonly used.

12.4.9 Pulse Modulation

Pulse modulation uses time averaging to convert digital signals to analog [6]. It has been called a "poor person's D/A converter." Figure 12-23 shows two common forms of pulse modulation, pulse-width modulation (PWM) and pulse-period modulation (PPM). PWM varies the duty cycle to vary the average output. PPM keeps the on-time (or off-time) constant and varies the period to vary the output. Both output pulses are smoothed by the output element so that a digital signal is converted to an analog signal. The smoothing is often the implicit integration of the plant. If an inductor is the

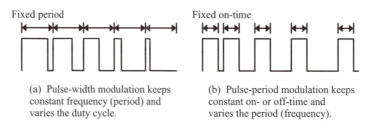

(a) Pulse-width modulation keeps constant frequency (period) and varies the duty cycle.

(b) Pulse-period modulation keeps constant on- or off-time and varies the period (frequency).

Figure 12-23. Pulse-width and pulse-period modulation.

output element, a pulsed voltage waveform is smoothed by the inductor to produce an analog current. Similarly, in an oven, pulses of heat (from pulsed current) are smoothed by the thermal mass inside the oven.

For example, consider an inductor driven by a pulse-width-modulated transistor stage. Here, current is controlled by varying the duty cycle of the transistors. The average current is driven by the average voltage; larger duty cycles imply higher voltage and thus generate more current. Pulse modulation is used because it greatly reduces the power losses in the transistor. Since the transistor is digital (either on or off), power losses in the transistor are minimized. That's because when the transistor is off, the current is low, and when it's on, the voltage is low; in either case, the conduction losses, which are equal to the product of voltage and current, remain low.

The primary disadvantage of pulse modulation techniques is the creation of undesirable harmonics in the output. For inductor-based current control, harmonics in the output voltage create harmonics in the current called *ripple*. These harmonics create problems such as audible noise. In addition, the fast voltage transients generated by the switching transistors cause electromagnetic interference (EMI). The amplitude of output harmonics is reduced by increasing the modulation frequency; however, this usually increases EMI and power losses.

Deadband is common in amplifiers where two transistors are connected in series from the positive to the negative voltage supply in a *totem pole* circuit, as shown in Figure 12-24. In this case, it must be ensured that both transistors are not on simultaneously. The time to turn off a power transistor is longer by a few microseconds than the time to turn one on. Because turn-off time is longer, if a transistor is commanded to turn on simultaneously with its opposite being commanded to turn off, both transistors will be on for a short time. Even though the time is brief, a large amount of current can be produced because the two transistors being on simultaneously will connect the positive and negative voltage supplies.

The normal means of ensuring that the transistors will not be on simultaneously is to have a short time period in which it is guaranteed that both transistors will be off. This time results in a small deadband. For example, if a PWM stage operated at 20 kHz (50 μsec period) and the dead time was 2 μsec, the deadband would be $2/50 \times 100\% = 4\%$. If the power transistor bridge controlled a 600-V bus, the resulting

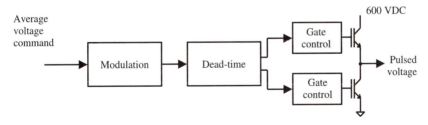

Figure 12-24. A "totem pole" transistor stage that usually uses dead-time to ensure that both transistors are never on simultaneously.

deadband would be approximately 4% of 600 V, or 24 V, a small but significant voltage. This deadband can be compensated by several methods, including using the technique provided earlier in this section.

For the control system, modeling modulation usually requires knowing only the relationship between the command to the modulator and the average output of the modulator; harmonics can be ignored in most cases because they have little effect on traditional measures of control systems. Most pulse-width modulators are approximately linear, with some deadband; if the deadband is small enough to ignore, only a constant of linearity need be derived. In the simplest case, the constant is the output span divided by the input span. For example, an analog voltage modulator that can output from 0 to 300 V with an input of 20 V (± 10 V) would have a constant of proportionality of 300/20, or 15. PPM usually has a nonlinearity because the average voltage does not vary proportionally with the variable parameter, such as the off-time in Figure 12-23b. In that case, gain scheduling can be used to compensate. Pulse-width modulation is discussed as it relates to motor control in Sections 15.5.7.2 and 15.6.3.1.

12.4.10 Hysteresis Controllers

Hysteresis, or *bang-bang*, control is perhaps the simplest control loop to implement [83]. With a hysteresis loop, the plant is operated in either of two states: off or on. A hysteresis band is defined, and if the feedback signal is above that band, the plant is operated in one state; if it is below that band it is operated in the other state. If the feedback is within the band, the operating state is left unchanged. Hysteresis control is widely employed. For example, most home heating and air conditioning systems use hysteresis control [6]. A thermostat is set to turn the furnace on if the temperature falls below 24 °C and turn it off when the temperature climbs above 25 °C. The operation of such a controller is shown in Figure 12-25. The home heating system demonstrates both of the key advantages of hysteresis control. It is simple and uniformly stable.

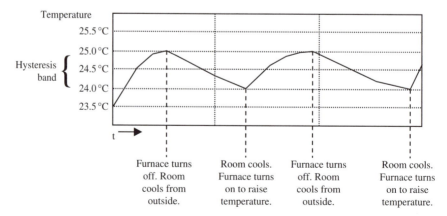

Figure 12-25. A heating system is an example of a hysteresis controller.

The hysteresis controller, being nonlinear, does not demonstrate the weaknesses of linear control; there is neither phase lag nor gain. Within the error of the hysteresis band and the ability of the power converter, the hysteresis controller follows the command and rejects disturbances perfectly, except that delays in the feedback device and power converter slow response. The weaknesses of the system are also evident. The loop never reaches a quiescent state; it is forever moving from full on to full off. The frequency of operation is not predictable; a furnace turns off and on at a higher frequency when the outside temperature is lower. Also, the error is never driven to zero but can always roam within the hysteresis band.

With hysteresis there is no process of tuning to achieve the optimal combination of stability and response. The key adjustment is setting the hysteresis band, which is a trade-off of how often the plant changes state against how much error can be tolerated. The tighter the hysteresis band, the higher the frequency of cycling power to the plant.

12.5 Questions

1. Assume a gain margin of 20 dB and a plant with a nominal gain of 50/sec. Assume also that the minimum GM of the system is specified to 10 dB. To what maximum value of gain can the plant climb under all operating conditions?
2. Assume that over a 1-second period, the inertia of a system falls according to the equation $J = 0.002 - 0.001t$. Further, assume the speed starts at zero and accelerates at a rate of 2000 rad/sec^2. At what maximum velocity will the variation be "slow"; here, slow variation is defined as when $\omega \times dJ/dt$ is no larger than 25% of $J \times d\omega/dt$.
3. Design the gain scheduling for the following variation of inductance with current:

$$L(i) = 1\,\text{mH} \quad i < 2\,\text{A}$$
$$= 2\,\text{mH}/i \quad 2\,\text{A} < i < 4\,\text{A}$$
$$= 0.5\,\text{mH} \quad 4\,\text{A} < i$$

Assume nominal gain is defined as the gain when $i = 0$.
4. Provide the ideal gain scheduling for Figure 12-10 as a function of Θ. Assume inertia varies with theta as follows:

$$J(\Theta) = J_\text{M} + J_\text{L}(1 + 2\sin(\Theta) \times R)^2$$

where J_M is the motor inertia, J_L is the load inertia, and R is the length of both arms connected to the elbow. Assume that the nominal gain is defined as the gain when $\Theta = 0$.
5. Correcting for reversal error correction and deadband compensation are special cases of what general technique?

Chapter 13

Seven Steps to Developing a Model

Developing a useful model for a control system is a complex task. It requires a thorough understanding of all components in the control loop: the controller, the power converter, the plant, and the feedback devices. Developing a model is usually tedious; frequent trips to the laboratory are required to measure and verify. For engineers who take on the job, the results can be well worth the effort. Those engineers can better predict the operation of their control system, more quickly troubleshoot problems, and more efficiently add improvements.

The principal function of a model is to improve your understanding of the system. The amount of detail necessary for a model depends on what behavior is being studied. Two pitfalls should be avoided. First, avoid making the model overly complex; the best model is the simplest one that demonstrates the behavior of interest. Including more detail than is necessary slows the modeling process, makes verification more difficult, and ultimately makes the model less useful.

Second, verify the model in the laboratory. There are two reasons: to check assumptions and to test for programming errors. All models include assumptions that, if incorrect, increase the likelihood that the model will provide unreliable results. Another reason to verify is that models are a form of software; like any software, models can have programming errors. The remainder of this chapter is a seven-step procedure for developing LTI and non-LTI models.

13.1 Determine the Purpose of the Model

The first step in modeling is to define the behaviors you want to understand better. There is no single best model of a control system; the detail in the model should correspond to the behaviors under study. Selecting a model is a little like choosing a car. If you need inexpensive transportation, choose a compact. If you taxi four children, choose an SUV. There is no single best car and you should not expect one model to predict everything about your system. There are at least four purposes for models in control systems: training, troubleshooting, testing, and prediction.

13.1.1 Training

Training models are the easiest models to develop. In a training environment, you know what results to expect. That means you can keep working with the model until it demonstrates the desired behavior. The *Visual ModelQ* models in this book are training models. They were developed to demonstrate a specific behavior. With training models, you know the answer before you start modeling.

13.1.2 Troubleshooting

Troubleshooting models are used to help correct a problem in a control system. They are used to identify the source of the problem and then to allow the engineer to simulate corrections without having to change the actual product or process. This can save time because a model is usually much easier to change than a working control system. In the end, the actual product or process must be modified, but, ideally, only once. A well-made troubleshooting model instills a high degree of confidence in a proposed solution.

With troubleshooting models, you are searching for some aspect of a component's behavior that, when combined with the rest of the system, results in undesirable operation. The key task is to include all the behaviors that contribute to the problem of interest. As with training models, you know what results to expect; when the model reproduces the problem, it is likely that the cause is properly modeled. Of course, troubleshooting models require verification of the model against the working system. To this point, troubleshooting and training models are similar.

After the model is verified, the next step in troubleshooting is modifying the model to eliminate the undesirable behavior. If the modification cures the problem in the model, it is likely that the same modification will correct the actual system. Trouble-shooting models illustrate the need to minimize model detail. A model that includes only enough component behaviors to reproduce the problem will be easier to build; also, by leaving out numerous behaviors that are unrelated to the problem under study, the list of possible corrections is reduced. Both aspects can speed the trouble-shooting process.

13.1.3 Testing

Testing models are designed to allow simulation of aspects of system operation that are difficult or dangerous to measure on the actual system. It is assumed that a partial set of tests can be run on the actual control system. Like troubleshooting models, testing models require that you verify the model against the operations that are known. The reliability of the model depends on how broadly it is verified.

Testing models are inherently less reliable than training or troubleshooting models. Although you can verify a model against the tests that can be run on the actual system, the tests being run on the model, by definition, cannot be verified. The closer the actual test is to the simulated test, the more reliable the model will be. For example, if you tested an actual temperature controller's behavior with respect to a 10 °C command change and then verified the model in the same conditions, you would have a high degree of confidence that the model could predict the reaction to a 5 °C, 15 °C, or 20 °C command change. However, your confidence in the model's ability to predict the temperature controller's rejection of changes in ambient temperature would probably be lower. The closer the verification conditions are to the simulation test, the more you will trust the simulation tests.

13.1.4 Predicting

Predictive models are used to predict behavior that has not been tested. For example, if a company were switching from an analog controller to a digital controller, a predictive model could be used to estimate the performance change. Confidence in such a model is difficult to establish because, by the nature of the model's function, the amount of verification that can be done is limited. Predictive models are similar to test models; the primary difference is that the predictive model is built before the product; the testing model is built after the product, so partial verification is possible.

The approach to modeling here emphasizes that you should define the use and verification of the model. If a model is used to make important decisions, as it often is in test and predictive models, the amount of verification should be in proportion to the magnitude of the decision. However, models, like any software, can have programming errors that escape detection from even the most thorough verification. Verification increases confidence, but some doubt should always remain. In the end, models provide evidence for decisions, but important decisions should be firmly grounded in broad technical expertise.

13.2 Model in SI Units

Write your model in *Système International* (*SI*) units: meters, seconds, kilograms, etc. Modeling in English units or other nonstandard units is more difficult because most processes must be scaled. For example, in SI units rotational velocity (rad/sec) is the unscaled integral of position (rad). By using SI units, you are more likely to avoid errors of scaling.

If you need to display results to the user in non-SI units (for example, RPM or °F), you can scale the output values just before displaying this. This is how *Visual ModelQ* works; all the internal calculations are done in SI units, and the scope and DSA scale the output results, where SI units are unfamiliar to the user.

13.3 Identify the System

The next stage of modeling is identifying the system. First address linear behaviors by building a transfer function for each component in the system. Every LTI transfer function can be written in the form of a gain, K, a series of zeros, $a_1, a_2, a_3, \ldots$, and a series of poles, $b_1, b_2, b_3, \ldots$, as shown in Equation 13.1:

$$K\frac{(s+a_1)(s+a_2)(s+a_3)\cdots}{(s+b_1)(s+b_2)(s+b_3)\cdots} \tag{13.1}$$

Sometimes the poles and zeros can be complex (as opposed to real) numbers. Anytime a complex pole or zero is present, its conjugate will also be present. Complex conjugates can be combined into a quadratic form so that the coefficients are always real. Assuming a_i and a_{i+1} are complex conjugates:

$$\begin{aligned}(s+a_i)(s+a_{i+1}) &= s^2 + (a_i + a_{i+1})s + a_i a_{i+1} \\ &= s^2 + c_j s + d_j\end{aligned} \tag{13.2}$$

where $c_j = a_i + a_{i+1}$ and $d_j = a_i \times a_{i+1}$ (c_i and d_i are guaranteed real). So Equation 13.1 can be rewritten with real coefficients as

$$K\frac{(s+a_1)(s+a_2)(s+a_3)\cdots(s^2+c_1s+d_1)(s^2+c_2s+d_2)\cdots}{(s+b_1)(s+b_2)(s+b_3)\cdots(s^2+e_1s+f_1)(s^2+e_2s+f_2)\cdots} \tag{13.3}$$

Although Equation 13.3 can appear intimidating, in practice most transfer functions will have a scale factor (K) and perhaps one or two poles.

After the linear transfer function is identified, nonlinear behaviors can be added to the transfer functions for time-domain models, as discussed in Chapter 12. This should include the nonlinear effect and any compensating algorithms provided in the controller. If the compensating algorithms work well enough over the frequency range of interest, the combined nonlinear effect and compensation can be ignored.

13.3.1 Identifying the Plant

The plant is usually understood through the study of reference material. For example, the linear behaviors for the majority of plants used in industry are described in Table 2-2. If you do not know the model for your plant, someone has probably written about it; it's usually a matter of locating the information.

If you purchase your plant, the vendor will probably have application notes that include information on the plant behavior in a control system. If your company manufactures the plant but does not provide this type of information, you can review application information from your competitors. Also, industry consultants are often able to help, and many can provide you with a complete model.

Also, look to companies that manufacture controllers for your plant or a plant similar to it. They will usually have application notes that provide detail on general models. Another good source of application notes is sensor manufacturers. Because sensor manufacturers frequently need to demonstrate the performance of their products, they often model typical plants in their industry. Start with the high-end manufacturers — they regularly need to prove their product's superior performance, and they usually have more technical resources.

You can also research the literature. Start with the trade press; trade journals are usually easy to read and easy to apply. Three trade journal publishing companies, Penton (www.penton.com), Cahners (www.cahners.com), and Intertec (www.intertec.com), publish trade journals for a wide range of industrial applications. Search their Web sites for the journals of interest. Once you locate a few journals, use the Web to search back issues that relate to your process.

There are numerous annual conferences that produce application-oriented papers, most of which include models of plants. One academic society, The Institute of Electrical and Electronic Engineers (IEEE), includes the Industrial Application Society (IAS), which addresses a broad range of applications. Each year the annual meeting of that society produces hundreds of papers in the *Conference Record of the IEEE IAS* (sometimes called *Proceedings of the IEEE IAS Annual Meeting*), which is available at most technical libraries (see www.ewh.ieee.org/soc/ias). The best of these papers is published in the bimonthly *IEEE Transactions on Industry Applications*. Also, the *IEEE Transactions on Industrial Electronics* prints application-oriented papers.

13.3.2 Identifying the Power Converter

Power converters are complicated, but they can often be simplified to low-pass filters. This is because when power conversion involves a complex process, the effects usually occur well above the range of frequencies interesting for the control system. Producing an analytical model of the power converter from basic principles often results in undue effort and excessive detail. For example, consider a pulse-width-modulator- (PWM)- based current controller (see Section 12.4.9). In many cases, the power transistors must be turned on and off within microseconds of the command. The process is nonlinear, and time-domain simulation with this level of detail requires that models run with time steps of perhaps 100 nsec. However, a well-built PWM-based power converter can be quite linear when viewed from a longer time step, say, 50 μsec.

When analytical models of power conversion overcomplicate the model, the best way to identify the power converter is to measure it. If the power conversion is not fully electrical, such as electrohydraulic or electropneumatic, sensors can be used in the laboratory to provide an electrical signal representing the power converter output.

As shown in Figure 13-1a, a two-channel scope can monitor the input and output of the power converter and a separate sine wave generator can provide commands. A preferred arrangement is shown in Figure 13-1b with a dynamic signal analyzer (DSA), which includes a waveform generator. The DSA produces Bode plots directly; the oscilloscope and a sine wave generator can be used to produce a Bode plot by measuring the phase and gain at many frequencies. This process is tedious and less accurate than a DSA but will produce useful results for most control problems. In either case, it is important to ensure that the power converter not saturate during the measuring process. Saturation invalidates Bode plots.

After you have measured the Bode plot, the next step is to find a filter that has a similar Bode plot over the frequency range of interest. For most controls applications, a good fit consists of being accurate to within a few degrees and a few decibels from low frequency to about five times the expected bandwidth of the control system.

Fitting the response to filters can be done by trial and error. First, determine the order of the filter. The phase lag of a filter at high frequency is 90° times the number of poles less the number of zeros. If the gain of the Bode plot does not peak and the phase lag is less than 90°, a single-pole low-pass filter should work; if the phase lag is less than 180°, a two-pole low-pass filter with damping <0.707 will often fit well. If there is peaking followed immediately by attenuation, a two-pole low-pass filter with damping <0.707 may work. If there is peaking, but the high-frequency phase is about −90°, a two-pole filter in line with a single zero may work well.

Remember to include a scaling constant. Fortunately, the scaling constant can be measured at DC, since power converters, unlike plants and controllers, usually have bounded DC gain (most plants and controllers have at least one order of integration). If a current converter were scaled so that 10 volts = 50 amps, the DC gain (K) of the transfer function would be 5 amps/volt.

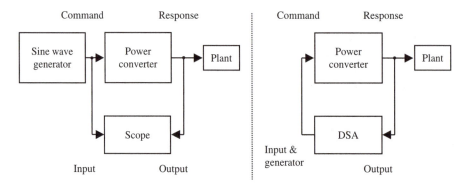

Figure 13-1. Measuring the power converter with (a) a scope and sine wave generator and (b) a dynamic signal analyzer (DSA).

13.3.3 Identifying the Feedback

Feedback identification may require measured results. If you purchase the feedback sensor, the vendor should be able to provide you with a measured transfer function. As with the power converter, expect a scaling constant and a one- or two-pole low-pass filter over the frequency range of interest. Don't forget to add the scaling constant to convert the units of the feedback device to SI. For example, a rotary encoder with 4000 counts per revolution would have an implicit DC scale of $4000/2\pi$ counts/rad.

13.3.4 Identifying the Controller

The controller is often the most difficult element to identify. It may incorporate many functions besides the control laws, including scaling and filtering. Controllers often have undocumented noise filters on the command and feedback signals and, for PID controllers, on the derivative term. The vendor may not disclose this information in the general literature if they judge it unnecessary for the majority of their customers. However, these filters can affect the Bode plot significantly.

For digital controllers, it is often hard to determine the scaling of constants. Many microprocessors and DSPs in industrial controllers are limited to integer representation. So if a typical gain spanned 0.0–0.01 m/sec, it would be scaled up (say, by 10,000) so that it could be represented in integer form without undue impact of quantization. Also, SI units of time are seconds, but many controllers have units scaled by the sample time of the controller. The z-domain representation of an integration gain, $K_I \times Tz/(z - 1)$, is not usually explicitly scaled by T; the scaling by T is implicitly part of the integral gain.

If you purchase your controller, you will in all likelihood need help from the vendor to get a model of the control system. This process often involves reading, phone calls, and even a visit to meet the vendor's control system experts. In most cases, the vendors are not intentionally hiding this information; only a minority of customers model their systems, and so vendors are reluctant to dedicate the resources to learning and documenting these details.

In a few cases, vendors judge that detailed information on the controller is proprietary and they provide only sparse information. If you determine that your vendor falls into this class, you may consider changing vendors. Your ability to evolve your system depends on how well you understand all the components. If you do not have enough information to build a model, your understanding will be limited. Conversely, avoid asking for detail that you don't need. Vendors will understandably become concerned or impatient with customers who request ever more detail about the operation of their products.

13.4 Build the Block Diagram

Now that you have the transfer functions of all the components of the system, you can build a block diagram. Be sure to show the integrals by themselves (without scaling constants); this simplifies conversion to state space, as discussed in Section 11.3.3 [32,60]. Review the system, and simplify where possible. Aspects of a plant that are unimportant should not be included in the model. The more effects you try to take into account, the more complex the model becomes; as a result, the model will take longer to develop, be more likely to have programming errors, and require more time to execute.

Two attributes of unimportant effects are small amplitude and high frequency. Quantization of D/A converters can be ignored if the D/A resolution is high enough, because the amplitude of the nonlinearity is small. And think of the control system in frequency zones. You can usually ignore effects that occur at frequencies well above the bandwidth of the system.

Simplify, but do not leave important detail out. Include all the important non-LTI relationships in the system. Include all the lag imposed by implicit filters. For example, you may need to include the effects of long transmission lines. Don't forget calculation delay, an effect that is often important but rarely documented. If appropriate, include imperfections of the feedback device: repeatable errors, noise, and resolution limitations.

A good strategy in modeling is to start with the simplest model you can test and build from there. By writing and verifying the simple model, you remove many opportunities for errors. Then add details, a few at a time. This method can be the fastest way to build a reliable model.

13.5 Select Frequency or Time Domain

After the block diagram is built, you can select the time or frequency domain. The frequency domain is better for classical controls issues: stability, command response, and disturbance response. The time domain is better for understanding nonlinear behavior. Also, most machines are rated in the time domain (overshoot, settling time, peak error). The ideal answer is to use a single model that can analyze the system in both domains.

13.6 Write the Model Equations

Now it's time to write the model equations from the block diagrams, as discussed in Chapter 11, for the time domain and for the frequency domain. For graphical modeling environments, draw the block diagram in the environment to write the model.

13.7 Verify the Model

You should verify your model carefully. For linear behaviors and slowly varying non-LTI behaviors, verify in the frequency domain. The preferred instrument is a DSA, but a scope can be used. For fast-varying nonlinear effects, the time domain must be used. A step command is the most popular command because it has a wide frequency spectrum. However, other signals, such as trapezoids, triangle waves, and swept sine waves, also have harmonics across a wide spectrum. The key is to excite the system with a broad range of frequencies.

The ability to verify the model goes up with the number of signals that are accessible. Consider Figure 13-2. The signals that would be available in most cases are the command and feedback. This will allow you to measure only the closed-loop Bode plot; if the Bode plots of the actual and modeled system do not agree, it is difficult to localize the cause. As discussed in Section 13.3.2, if you have access to the commanded and actual power, the power converter can be verified independently (*Actual power vs. Commanded power*). If you have access to the error signal, the control laws can be verified (*Commanded power* vs. *Error*). By using the system feedback device, you can verify the combined plant and feedback (*Feedback* vs. *Actual power*). If the feedback device used in the production machine or process is in question, you can use a laboratory instrument for an accurate measurement of *Response*, allowing you to verify independently the plant (*Response* vs. *Actual power*) and feedback (*Feedback* vs. *Response*).

The more you can exercise the system, the better you can verify operation. Generally, verification is better executed when the control law gains are high, so tune the control system for maximum response during verification. Also, vary individual controller gains over a wide range of values. If the digital controller contains other parameters, such as selectable filters, run the system in as many modes as possible to verify operation across the widest range of configurations. Of course, no model should be considered 100% verified, but the more you verify your model, the more confidence you will be able to place in it.

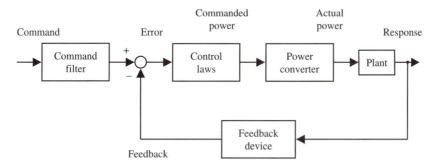

Figure 13-2. Measuring the loop.

Section III
Motion Control

Chapter 14

Encoders and Resolvers

Motor feedback sensors determine, in large measure, the capability of a motion-control system. Sensor error limits the steady-state accuracy of position or velocity. Also, imperfections in the feedback sensor such as coarse resolution and cyclical error cause torque perturbations. There are a number of feedback devices used on servomotors. This chapter will present detailed discussions of the two most common ones, encoders and resolvers, and the limitations they impose on a servo system.

The most common motion sensors read velocity or position. When electronic motion control was first widely used in industrial applications, high-performance motion systems relied on two motion sensors: a tachometer for velocity and either an encoder or a resolver for position. The velocity of each axis of motion was controlled through independent motor controllers called *drives*; drives used a tachometer as their only motion sensor. A multiaxis motion controller coordinated the many axes on the machine; it was configured as a position controller and used one position sensor (either resolver or encoder) for each axis to close the position loops. This configuration is shown for a two-axis system in Figure 14-1.

The position and velocity signals were segregated in early systems. The motor was a brush DC motor, the velocity of which could be controlled without knowledge of motor position; thus the drive required only the tachometer. The motion controller did not have the computational resources to close the velocity loop; it was limited to closing the position loop and thus required only a position signal. In the succeeding decades, two important changes occurred in motion systems. First, the AC (brushless-DC, or induction) motor has largely replaced the DC (brush) motor in high-end applications. Second, the increased computational resources of drives and controllers have allowed the estimation of velocity from the position sensor.

The change to AC motors has affected motion feedback because *commutation*, the process of channeling motor current as a function of position to produce torque (see Section 15.6.2), is carried out electronically by the drive; this implies that the drive must have access to position feedback. The estimation of velocity from a position

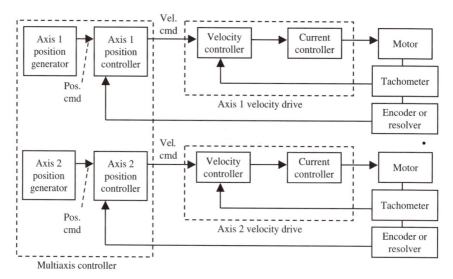

Figure 14-1. Two-axis motion-control system with separate position and velocity sensors.

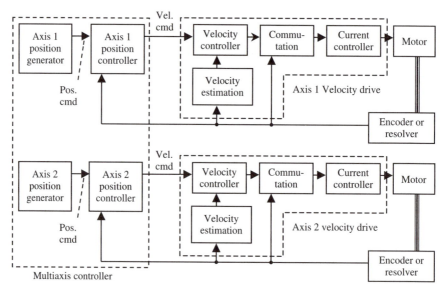

Figure 14-2. Two-axis motion-control system with a single position feedback sensor per axis.

sensor allows the elimination of the tachometer, reducing component cost and wiring complexity. Today, the majority of motion-control applications use a position sensor as the sole motion sensor. This implies that the drive and the position controller share a single sensor, as shown in Figure 14-2.

Because of the trends just discussed, this chapter will focus on position sensors. In industrial applications, the two dominant position sensor technologies are encoders and resolvers. Within these broad classifications, there are several subclasses. Encoders can be based on optical or magnetic transducers; they can be rotary or linear. Resolvers can be designed for low or high resolution and accuracy. Typically, encoders are the more accurate sensor but resolvers the more reliable, especially in harsh environments. Finally, because most encoders transmit square waves and resolvers output sine waves, encoders often generate more electrical noise when the cable between motor and encoder is long.

14.1 Accuracy, Resolution, and Response

The three main factors in feedback devices that affect the performance of the control system are accuracy, resolution, and response. Response is the easiest to understand. A sluggish feedback sensor will behave like an ideal (fast) sensor in series with a low-pass filter. The sensor will generate phase lag that will limit the stability of the system according to the discussions in Chapters 3 and 4. The implication is that fast systems require fast sensors. In most applications, the feedback from encoders is so fast that it can be considered ideal. The signal processing for resolvers inserts a two-pole low-pass filter in series with the velocity loop; the bandwidth of that filter is typically between 200 Hz and 1200 Hz.

Accuracy and resolution of the position sensor as applied to the position loop are also readily understood. The position of the motor can be controlled no more accurately than the accuracy of the position sensor and no better than to one or two counts of the sensor resolution. The resolution of feedback sensors varies from hundreds of counts per revolution in low-end encoders up to over 1 million counts per revolution for *sine encoders* and *multispeed resolvers*. The accuracy of typical devices varies from about $30 \min^{-1}$ ($1 \min^{-1}$, or 1 *arc-minute*, is $(1/60)°$) to under $1 \sec^{-1}$ ($1 \sec^{-1}$, or 1 *arc-second*, is $1/60 \min^{-1}$).

The effects of accuracy and resolution on the velocity loop are more difficult to understand; much of this chapter will discuss these effects. The velocity estimation algorithm used most often is the simple difference (dividing the difference of the two most recent samples by the sample time), which was discussed in Section 5.6.2. As an alternative, analog drive systems sometimes use a frequency-to-voltage converter, a process that generates a pulse at each encoder edge, to perform the equivalent of simple differences in analog electronics.

14.2 Encoders

The position feedback sensor in modern motion controllers is usually either an encoder or a resolver coupled with a *resolver-to-digital converter*, or RDC. In both cases, the servo controller knows the position only to within a specified resolution. For the two-channel encoders used in most servo applications, the resolution is four times

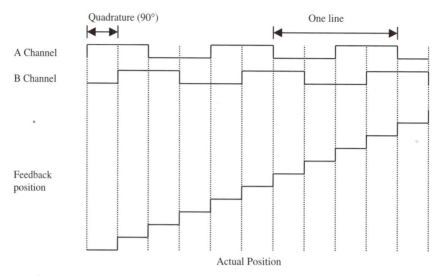

Figure 14-3. Quadrature decoding: two encoder channels generate four counts per line.

the line count; encoder interface circuitry uses a technique called *quadrature decoding* to generate four pulses for each encoder line. Figure 14-3 shows the technique, which is based on having two encoder signals, named Channel A and Channel B, separated by 90°. Because of quadrature, there are four edges for each encoder line, one positive and one negative for each channel. Thus a 1000-line encoder will generate 4000 counts for each encoder revolution. It should be noted that single-channel encoders are used occasionally in servo applications. This is rare because two channels are necessary to determine the direction of motion.

14.3 Resolvers

Resolvers are also commonly used to sense position. A resolver is a rotating transformer that generates two signals: $\sin(p(t))$ and $\cos(p(t))$, where $p(t)$ is the rotor position. A free-running oscillator drives the excitation winding, which is magnetically coupled to two sense windings, sin and cos: This coupling is a function of the sine and cosine of the resolver rotor position. The sin and cos signals are fed to an RDC, an integrated circuit that generates the motor position. This is shown in Figure 14-4.

14.3.1 Converting Resolver Signals

Position is sensed only to within a specified resolution. For example, if the RDC is configured for 14 bits/revolution, there are 2^{14}, or 16,384, counts for each revolution.

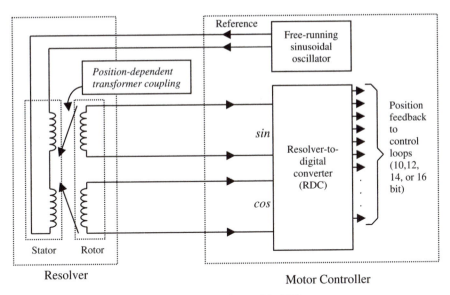

Figure 14-4. Resolver and the RDC.

The traditional RDC is a monolithic chip that implements a variety of digital and analog functions [13, 89]. As depicted in the conceptual diagram of Figure 14-5, the *SIN* and *COS* signals are demodulated to create $\sin(P_{RES})$ and $\cos(P_{RES})$, where P_{RES} is the actual position of the resolver. Simultaneously, the estimated position, P_{RD}, which is stored in an up/down counter, is fed to specialized D/A converters that produce the signals $\sin(P_{RD})$ and $\cos(P_{RD})$. These signals are multiplied to produce the signal

$$\sin(P_{RES}) \times \cos(P_{RD}) - \sin(P_{RD}) \times \cos(P_{RES}) = \sin(P_{RES} - P_{RD})$$

Assuming that the position from the RDC is fairly close to that of the resolver,

$$\sin(P_{RES} - P_{RD}) \approx P_{RES} - P_{RD}$$

This approximation is based on $\sin(\theta) \approx \theta$ if θ is small.

A PI compensator is applied to drive the error signal, $\sin(P_{RES} - P_{RD})$, to zero. The compensator is equivalent to an op-amp circuit with gains set by discrete resistors and capacitors. The compensator output, which is an analog signal, is converted to a pulse train through a voltage-controlled oscillator, or VCO. The output of the VCO is fed to the up/down counter, which acts like an integrator, summing the VCO pulses over time.

The RDC of Figure 14-5 is redrawn in Figure 14-6 to emphasize the effects of the conversion process on the servo system. In Figure 14-6, the demodulation and

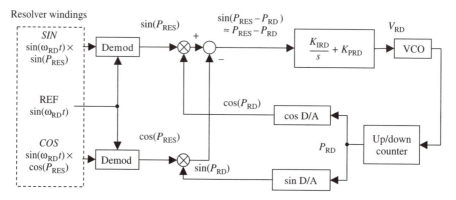

Figure 14-5. Simplified RDC.

trigonometry are combined to create a signal representing the actual position of the resolver, P_{RES}. This signal is compared to RDC position, P_{RD}, to create an error. That error is PI compensated and fed to an integrator to create P_{RD}. Note that some of the intermediate signals in Figure 14-6 do not explicitly exist, but RDC dynamics are represented reasonably accurately.

Figure 14-6 can be used to derive a transfer function of the RDC using the $G/(1 + GH)$ rule. Assuming that demodulation and trigonometric functions do not significantly affect the dynamics, the relationship between the actual resolver position and the output of the RDC is

$$\frac{P_{RD}(s)}{P_{RES}(s)} = \frac{s \times K_{PRD} + K_{IRD}}{s^2 + s \times K_{PRD} + K_{IRD}} \qquad (14.1)$$

Equation 14.1 shows that the RDC behaves like a second-order low-pass filter. At low frequencies, where the s^2 denominator term is overwhelmed by the other terms, Equation 14.1 reduces to 1; so at low frequency, the converter generates no significant effects. However, at high frequencies, the s^2 term will become larger, inducing attenuation and phase lag. Drive manufacturers are responsible for selecting and installing the components that set the PI compensator gains. They normally try to maximize the gains in order to raise the effective bandwidth of the RDC, which minimizes the phase lag induced by the process. Stability margins and noise combine to limit the bandwidth, so the typical RDC has a bandwidth of about 300–600 Hz.

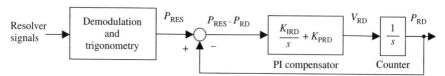

Figure 14-6. Idealized structure of R/D conversion.

Because of the techniques used in hardware RDCs, the position must be interpreted count by count, and the top frequency of the counts is limited, typically to a few megahertz. This limits the top velocity of the motor in the application. For example, for an RDC limited to 2 MHz and configured for 14-bit resolution, the top speed would be

$$V_{\text{MAX}} = 2 \times 10^6 \frac{\text{counts}}{\text{second}} \times \frac{1 \text{ rev}}{2^{14} \text{ counts}} \times 60 \frac{\text{seconds}}{\text{minute}} = 7324 \text{ RPM} \qquad (14.2)$$

Most hardware RDCs can be configured for multiple resolutions, typically 10, 12, 14, and 16 bits/revolution. For the example just given, if the application required more than 7324 RPM, the RDC would need to be configured for a resolution lower than 14 bits; reducing the resolution from 14 bits to 12 bits (four times) would allow the RDC to work at 29,296 RPM (4×7324 RPM). Normally, the resolution should be maximized for the required top velocity of the application.

14.3.2 Software Resolver-to-Digital Converters

Software RDCs perform the conversion process in a programmable chip [52]. The methods used are often similar to those of hardware RDCs. Software RDCs have numerous advantages over their hardware counterparts. First, they reduce cost by eliminating the hardware RDC, which is often the single most expensive component in the control section of a drive. Second, the conversion in a software RDC is not executed count by count, so the maximum speed is not limited by the count frequency as it was with the hardware RDC. A third advantage of software RDC is that it becomes practical to implement an observer to reduce phase lag from the conversion process. This is discussed in Ref. 26 and 28.

The fourth advantage of software RDCs is that the bandwidth of the conversion can be modified in the software. The bandwidth of hardware RDCs is difficult to change because it is set by several passive components. For off-the-shelf drives, this flexibility is rarely provided, and users must accept the RDC conversion bandwidth provided by the drive manufacturer. Manufacturers must balance the needs of many applications when setting the RDC bandwidth. When the bandwidth is set too high, the system generates excessive noise. This noise is most easily observed when the motor is commanded to zero speed; the average speed may be zero, but small torque perturbations can cause the shaft to be active. When the conversion bandwidth is set too low, the RDC causes excessive phase lag, limiting the bandwidth of the system.

The software RDC allows the user to set the bandwidth according to the application. If the application does not have demanding response requirements but does require very low noise, the user can reduce the conversion bandwidth to perhaps as low as 100 or 200 Hz. This greatly reduces noise. If the application is insensitive to noise but requires the highest response rate, the RDC bandwidth can be raised. For example, Ref. 52 shows the behavior of a software RDC configured for 400, 800, and 1200 Hz.

Hardware RDCs convert the resolver signals to a position signal one count at a time. The disadvantage of this, as discussed, is the maximum speed limitation for a given resolution. However, the hardware RDC signal does readily provide a digital, serial output of the position. Many RDC-based drives format this serial signal to an *encoder-equivalent output*, which acts like a two-channel encoder signal for other system components, chiefly the motion controller. This implies that only one RDC (in the drive) is required even though two devices (drive and controller) require information from the resolver. Because software RDCs do not convert the signal count by count, they must synthetically generate the encoder-equivalent output.

Synthetic generation of the encoder-equivalent signal requires three steps: measuring the distance that the resolver has turned since the previous sample, converting that distance to an equivalent number of encoder pulses, and feeding that pulse train into a component that can transmit the pulses in encoder format. Because the output frequency can be high, the counter and conversion to encoder format are usually done in a programmable logic chip, such as an FPGA or CPLD. One disadvantage of synthetic pulse generation is the implied phase lag. Because of the processing, the encoder pulses lag the position sensor by approximately one sample. This is a minor concern if the velocity loop is closed in the drive and the encoder pulses are used by the position loop. The position loop runs at a relatively low bandwidth, and the lag of one sample is less important, especially if the sample rate is high. However, if the drive is a torque controller and the velocity loop is closed in the motion controller using the encoder-equivalent output, the lag of perhaps a few hundred microseconds will be significant in high-response systems.

14.3.3 Resolver Error and Multispeed Resolvers

Resolver feedback signals usually contain more position error than do encoder signals. Resolvers are constructed by winding copper wire through slots in steel rotor and stator cores. Small imperfections in the rotor and stator geometry and in the winding placement generate typically between 4 and $20\,\text{min}^{-1}$ in mass-produced resolvers. Specially manufactured resolvers can be much more accurate, although they are usually expensive.

One technique that improves accuracy is to design the windings so there are multiple electrical revolutions of the resolver for each mechanical revolution. These are called *multispeed* resolvers, and they generate $\sin(Np(t))$ and $\cos(Np(t))$; N is called the *speed* of the resolver. Note that sometimes resolvers are described as having *poles* rather than a *speed*, where *poles* $= 2 \times speed$. Multispeed resolvers enhance accuracy in two ways. First, because of construction techniques in manufacturing, the resolver error drops almost in proportion to the speed, N. Second, because RDCs contribute error to the electrical position (typically between 2 and $10\,\text{min}^{-1}$) and N scales the electrical position down, mechanical RDC error is also scaled down by N. The end result is that multispeed resolvers scale down positional error roughly by a factor of the speed of the resolver.

With hardware RDCs, increasing the speed of the resolver may increase or decrease the resolution of the feedback, depending on the motor's top velocity. The resolution of multispeed resolvers increases by N times, but the top velocity supported by the RDC falls by N. Equation 14.2 can be rewritten to take the resolver speed (N) into account for a 14-bit, 2-MHz RDC:

$$V_{MAX} = \frac{2 \times 10^6}{N} \frac{counts}{second} \times \frac{1 \, rev}{2^{14} \, counts} \times 60 \frac{seconds}{minute} \tag{14.3}$$

As an example of how maximum motor velocity may increase or decrease with *resolver* speed, consider a two-speed resolver. For Equation 14.3 set $N = 2$:

$$V_{MAX} = \frac{2 \times 10^6}{2} \times 2^{-14} \times 60 = 3662 \, RPM \tag{14.4}$$

If the application required only 3500 RPM, then a two-speed resolver could be used with 14-bit resolution without violating the 2-MHz maximum in Equation 14.3. This effectively doubles the mechanical resolution from Equation 14.2. However, if the top velocity of the system were instead 4000 RPM, the two-speed resolver would not allow 14-bit resolution. The hardware RDC resolution would have to be dropped to 12 bits, and the mechanical resolution would halve (it doubles because of N but falls by 4 because of reduced RDC resolution). If a 13-bit RDC were available, it could be used to keep the mechanical resolution the same. However, hardware RDCs are typically available only in resolutions of 10, 12, 14, and 16 bits/revolution. The impact on resolution of increasing the resolver speed must be evaluated based on maximum motor velocity.

14.4 Position Resolution, Velocity Estimation, and Noise

All digital position sensors provide the position quantized to some limited resolution. Figure 14-7 shows the effect of quantization, comparing true position (the nearly-straight line) with feedback position (the stair step). Digital controllers sample the position at regular intervals, and at each sample most estimate velocity from position using simple differences: $V_N = (P_N - P_{N-1})/T$.

Estimating velocity from the two positions causes resolution noise. As Figure 14-7 shows, the simple-differences algorithm induces a pulse that is one sample time wide at each step of the position sensor. The noise from velocity estimation travels through the control law and produces current spikes. These current spikes are responsible for most of the noise in many servo systems.

Low-pass filters are commonly used to reduce resolution noise. Such filters attenuate high-frequency components and thus remove what is often the most objectionable result of limited resolution. Unfortunately, these filters also inject phase lag into the loop, often in substantial portions. Generally, the more coarse the resolution, the more

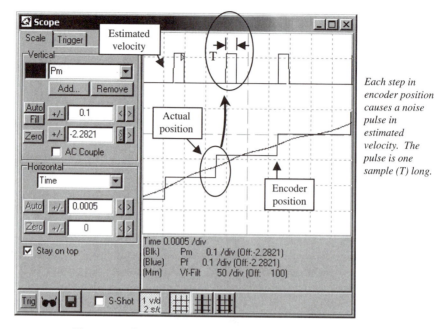

Each step in encoder position causes a noise pulse in estimated velocity. The pulse is one sample (T) long.

Figure 14-7. Quantized position and the effects on velocity estimation.

severe the required filter. An indirect relationship between resolution and control law gains results: (1) Limited resolution demands the use of low-pass filters to reduce noise; (2) low-pass filters inject phase lag in the loop; (3) increased phase lag reduces margins of stability; (4) lower margins of stability force a reduction in control law gains.

14.4.1 Experiment 14A: Resolution Noise

Experiment 14A, shown in Figure 14-8, demonstrates the relationships of resolution, filtering, and noise. This system is a PI velocity loop using a resolution-limited encoder for feedback. The command is generated from a waveform generator (Command), which is a constant 10 RPM by default. The resolution can be set with the *Live Constant* Resolution, which defaults to 4096 counts/revolution (equivalent to 1024 lines). A *Live Button* by the same name removes the resolution limitation when off. There are two filters, both of which reduce noise generation; one is in the forward path (FwdFilt), and a second is in the feedback path (FbFilt). Both filters can be enabled with *Live Switches*, which have the same name as the filter they control. There are two *Live Scopes*. On the left, commanded current (above) is compared to actual current (below); the smoothing demonstrates the implicit filtering provided by the current loop. The *Live Scope* on the right shows velocity, both filtered (above) and unfiltered (below).

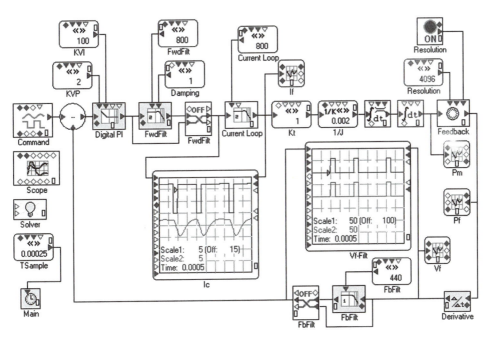

Figure 14-8. Experiment 14A demonstrates smoothing from filters to reduce perturbations in velocity feedback and current.

14.4.2 Higher Gain Generates More Noise

The most responsive feedback systems require the largest gains. That means that when a servo system must move quickly, K_{VP} will be relatively large. Note that all control loops have an equivalent to a proportional gain. For example, PID position loops use the "D" gain for the same purpose as the velocity loop K_{VP}. So responsive systems have a large K_{VP} (or its equivalent), and K_{VP} amplifies noise. That means quick servos are more sensitive to resolution problems. (That is one of the reasons the most responsive servos need the highest feedback resolution.) This is demonstrated in Figure 14-9, which compares resolution noise in Experiment 14A, with K_{VP} set to 0.5 and 2.0.

Large load inertia also implies that the system may be noisier. Consider the motor as a control element: When the inertia becomes large, the motor gain is low. A low motor gain requires a large electronic gain to provide reasonable total loop gain. For example, if the inertia doubles, the gain, K_{VP}, must double to maintain loop performance. In practice, the one-to-one relationship of gain and inertia often cannot be maintained, so increases in inertia may not be fully compensated by higher gains. This is because mechanical resonance will often prevent the gain from being raised that high. Still, when the load inertia is increased, the loop gain (K_{VP}) will almost always

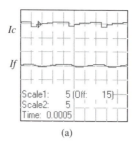 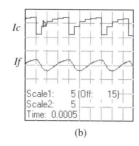

(a)   (b)

Figure 14-9. Noise in commanded (above) and actual current increases with control law gains: (a) $K_{VP} = 0.5$; (b) $K_{VP} = 2.0$.

increase, even if the increase is less than proportional. This implies that raising inertia forces larger values of K_{VP}, which makes the system more susceptible to resolution noise. The interested reader can simulate this using Experiment 14A.

14.4.3 Filtering the Noise

Resolution noise can be reduced by adding filtering in one of three places: in the feedback path, in the forward path, and using the current loop as an implicit filter. These filters are shown in Figure 14-8. Filters reduce noise but at the expense of adding phase lag, which causes overshoot and instability, as discussed in Chapters 3, 4, and 6.

Filtering the feedback signal reduces noise by smoothing the velocity noise pulse. This filter is usually a single- or double-pole low-pass filter with a bandwidth that can sometimes be set according to the application. When possible, the bandwidth of the low-pass filter should be set lower for less responsive applications; for the most responsive applications, it should be set higher or left off all together. Figure 14-8 (right-hand *Live Scope*) can show the effect of a 440-Hz single-pole filter applied to the velocity estimation run at a 250-μsec sample time; the effect of filtering here is to halve the amplitude of the noise pulse. This comes at the expense of the phase lag from a 440-Hz filter, which is acceptable for most applications. For applications where responsiveness is not a primary concern, a lower-bandwidth filter will provide greater noise reduction, perhaps allowing the use of a coarser feedback device.

The noise from the filtered velocity signal proceeds to the velocity-loop summing junction, where it is inverted and then processed by the velocity controller, which for this example is a PI controller. The proportional gain amplifies the noise; the effect of the integrator is small because the frequency of resolution noise is high and is smoothed by the integrator. The noise is then attenuated by the filter in the forward path (FwdFilt). The forward filter here is a two-pole low-pass filter with an 800-Hz break frequency; this frequency is relatively high, so the minimal smoothing is provided.

The current command is fed to the current loop, which generates the actual current. The current loop here is represented by an analog current controller. Most digital controllers today use digital current loops. These loops run at much higher sample rates than the digital velocity loop, typically four to eight times higher. The differences in sample rates make the current loop so much faster than the velocity loop that, for the purposes of studying the velocity loop, the analog model is a reasonable simplification. The current loops provide smoothing according to the bandwidth. Since the current loop bandwidth is relatively high (800 Hz), the smoothing is minimal.

The assumption for this example has been that the loop must be responsive. Thus, the filters all had relatively high bandwidths, the lowest being the 440-Hz feedback filter. As a result, the combined smoothing was small, and the current ripple from each position step is about 3 amps. The velocity loop is fairly responsive for modern servo systems. The 3-amp spikes may or may not be a problem for a given application. If not, and if the response is high enough, the position feedback sensor would have appropriate resolution. For this example, the resolution was 4096 counts/revolution, equivalent to a 1024-line encoder or to a 12-bit RDC. This resolution is relatively coarse and will not work in many applications. Note that had the inertia been larger, the gain K_{VP} would have been raised and the current spikes would have gone up proportionally. Most industrial applications require between 1000- and 8000-line encoders.

14.5 Alternatives for Increasing Resolution

The most straightforward way to increase resolution is to use a feedback device with higher line count. Beyond that, there are several other means of enhancing resolution.

14.5.1 The 1/T Interpolation, or Clock Pulse Counting Method

The 1/T *interpolation*, or *clock pulse counting*, method enhances resolution by measuring precise times in which encoder transitions occur rather than simply counting encoder transitions per sample time. Using simple differences, the resolution of the time measurement is the sample interval — perhaps several hundred microseconds; by comparison, using 1/T interpolation, the resolution can be 1 microsecond. See the example in Figure 14-10.

The average speed in Figure 14-10 is 10 rad/sec. However, during the 1.0 msec between samples, the encoder moved 3 counts, or 0.009 rad, or at an apparent rate of 9 rad/sec (for convenience, the resolution in this example is 0.003 rad and the sample time is 0.001 sec). The random relationship between the sample time and the encoder steps creates resolution noise. Over a longer period of time, the average sampled velocity would indeed be 10 rad/sec, but the noise for any given sample would be as much as ± 1 count/sample, equivalent here to ± 3 rad/sec, or 28.7 RPM.

The 1/T interpolation method works by using hardware to measure the time that the last encoder transition took place. Most microprocessors with onboard hardware timers can measure time to an accuracy of a microsecond or so. The 1/T interpolation

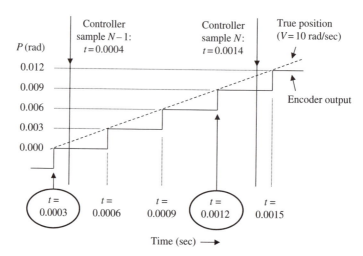

Figure 14-10. Position vs. time for $V(t) = 10$ rad/sec, 1-kHz sampling.

method does not rely on a fixed sample time but instead enhances the velocity-estimation formula for simple differences:

$$V_N = (P_N - P_{N-1})/(t_N - t_{N-1}) \tag{14.5}$$

For Figure 14-8, t_{N-1} (when the transition for P_{N-1} took place) is 0.0003 sec and t_N is 0.0012 sec, so the $1/T$ velocity is correct:

$$(0.012 - 0.003)/(0.0012 - 0.0003) = 10 \, \text{rad/sec}$$

The $1/T$ method also helps the noise sensitivity of the *inverse trapezoidal method*, discussed in Section 5.6.2.1. However, $1/T$ does add phase lag, especially when the motor is rotating slowly, so the overall improvement in phase lag offered by the inverse trapezoidal method will be mitigated at low speed.

14.5.2 Sine Encoders

Sine encoders provide sinusoidal A-Channel and B-Channel feedback [38,56] rather than the digital signals shown in Figure 14-3. Those channels are measured as analog signals and can be interpolated 250–1000 times more finely than the equivalent digital encoder signals. Typical line counts for sine encoders are 500–2000 lines per revolution; that is, there are about 500–2000 sinusoidal cycles of channels A and B for each mechanical revolution of the encoder. However, because of interpolation, it is common for sine encoders to provide more than 1 million counts per revolution, well beyond the resolution available to digital encoders. Also, sine encoders are relatively accurate; most rotary sine encoders have cyclic errors of less than $1\,\mathrm{min}^{-1}$.

There are two reasons why sine-encoder techniques are employed: resolution and maximum velocity. Resolution of digital encoders is limited by the dimensions of the slots through which encoder light travels [76]. Encoders work by sourcing light through two masks: one stationary and the other attached to the encoder rotor. Light travels through slots formed by the two masks and is alternately passed and blocked as the rotating mask moves. Higher resolution requires smaller slots. Smaller slots increase the opportunity for light to fringe and for crosstalk. Slots for very high-resolution incremental encoders would need to be so small that they would be impractical to build. Because sine encoders interpolate within a single slot, they achieve very high resolution but require only slot widths approximately equivalent to those used for standard encoders.

The maximum velocity of digital encoders is often limited by the frequency of the output signal. The frequency output for high-resolution encoders can be quite high. For example, a 20,000-line encoder rotating at 3000 RPM outputs a frequency of 1 MHz. Many encoder output stages cannot generate such a high frequency, and many drives could not accept it. However, with a sine encoder this limit is rarely a factor because the sine waves are at a low frequency; a 2000-line sine encoder rotating at 3000 RPM generates only 100 kHz but provides an interpolated resolution of perhaps 512×2000, or more than 1 million lines/revolution.

Note that many high-resolution digital encoders use sine-encoder techniques internally but convert the signal to the digital format at the output stage. For example, a 25,000-line encoder may be a 2500-line sine encoder with a 10-times interpolation circuit creating a digitized output. This is done inside the encoder and is transparent to the user. This technique provides high accuracy but still suffers from the need to produce high-frequency output for high-resolution signals. The natural question is: Why not output the sine signal directly to the drive? One answer is that many modern motion controllers' drives cannot accept sine-encoder signals directly, although the number that can is increasing. Also, many sine encoders are sold with a *sine interpolation box*, a separate device that interpolates sine-encoder signals and converts them to a digital pulse train to be fed to the drive. As more devices become able to accept sine-encoder signals directly, the need for sine interpolation boxes should decline.

One interesting facet of sine encoders is that the signals are often processed in a manner similar to the way resolver signals are processed. Because the sine encoder feeds back analog signals that are similar in form to demodulated resolver signals, a tracking loop is often applied to process those signals. As was the case with the resolver, the tracking loop can inject phase lag in the sine encoder signals [10].

14.6 Cyclic Error and Torque/Velocity Ripple

Cyclic error is low-frequency error that repeats with each revolution of the feedback device. It causes low-frequency torque ripple in motion systems. The resolution errors discussed earlier generate high-frequency perturbations; the undesirable effects are usually audible noise and vibration. By contrast, cyclic errors cause low-frequency torque ripple on the motor shaft. A typical low-frequency error might occur two or four times per motor revolution, a fraction of the frequency of resolution errors. Effects of cyclical errors cannot be filtered when they are low enough to be at or below the velocity loop bandwidth. For example, at 60 RPM, a two-times-per-rev cyclic error would generate 2-Hz ripple, well below the reach of a low-pass filter in almost any servo system.

Cyclic errors are caused by imperfections in the feedback device and in the device mounting. For resolvers these imperfections are in the winding inaccuracies and component tolerances; also, the RDC contributes inaccuracy because of errors such as those in the D/A converters that RDCs rely on to produce the sine and cosine of the input signals. For a typical single-speed resolver system, the resolver error is between 4 and $10 \min^{-1}$ and the RDC error is between 2 and $20 \min^{-1}$. Resolver-mounting inaccuracy can contribute some error as well. Resolvers are commonly built in two styles, housed and frameless. Housed resolvers (Figure 14-11) have their own bearing system. The resolver housing is mounted to the motor housing; the motor and resolver shafts are mechanically connected with a coupling that allows small alignment inaccuracies between the two rotors. Housed resolvers have little mounting inaccuracy.

Figure 14-11. Housed resolver.

Figure 14-12. Frameless resolver.

Frameless resolvers (Figure 14-12) rely on the motor bearing system. The motor shaft extends through the resolver; the resolver rotor is fixed directly to the motor shaft. This reduces motor length by virtue of eliminating the coupling. Frameless resolvers can reduce cost by simplifying the feedback assembly and reducing material. However, frameless resolvers often have additional error because of imperfections in the mounting of the resolver stator with respect to the resolver rotor. Frameless resolvers rely on the motor shaft; they cannot be aligned as well as the shaft of a housed resolver, which relies on instrument-grade bearings and machining. Typically, frameless resolvers contribute an additional mounting inaccuracy of 3 to 5 min^{-1}.

Encoders also have low-frequency inaccuracy, although usually smaller than that of resolvers. Industrial encoders are commonly available with low-frequency inaccuracies under 1 min^{-1}. Also, industrial encoders are usually housed so that mounting inaccuracy is not an issue. However, inexpensive encoders are commonly provided in "kit" or frameless form. These encoders often have inaccuracies as high as 20 min^{-1}, and, being frameless, they also suffer from higher mounting inaccuracy.

So the total cyclic error from the feedback device can be in excess of 40 min^{-1}, although 15–20 min^{-1} for resolvers and 1–10 min^{-1} for encoders are more typical in industrial systems. This may sound small, being only a fraction of a degree, but this error can generate a surprisingly large amount of torque ripple.

14.6.1 Velocity Ripple

Velocity error can be calculated directly from position error. Position error versus time for any component of cyclical error is given as

$$P_E(t) \equiv P_{E-PEAK} \times \cos(N \times P(t))$$

where P_{E-PEAK} is the peak of the error signal, N is the harmonic, and $P(t)$ is motor position as a function of time. Velocity error is the derivative of position error with respect to time of position error. Using the chain rule of differentiation:

$$V_E(t) \equiv dP_E(t)/dt = P_{E-PEAK} \times N \times -\sin(N \times P(t)) \times dP(t)/dt$$

Noting that $dP(t)/dt$ is velocity, velocity ripple as a fraction of velocity is related to the peak of position error by a simple formula:

$$V_{E\ PEAK}/V = P_{E-PEAK} \times N$$

with all terms in SI units. For example, the velocity error as a fraction of motor velocity of a system with 15-min^{-1} (0.0044 rad) error at the 4th harmonic would have 0.0044×4, or 1.7%, velocity error due to this component.

The velocity error is reflected directly in the velocity feedback signal when the actual motor speed is constant. Of course, the servo loops work to reduce the ripple in the velocity signal; unfortunately, the only way to smooth ripple in the feedback signal is to create torque ripple that induces ripple in the actual speed. This results in an improvement in the feedback signal while actually worsening the actual velocity. In fact, if the servo-loop bandwidth is high enough in comparison to the ripple frequency, it can induce severe torque ripple to cancel an error that exists only in a feedback signal. On the other hand, if the ripple frequency is well above the servo bandwidth, the servo loop does not correct; the motor speed is relatively smooth, but the feedback signal indicates ripple.

14.6.2 Torque Ripple

Torque ripple is a function of velocity ripple and the servo response at the ripple frequency. Torque ripple from cyclical error is easier to analyze than the resolution noise because the frequencies are low, so filters described in Figure 14-8 have little effect on the magnitude. Typical resolver cyclic errors in single-speed resolvers are at two and four times the motor speed. So a motor rotating at 1200 RPM (20 revolutions/second) would have its major error components at 40 and 80 Hz. Usually, the effects of current loops and low-pass filters are minimal at such low frequencies. Also, the sampling process can be ignored, again because the low frequency allows the assumption of ideal differentiation.

The model in Figure 14-13 ignores all loop gains except the proportional velocity gain, because this gain causes the majority of ripple. It also ignores the low-pass filters

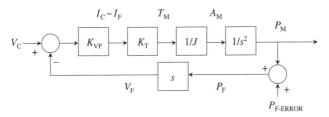

Figure 14-13. Simplified model for analyzing ripple caused by sensor error.

and the sampling process. Following Figure 14-13, velocity error is fed into a proportional controller, which generates current command. Current command is assumed equal to feedback because only low-frequency effects are being considered. Current is scaled by the motor torque constant to create torque, which is scaled through inertia to create acceleration. Acceleration is double-integrated to produce true position, which is summed with feedback error to produce feedback position.

Using the $G/(1 + GH)$ rule on Figure 14-13, the transfer function $G_{\text{RIPPLE}}(s)$ from resolver error ($P_{\text{F-ERROR}}$) to torque (T_{M}) is shown in Equations 14.6 and 14.7.

$$G_{\text{RIPPLE}}(s) = \frac{T_{\text{M}}(s)}{P_{\text{F-ERROR}}(s)} \tag{14.6}$$

$$= \frac{s^2 K_{\text{VP}} K_{\text{T}}}{s + K_{\text{VP}} K_{\text{T}}/J} \tag{14.7}$$

The gain $K_{\text{VP}} K_{\text{T}}/J$ is approximately the bandwidth of this proportional velocity controller. This can be seen by using the $G/(1 + GH)$ rule to form the transfer function from command to feedback:

$$G_{\text{CLOSED}}(s) = \frac{K_{\text{VP}} K_{\text{T}}/Js}{1 + K_{\text{VP}} K_{\text{T}}/Js} \tag{14.8}$$

$$= \frac{K_{\text{VP}} K_{\text{T}}/J}{s + K_{\text{VP}} K_{\text{T}}/J} \tag{14.9}$$

The bandwidth of this simple transfer function is easily seen to be $K_{\text{VP}} K_{\text{T}}/J$ rad/sec by setting $s = jK_{\text{VP}} K_{\text{T}}/J(j = \sqrt{-1})$ and noting that the magnitude of the result is 0.707, or −3 dB:

$$F_{\text{BW}} \approx \frac{K_{\text{VP}} K_{\text{T}}}{J} \text{ rad/sec} = \frac{K_{\text{VP}} K_{\text{T}}}{J2\pi} \text{ Hz} \tag{14.10}$$

Combining Equations 14.9 and 14.10, we get

$$G_{\text{RIPPLE}}(s) \approx \frac{s^2 K_{\text{VP}} K_{\text{T}}}{s + 2\pi F_{\text{BW}}} \tag{14.11}$$

Now two observations can be made. When the ripple frequency is high, the denominator is dominated by the s term, so Equation 14.11 reduces to

$$G_{\text{RIPPLE-HIGH}}(s) \approx s K_{\text{VP}} K_{\text{T}} \qquad (14.12)$$

This indicates that when the ripple frequency is high, the torque ripple will be proportional to the frequency (via s) and proportional to K_{VP}, whether K_{VP} was raised to accommodate higher inertia or to raise the bandwidth. This is why systems with large inertia or high response are more susceptible to problems caused by cyclic error.

The second observation relates to ripple frequency well below the velocity loop bandwidth. Here, the denominator of Equation 14.11 reduces to $2\pi F_{\text{BW}} = K_{\text{VP}} K_{\text{T}}/J$, so Equation 14.11 becomes

$$G_{\text{RIPPLE-LOW}}(s) \approx s^2 J \qquad (14.13)$$

In Equation 14.13, the torque ripple increases with the square of the frequency and proportionally with the inertia. Note that the error reduction provided by multispeed resolvers does not improve the torque ripple; in fact, it enlarges it at lower ripple frequencies. Multispeed resolvers typically reduce position error approximately in proportion to the resolver speed; but the electrical frequency increases by that same constant. So, although position error is reduced, the s^2 term more than offsets the improvement and torque ripple increases. This will be studied in *Visual ModelQ* Experiment 14B.

Bear in mind that resolver errors are rarely pure sinusoids. Instead the errors are combinations of several harmonics, with the harmonic at two and four times the electrical frequency dominating. If a resolver is guaranteed to have $20\,\text{min}^{-1}$ of errors, that will be a combination of multiple error harmonics, where the peak magnitude is within that error margin. Manufacturers usually will not provide the individual harmonics. Here, the focus is on the second and fourth harmonics. Since the ripple is worse for higher frequencies, this is an optimistic assumption: If the entire ripple were at a higher harmonic, the amount of torque ripple could be much worse.

14.7 Experiment 14B: Cyclical Errors and Torque Ripple

Experiment 14B, shown in Figure 14-14, demonstrates the relationship between cyclical error and torque ripple. This system is a PI velocity loop, with the default tuning yielding a bandwidth of 92 Hz. The command is generated from a waveform generator (Command), which creates a 1000-RPM DC signal. Note that the inertia, J, is large ($J = 0.02\,\text{kg-m}^2$) because larger inertia makes torque ripple more apparent. A 2-arc-minute cyclical position error at the second harmonic is injected via a *Visual ModelQ* harmonic block (Pos Err); an analog switch at far right allows the DSA to temporarily disable position error because this signal corrupts Bode plots. There are two *Live Scopes*. On the left, feedback velocity is shown above actual velocity; both signals are scaled at 2 RPM/division and AC coupled to remove the DC command; note that the

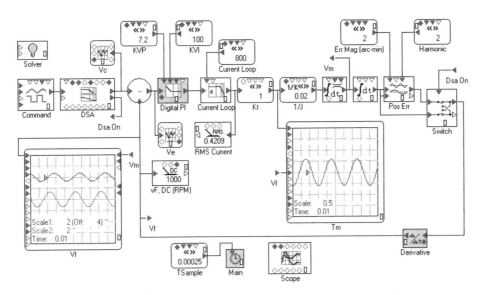

Figure 14-14. Experiment 14B demonstrates cyclical error and torque ripple.

model must execute about 2 seconds of simulation time for the signals to couple through the scope's AC coupling filter and come into view. On the right, torque ripple is shown. Both *Live Scopes* are triggered by the V_f, the right scope through an explicit connection to the trigger node (the 8th node from the top on the left), and the left scope implicitly because V_f is Channel 1 and the trigger node is left unconnected.

14.7.1 Relationship Between Error Magnitude and Ripple

The ripple is proportional to the magnitude of cyclic error. This applies to all frequencies of ripple. The only exception is that the ripple magnitude cannot exceed the drive peak current. Change the error magnitude (Err Mag) and note the proportional relationship.

14.7.2 Relationship Between Velocity and Ripple

The frequency of the ripple is proportional to the velocity. You can change the commanded speed by changing the offset in the waveform generator. For example, changing the offset from 1000 RPM to 500 RPM will cut the amplitude by four times. This is expected because the ripple frequency (33 Hz) is below the bandwidth of the velocity loop (92 Hz), so the amplitude should be approximated by Equation 14.13, which predicts a square relationship. When the frequency exceeds the bandwidth, switch to Equation 14.12, which predicts a proportional relationship. In fact, this behavior is demonstrated above 125 Hz (velocity = 4000 RPM).

14.7.3 Relationship Between Bandwidth and Ripple

The bandwidth of the default system is 92 Hz. You can verify this with a Bode plot. Now, reduce the bandwidth to investigate the effect on the ripple magnitude. Since the bandwidth is well above the ripple frequency, the ripple amplitude is approximated by Equation 14.13. Reducing the bandwidth has little effect as long as it remains above the ripple frequency. Set the velocity loop bandwidth to approximately 45 Hz by cutting K_{VP} from 7.2 to 4 and reducing K_{VI} from 100 to 55. Notice that the magnitude of ripple is affected very little, as predicted.

Continue with this experiment. When the ripple frequency is below the bandwidth, the amplitude is little affected by the bandwidth. When the ripple frequency is above the bandwidth, the velocity loop acts to attenuate the ripple. For example, cutting bandwidth to 7 Hz ($K_{VP} = 0.72$, $K_{VI} = 10$) has a large impact on the ripple because the ripple frequency is now well above the bandwidth.

14.7.4 Relationship Between Inertia and Ripple

If you adjust inertia and K_{VP} to accommodate the change, you will see the ripple change according to Equations 14.12 and 14.13, both of which predict a proportional relationship assuming K_{VP} is adjusted to maintain the original bandwidth. Note that the peak of the simulated drive is 20 amps, so these equations do not work when the inertia increases so much that the ripple reaches that limit.

14.7.5 Effect of Changing the Error Harmonic

Experiment with the error harmonic. The default is "2," indicating that the error is the second harmonic. If all the error were at the fourth harmonic, the amplitude would increase by about four times. The reason for the square relationship is that the ripple frequency is below the bandwidth, so Equation 14.13 approximates the behavior of the system; this equation has an s^2 in the numerator. When the ripple frequency exceeds the bandwidth, the behavior is approximated by Equation 14.12, which increases only proportionally to the frequency. Changing the harmonic of the error changes the frequency in much the same way as changing the speed of the motor did.

14.7.6 Effect of Raising Resolver Speed

Raising the *resolver* speed for mass-produced resolvers changes two things. The position error of higher speed resolvers declines approximately in proportion to the resolver speed. In other words, a single-speed resolver with 20 min^{-1} of errors can often be rewound as a two-speed resolver with just 10 min^{-1} of errors. At first it seems that this would reduce torque ripple. However, raising the speed also increases the frequency of the ripple. For low-frequency ripple, often the area of most concern, the ripple increases as the square of the ripple frequency (Equation 14.13). The overall effect is that low-frequency torque ripple is not reduced but instead increases in proportion to resolver speed. Even when the ripple is at high frequency (Equation 14.11), the increase in frequency offsets the reduction in magnitude. Increasing the resolver speed does reduce ripple at high frequencies, where the filters and the current loop can provide attention. For our example, that benefit would not be enjoyed until the ripple frequency was above the current loop bandwidth (800 Hz); this is so high that it is usually not important for the study of cyclical error.

14.7.7 Relationship Between Ripple in the Actual and Feedback Velocities

The relationship between ripple in the actual motor velocity and that in the feedback signal is counterintuitive. Higher gains in the control law increase the reaction to cyclical error, smoothing the velocity feedback signal while actually increasing the ripple in the motor. This is demonstrated in Figure 14-15; in Figure 14-15a, the system with relatively low gains (7-Hz bandwidth, $K_{VP} = 0.72$, $K_{VI} = 10$) responds little to the ripple in the velocity feedback signal (above) so the actual ripple (below) remains low. However, in Figure 14-15b, the system with high gains is shown (92-Hz bandwidth, $K_{VP} = 7.2$, $K_{VI} = 100$). Here, the system responds to the erroneous position harmonic, reducing the ripple in the feedback signal. However, it does so by injecting torque ripple, which actually increases the ripple in the actual signal. Of course, in an actual application, only the measured velocity (top) signal is normally available; relying wholly on that signal, you might conclude that ripple improves with higher control law gains.

In the case of Experiment 14B, the torque signal (or, equivalently, the current feedback) is a better indicator of ripple in the control system than is the velocity feedback signal. This is demonstrated in Figure 14-16. Figure 14-16a shows low-torque ripple

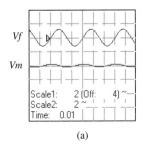 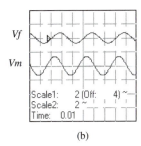

(a)   (b)

Figure 14-15. Ripple in actual velocity (above) vs. the feedback signal. (a) Low gains produce high ripple in the feedback, but the actual ripple is low; (b) high gains reduce apparent ripple but increase actual ripple.

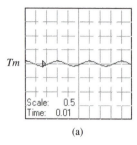
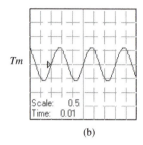

(a)   (b)

Figure 14-16. Torque ripple with (a) low and (b) high control law gains.

generated low-control-law gains ($K_{VP} = 0.2$, $K_{VI} = 5$); Figure 14-16b shows high ripple from high gains ($K_{VP} = 7.2$, $K_{VI} = 100$). In an application where the only velocity signal available is the one derived from a position signal, which is influenced by error harmonics, the current feedback signal is often the best indicator of actual ripple.

14.8 Choosing a Feedback Device

There are a number of feedback devices from which to choose. This section will provide an overview of some of the more popular feedback devices. The most popular rotary position feedback device is certainly the optical encoder. In the past, these sensors suffered from reliability problems, especially in the heat sensitivity of the optical electronics. This caused motor manufacturers to *derate* motors when using optical encoders, reducing the output torque of the motor to reduce the housing temperature. These problems are largely cured, and many industrial encoders can run at high enough temperatures that they have little or no effect on motor ratings. Optical encoders are available in a wide range, from kit encoders, which are inexpensive encoders that share the shaft of the motor and have a resolution of a few hundred lines per revolution and a price of about $10 in volume, to fully housed encoders with 50,000 or 100,000 lines per revolution costing several hundred dollars. As discussed before, sine encoders raise the resolution to the equivalent of 500,000 lines per revolution or more. Encoders are available in the resolution, accuracy, and temperature range required by most servo applications.

One shortcoming of optical encoders is that they rely on a masked transparent disk, which can be clouded by contaminants. In environments where the feedback device is exposed to contamination, magnetic techniques offer higher reliability. A class of encoders called *magnetic encoders* operates on principles similar to those of optical encoders but with less sensitivity to contaminants.

Resolvers are generally thought to be among the most reliable motion sensors. First, the sensing is based on magnetic techniques and so is insensitive to most contaminants. For example, in aerospace applications, resolvers are sometimes immersed in jet fuel as part of normal operation. Because resolvers have no internal electronic components, they can be exposed to extremely high temperatures. Also, because resolvers consist of copper wound around steel laminations, they can be built to be quite rugged, able to absorb high levels of shock and vibration compared with encoders. Resolvers are also inexpensive compared with encoders with equivalent resolution (that is, between 1000 and 16,384 lines), especially when software RDC is used.

For applications with a long distance between the motor and drive, the maximum cable length of the feedback device must be considered. For resolvers, the problem is that cable capacitance can create a phase lag for the modulated signal. Resolver signals are usually modulated at between 4 and 8 kHz. For cables that are in excess of 100 meters, the phase shift of the resolver's signals can degrade the RDC performance. Some drive manufacturers provide adjustments for the RDC circuit so that it can accommodate the phase shift generated by the long cables.

For encoders, cable length is limited by IR drop in the cable. Encoders usually draw between 100 and 250 mA. If the supply voltage is 5 V (as it usually is), the encoder may allow only a few hundred millivolts drop in the cable. This limits the maximum resistance of the cable. At least one company, Stegmann, has addressed this problem by allowing users to apply an unregulated voltage to the encoder. This voltage can vary by several volts, allowing the cable to induce a much larger voltage drop. In other cases, users can employ power supplies with *sense* inputs; these power supplies regulate their output voltage based on the voltage at a remote location using non-current-carrying leads to sense the voltage. This allows very long encoder cables but at the expense of convenience. Few drive manufacturers provide such power supplies onboard drives. Encoder cable length is also limited by transmission-line effects, especially when the encoder output frequency is more than a few hundred kilohertz.

Tachometers are still used in servo applications. A DC motor, an analog drive, and a DC tachometer constitute a well-tested technology that is still cost effective, especially in low-power applications. Also, tachometers are used in some very high-end applications, especially when the speed is very slow, such as when controlling a radar dish or telescope. Because a tachometer feeding an analog drive has no explicit resolution limitation, it can control motors rotating at small fractions of an RPM.

14.8.1 Suppliers

There are numerous suppliers of encoders. A recent search of the Thomas Register (www.thomasregister.com) showed over 100 companies that supply encoders, although only some are for servo motors. Suppliers of encoders include

BEI (www.beiied.com)
Canon Components (www.canon.com)
Computer Optical Products (www.opticalencoder.com)
Danaher Controls (www.dancon.com)
Dynamics Research (www.drc.com)
Encoder Products (www.encoderprod.com)

Gurley Precision Instruments (www.gurley.com)
Hengstler (www.hengstler.de)
Heidenhain (www.heidenhain.com)
Ono Sokki (www.onosokki.co.jp)
Renco (www.renco.com)
Sony (www.sony.com)
Stegmann (www.stegmann.com)
Sumtak (www.sumtak.co.jp)
Tamagawa (www.tamagawa-seiki.co.jp)
US Digital (www.usdigital.com)

Sine encoders are supplied by a few manufacturers, including

Heidenhain (www.heidenhain.com)
Hengstler (www.hengstler.de)
Stegmann (www.stegmann.com)

Resolvers are supplied by several manufacturers, including

Artus (www.psartus.com)
Axsys (www.axsys.com)
Harowe (www.danahermotion.com)
Litton Polyscientific (www.polysci.com)
Transicoil (www.transicoil.com)
Tamagawa (www.tamagawa-seiki.co.jp)

Hardware R/D converters are manufactured by

Analog Devices (www.analog.com)
Control Sciences Inc. or CSI (www.controlsciences.com)
Data Device Corporation (www.datadevicecorp.com)
North Atlantic Instruments, Inc. (www.naii.com)

14.9 Questions

1. What is the resolution of a 5000-line encoder in degrees or arc-minutes?
2. Name three benefits of using software R/D conversion compared to hardware R/D converters.
3. What are two benefits of multispeed resolvers compared to single-speed resolvers?
4. For a 2000-line encoder sampled at 5 kHz, what is the speed resolution absent filtering?

5. a. For a resolver with a $3\,\text{min}^{-1}$ ($1\,\text{min}^{-1} = 1/60$ degree) error at four times per revolution, what is the magnitude of velocity ripple on the sensed signal as a percentage of motor speed?
   b. Use Experiment 14B to verify.

6. a. Run Experiment 14B. Measure the torque ripple as a function of velocity loop bandwidth for $K_{VP} = 1, 2, 5, 10$, and 20. Note that you can set $K_{VI} = 0$ to ensure stability.
   b. Compare this to the ripple estimated by Equation 14.13.
       What conclusions can you draw? (*Hint: ripple frequency is* $1000\,RPM/60\times$
       $2 = 33.33\,Hz$)

7. a. What are two advantages of resolvers over encoders?
   b. What are two advantages of encoders over resolvers?

Chapter 15

Basics of the Electric Servomotor and Drive

Control systems span four major areas: temperature, pressure and flow, voltage and current, and motion. Motion control is implemented with three major prime movers: hydraulic [100], pneumatic [96], and electric motors. This chapter will provide an overview of electric motors and their associated drives.

Motor control requires three basic elements: a motor, a drive, and one or more feedback devices. The drive controls current in order to produce torque; drives also commonly control velocity and sometimes control position. In many control systems, multiple axes of motion must be controlled in synchronization. Usually a separate multiaxis controller provides coordination for the machine while the drives are left to implement the control associated with individual motors [100]. A typical configuration is shown in Figure 15-1. There are many variations of this configuration, including systems that have only one axis and thus have no need of a multiaxis controller and systems where the application is so well defined that the drive and controller are integrated into one device. An example of an integrated system is a computer hard-disk controller.

There are many control functions to be implemented for each axis of motion. First, a position *profile*, which is a sequence of position commands versus time, must be generated. Second, a position loop must be closed, and usually within that, a velocity loop is closed. The output of the position/velocity loop is a torque command. In brushless motors (including permanent-magnet brushless DC and induction motors), the torque command and the motor position are used to calculate multiple current commands in a process called *commutation*; for brush motors, commutation is mechanical. Then the current loop(s) must be closed. A power stage delivers power to the motor and returns current feedback. These functions are shown in Figure 15-2.

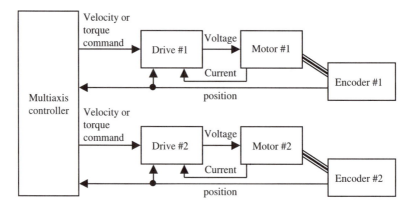

Figure 15-1. Typical multiaxis motion-control system.

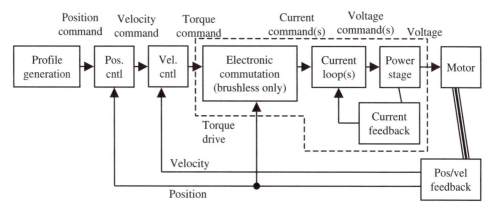

Figure 15-2. Functions of a single axis of motion.

15.1 Definition of a Drive

Defining the term *drive* can be difficult because most of the functions of an axis of motion can be implemented either in a drive or in a multiaxis controller connected to the drive. At one end of the spectrum, the drive is highly intelligent; a single-axis controller (SAC) performs all the functions in Figure 15-2. At the other end, a *power block* provides just the power stage; the remaining functions are performed by the multiaxis controller. Other configurations fall within this spectrum. For example, the *torque drive*, shown in Figure 15-2 in dashed lines, provides all functions after the torque command. In this chapter the term *drive* will imply *torque drive*, because the focus here is on the control of servomotors, in which the goal is to produce well-regulated torque; the torque command is a natural dividing point for that discussion.

15.2 Definition of a Servo System

Electronic motion control is a multi-billion-dollar industry. Servo motion is a fraction of that industry, sharing it with non-servo motion, such as stepper motors [49] and variable-frequency systems. A servo system is defined here as the drive, motor, and feedback device that allow precise control of position, velocity, or torque using feedback loops. Examples of servomotors include motors used in machine tools and automation robots. Stepper motors allow precise control of motion as well, but they are not servos because they are run "open-loop," without tuning and without the need for stability analysis.

The most easily recognized characteristic of servo motion is the ability to control position with rapid response to changing commands and disturbances. Servo applications commonly cycle a motor from one position to another at high rates. However, there are servo applications that do not need fast acceleration. For example, web-handling applications, which process rolled material such as tape, do not command large accelerations during normal operation; usually, they attempt to hold velocity constant in the presence of torque disturbances.

Servo systems must have feedback signals to close control loops. Often, these feedback devices are independent physical components mechanically coupled to the motor; for example, encoders and resolvers are commonly used in this role. However, the lack of a separate feedback device does not mean the system is not a servo. This is because the feedback device may be present but may not be easily identified. For example, head-positioning servos of a hard-disk drive use feedback signals built into the disk platter rather than a separate feedback sensor. Also, some applications use electrical signals from the motor itself to indicate speed. This technology is often called *sensorless* [12,66] although the name is misleading; the position is still sensed but using intrinsic motor properties rather than a separate feedback device.

15.3 Basic Magnetics

This section provides an overview of magnetic design, including permanent magnets and electromagnetism. The goal of the discussion is to provide a review so that you can better comprehend the tasks a drive must implement to regulate torque. The physical behavior of magnetic interactions closely parallels that of electric interactions. For example, Ohm's law, $V = IR$, has the magnetic parallel

$$\mathcal{F} = \Phi \mathfrak{R} \tag{15.1}$$

where $\mathcal{F}$ is MMF, the magnetomotive force, Φ is magnetic flux, and $\mathfrak{R}$ is reluctance, which is the magnetic impedance of an object. Recall that voltage is sometimes termed EMF (electromotive force) and the parallel becomes clearer. In both electric and magnetic circuits, a force (V or $\mathcal{F}$) moves a flow (I or Φ) through an impedance (R or $\mathfrak{R}$) [9].

In the simple magnetic circuit of Figure 15-3, there is a permanent magnet placed inside a steel core. The magnet is a source of constant MMF, equivalent to a battery in an electric circuit. The amount of flux flowing through the magnetic circuit is inversely proportional to the reluctance of the circuit. Flux, like current, always flows in a complete path. By convention, flux travels out of the north pole and returns to the south pole. This convention is due to the fact that when a magnet is affixed to a freely rotating structure such as a compass needle, the magnet's north pole points approximately to the earth's north pole.

The reluctance of the magnetic circuit shown in Figure 15-3 is the sum of the reluctances of the core and the magnet. The reluctance of an object depends on its material and on its dimensions. Every material has a permeability, μ. Higher permeability in a material implies that objects formed of that material will present lower reluctance to a magnetic field. In the design of motors, the material of choice for the core is most often steel because it is readily available and because the permeability of steel is quite high compared to most other materials. The permeability of a vacuum is defined as μ_0, which, in SI units, is equal to $4\pi 10^{-7}$ henrys/meter. The permeability of steel used in motors ranges from 2000 to 80,000 times higher than that of a vacuum (2.5×10^{-3} to 0.1). Often the magnetic permeability of a material is defined as μ_R, permeability relative to that of a vacuum:

$$\mu_R = \mu/\mu_0 \tag{15.2}$$

In modern permanent-magnet (PM) servomotors, the magnets are often made of neodymium-boron-iron (NdBFe), a *rare earth* magnetic material that exhibits much stronger magnetic properties than older motor-magnet materials. NdBFe magnets generate intense magnetic fields, allowing large amounts of torque to be generated in a small motor. In brushless (PM) motors, where the magnets are fixed to the rotor and thus must be accelerated, NdBFe magnets provide the additional advantage that they allow smaller rotor dimensions. This reduces rotor inertia, supporting higher acceleration rates [99]. The permeability of NdBFe is close to that of air; in fact, the magnet is usually the highest-reluctance component in the magnetic path.

The dimensions of an object also affect its reluctance. For an object with a uniform cross-sectional area, reluctance falls with area and grows with the object's length:

$$\Re = 1/\mu \times L/A \text{ henrys}^{-1} \tag{15.3}$$

For instance, a steel rod of length 50 mm with a cross section of 10 mm square (Figure 15-4) and with a permeability of 0.01 henrys/meter ($\mu_R \cong 8000$) would have a reluctance along the long axis of

$$R = (1.0/0.01 \times 0.050)/0.010^2 \text{ henrys}^{-1}$$
$$= 50,000 \text{ henrys}^{-1}$$

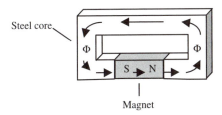

Figure 15-3. Flux path in a simple magnetic circuit.

Computing the reluctance of more complex shapes requires that the object be divided into multiple shapes of uniform area and permeability; the reluctance of the object is the sum of the reluctances of the shapes.

Magnetic properties can also be described based on material properties, which are independent of the shape of an object. In this case, the force is defined as a field intensity (H) and the result is a magnetic flux density (B). The equation relating B and H is

$$H = B/\mu \tag{15.4}$$

The parallel in electric fields is

$$E = J \times \rho \tag{15.5}$$

where E is the electric field intensity across a material, J is the current density in the material, and ρ is the resistivity of the material. Like μ, ρ depends only on an object's material and not on its dimensions. Table 15-1 compares the four equations of magnetic and electric interaction.

15.3.1 Electromagnetism

Electromagnetism is the phenomenon in which magnetic flux is generated with electric current. Multiple turns of wire are wrapped around a magnetic core; the MMF created by electromagnetism is in proportion to the amount of current and the number of turns in the coil:

$$\mathcal{F} = n \times I \tag{15.6}$$

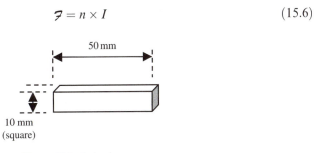

Figure 15-4. A simple core.

TABLE 15-1 EQUATIONS DESCRIBING MAGNETIC AND ELECTRIC INTERACTION

	Magnetic interaction	**Electric interaction**
Equations for objects	$\mathcal{F} = \Phi\mathfrak{R}$	$V = IR$
	$\mathcal{F}$ Magnetomotive force (MMF)	V Voltage (EMF)
	Φ Magnetic flux	I Current
	$\mathfrak{R}$ Reluctance	R Resistance
Equations for materials	$H = B/\mu$	$E = J\rho$
	H Magnetic field intensity	E Electric field strength
	B Magnetic flux density	J Electric current density
	μ Permeability	ρ Resistivity

In Figure 15-5, the core is only part of the flux path; the flux must return to the core through air. The total reluctance of the magnetic circuit is the sum of the reluctance in the core and the reluctance of the air, which constitutes the return path. The return path dominates the magnetic path because the permeability of air is so much smaller than that of steel.

15.3.2 The Right-Hand Rule

A convention of electromagnetic circuits, the right-hand rule, describes the direction of flux produced by electric current. Imagine a right hand grasping the core of Figure 15-5, with the thumb extending parallel with the core. If the current is traveling around the core in the direction in which the fingers point, by convention flux travels parallel with the thumb. This is illustrated in Figure 15-6.

15.3.3 Completing the Magnetic Path

In order to reduce reluctance, material can be added to the core in Figure 15-5 that substitutes steel for air in the return path; this material is called *back iron*. For example, Figure 15-7 shows a C-shaped back-iron piece. The reluctance of the magnetic circuit

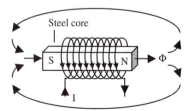

Figure 15-5. A core with flux.

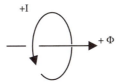

Figure 15-6. The right-hand rule for electromagnetism.

may be reduced thousands of times compared to the core in Figure 15-5, where the flux return path was air.

It should be stated that this discussion is for ideal magnetic circuits. In all magnetic circuits, some flux escapes the core in a phenomenon called *fringing*. Also, the permeability of steel and other magnetic materials declines as the applied field increases. This is called *saturation*. These factors are significant and must be accounted for in the design of electric motors. However, they are of limited importance for understanding control issues related to motors. For more information, see Refs. 15, 31, 40, and 71.

When an object is formed as a complex shape, the effect of the geometry on reluctance is difficult to calculate. Most products that use magnetic circuits (motors, transformers, and inductors) are composed of such complex shapes. Further complicating analysis, magnetic circuits must include the effects of fringing. In addition, saturation is a local effect — part of a core may be saturated while other parts are not. As a result of these complexities, designers often use an iterative method called "*finite element analysis*" (FEA) to analyze magnetic objects [18]; with FEA, a complex object is divided into many triangles (see Figure 15-8), and each triangle is assumed to be homogeneous (either air or steel and, if steel, a homogeneous level of saturation).

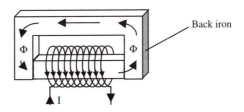

Figure 15-7. Wound core with back iron.

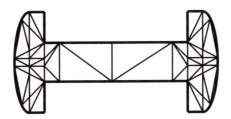

Figure 15-8. An object divided into triangles for FEA.

An FEA program iteratively solves the magnetic circuit. Manufacturers of such programs include Ansoft (www.ansoft.com), Ansys (www.ansys.com), and Cedrat (www.cedrat.com).

15.4 Electric Servomotors

The key characteristic of a servomotor is the ability to provide precise torque control. Ideally, the output torque of a servo system should be highly responsive and independent of motor position and of speed across the system's entire operating speed range. Although it is impractical for torque and speed to be completely independent, most servomotors are close enough to this ideal that simple models for servo systems can be based on that assumption. This is especially true when the purpose of the model is to predict stability, which was the case in earlier chapters of this book. In fact, the models of motors in those chapters made this assumption: The electromagnetic torque production was simply the product of current and the torque constant K_T (see Section 15.5.5).

More accurate models of servomotors show torque declining as speed increases. This reduction comes from numerous effects, including increased losses from wind friction (*windage*) and bearing friction. In brush motors, the brush commutator often limits the amount of current that can be produced in the motor at high speed. In brushless motors, a similar effect often occurs because of the current controllers. All these factors cause the torque to decline as speed increases.

15.4.1 Torque Ratings

Electromagnetic torque output, T_E, from servomotors is rated two ways: peak and continuous. A motor's peak torque is the maximum torque it can generate for a short period of time, usually one or two minutes. The continuous torque indicates how much torque the motor can generate over an indefinite period of time. These ratings represent thermal limits in the motor. When a motor outputs power, it does so with less than 100% efficiency. The power that is lost in the motor translates to heat. More power out means more losses; more losses inside the motor drive up the temperature. Excessive temperature in the motor will degrade lubricants and winding insulation. Limiting torque output (both peak and continuous) protects the motor by limiting its internal temperature. In addition, exceeding a motor's peak torque can permanently demagnetize the magnets.

Manufacturers of servomotors provide torque–speed curves to users. These curves show the envelope of motor torque available across the speed range. Figure 15-9 shows an example of a torque–speed curve. The ratings may be for the motor alone or for a system (i.e., a motor and a drive). If the rating is for a system, the thermal limits protect both the motor and the drive. Brush motors, as we will discuss, are easy to control, and their torque–speed curves are often for the motor only. Brushless motors are more complex and are often documented as part of a system. This is because

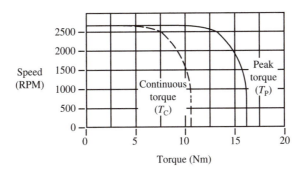

Figure 15-9. Typical torque–speed curve.

brushless drives are difficult to match to a motor, and manufacturers are better able to guarantee the performance of a motor–drive combination than a motor that will be mated to an unspecified drive.

An important area of expertise in applying servomotors is motor (or motor–drive) sizing. Here the application must be analyzed to select a motor (or motor–drive). There are many factors to consider, such as acceleration torque, machine cycle time, optimal motor inertia, ratios of transmission components such as belts or gears, and external loads. Many manufacturers provide software sizing programs, such as Motioneering by Kollmorgen (www.motionvillage.com). These packages aid users by selecting appropriate motors while requiring a minimal number of calculations on the part of the user.

The torque–speed curve is static; it does not specify the length of time a motor–drive system requires to produce that torque. In most servo systems, the limit to torque responsiveness is the responsiveness of the current loop. Current loops, like all control loops, are bandwidth limited by stability requirements. The bandwidth of current loops varies from about 300 Hz to 2500 Hz in servo systems. Use caution when reading torque–speed curves of induction motors. Some induction motors require long periods of time to generate torque (perhaps several seconds) when operating at high speeds. These drives are inappropriate for many servo applications.

15.4.2 Rotary and Linear Motion

Servomotors can create two types of motion: rotary and linear. Most servomotors are rotary and have a cylindrical rotor inside a cylindrical stator; the magnetic forces act along the circumference of the interface. In lower-power applications, disk motors [49] are sometimes applied; here, both the rotor and the stator are disks, with the magnetic forces acting between the ends of the disks. With inside-out motors, the rotor is moved outside the stator to increase rotor inertia and smooth rotation; these motors are commonly used in nonservo constant-speed applications, such as CD players and tape drives. The rotary configurations are shown in Figure 15-10.

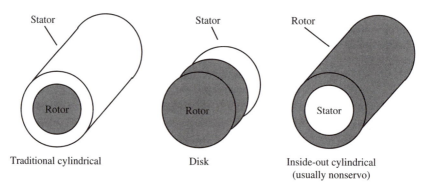

Figure 15-10. Three rotary motor geometries.

The terms *rotor* and *stator* imply mechanical, not magnetic, attributes: Rotors rotate and stators are stationary. In permanent-magnet brush motors, the field (magnet set) is on the stator and the armature (winding set) is on the rotor. For brushless PM motors, this is reversed: The magnets rotate and the windings are stationary.

15.4.3 Linear Motors

Recently, the linear motor has become important in industrial applications. Here, the magnets are laid out in line and an armature is carried on a table that moves along the magnet path, as shown in Figure 15-11. Usually, the armature is much shorter than the magnet way. Linear motors have many advantages in servo applications where the final motion is linear. Many servo applications generate linear motion through rotary motors, using lead screws or other mechanisms to translate the motion. Linear motors cure the shortcomings of rotary-to-linear translation, such as positional error caused by imperfections in lead screw threads, backlash caused by allowing clearance between gear teeth, and reduced mechanical stiffness due to compliance of transmission components. This last area will be discussed in detail in Chapter 16. Linear motors also allow higher acceleration and linear speeds than can be supported

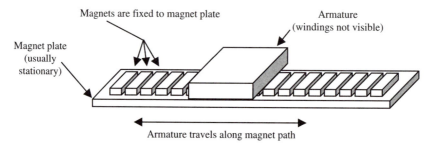

Figure 15-11. The linear motor.

by a rotary motor driving a lead screw. Also, they generate less noise and have greater reliability owing to fewer moving parts. When properly applied, linear motors can improve accuracy and response by an order of magnitude.

Linear motors are often more expensive than their rotary counterparts. One problem is that not all the magnets can be used in any one position. In Figure 15-11, the windings are opposite just a few magnets; the force created is just a fraction of what it would be were all the magnets used. Compare this to rotary motors, which can maintain interaction between all magnets and the windings in all positions.

For position feedback, linear motors usually require highly resolved scales that run the length of the motor; a linear encoder is usually many times more expensive than a rotary encoder. Linear motors also lack the mechanical advantage that a gearbox or belt transmission can provide a rotary motor. Linear motors can mitigate the price differences by removing transmission components; however, in most cases, a linear motor will be more expensive than a rotary motor. The remainder of this chapter will focus on rotary motors, but the discussion does apply to linear motors, except linear motors create force, not torque.

15.5 Permanent-Magnet (PM) Brush Motors

Electric motors create torque with flux from two sources: the armature and the field. In permanent-magnet motors [39], the field flux (Φ_F) is created by magnets, as shown in Figure 15-12 for a four-pole brush motor. The magnetic path for the field flux is the circular equivalent of the magnetic circuit of Figure 15-3, where the magnets provide F, the field strength, and the circuit reluctance is the sum of the reluctances of the magnets, the back iron (steel on the outside of the stator), the rotor steel, and the air

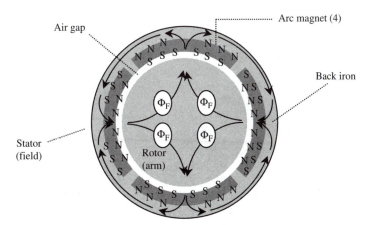

Figure 15-12. Field flux path for four-pole permanent-magnet brush motor (armature windings, which are on the rotor, are not shown).

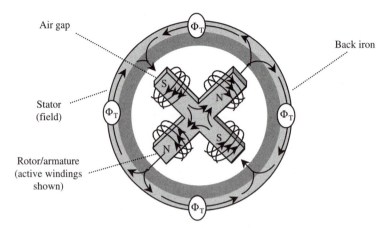

Figure 15-13. Armature flux path for a four-pole brush motor.

gap; it is necessary to have an air gap between rotor and stator to allow mechanical clearance for the moving rotor. Note that the rotor windings are not shown here, to simplify the drawing.

15.5.1 Creating the Winding Flux

The second flux that must be created for torque is that from the armature windings. Figure 15-13 shows the flux created from the armature windings (Φ_T)[1] in a four-pole brush motor. Follow the flux path and notice that flux travels from the rotor between the magnets, through the back iron, and then again between the magnets to return to the rotor. This is the proper path for flux to generate torque from the windings.

15.5.2 Commutation

The armature of a brush motor has many windings, and only some portion of those windings is excited with current when the motor is in any one position; the process of selecting the proper windings in a given rotor position is called *commutation*. Brush DC motors use mechanical commutation, so the drive needs no knowledge of motor position to regulate torque.

In brush motors, a mechanism of brushes contacting a commutator (see Figure 15-14) controls which windings are excited in any given position. Figure 15-13 shows only the windings that are excited in that position, assuming the commutation mechanism has selected the winding that yields maximum torque.

---

[1]The armature flux is often referred to as the "torque producing flux" and so is named Φ_T. This is a misnomer, in that torque is produced by the interaction of the two fluxes Φ_T and Φ_F.

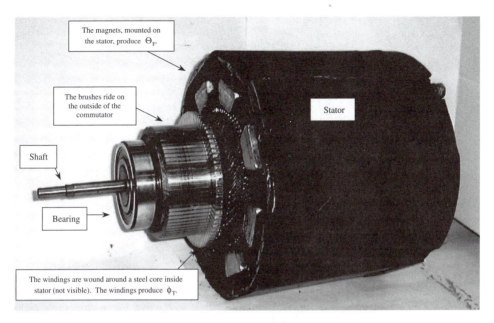

Figure 15-14. Partially disassembled brush motor showing commutator.

15.5.3 Torque Production

Electromagnetic torque is created by the interaction of the field flux and the armature flux. It is proportional to both according to Equation 15.7:

$$T_E \propto \Phi_T \times \Phi_F \times \sin(\theta_E) \qquad (15.7)$$

where θ_E is the electrical angle between the field and armature flux. The four-pole motor of Figure 15-15 has the flux from Figures 15-12 and 15-13 overlain on one cross-sectional view. Only the flux in the air gap is shown. The angle between the two flux vectors in Figure 15-15 is 90° (electrical). This is equivalent to 45° (mechanical) for this four-pole motor.

15.5.4 Electrical Angle Versus Mechanical Angle

The magnetic circuit of a motor can rotate faster than the rotor itself. Consider the four-pole motor of Figure 15-15; follow the path along the air gap all the way around and notice that the flux from the magnets cycles twice, once for each north–south magnetic pole pair. There are 720° electrical for each 360° mechanical, or $\theta_E = 2\theta_M$. In general,

$$\theta_E = \text{Poles}/2 \times \theta_M \qquad (15.8)$$

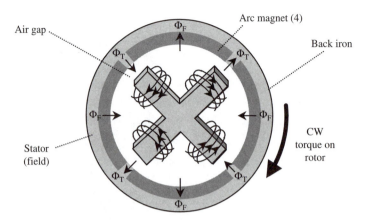

Figure 15-15. Four-pole brush motor with field and armature flux.

Most servomotors have more than two poles, so the electrical angle will usually rotate faster than the mechanical angle.

In any given position, torque is created by producing current in one or more of the windings. The winding must be selected so that θ_E in Figure 15-16 remains at or near 90° (electrical) because of Equation 15.7. For brush PM motors, the commutation angle (the difference between Φ_F and Φ_T) is fixed by the placement of the brushes on the commutator. As the rotor rotates, the commutator, which is fixed to the rotor, rotates and switches in the winding set that maintains the commutation angle at about 90°.

15.5.5 K_T, the Motor Torque Constant

For a brush PM motor, the torque output is approximately proportional to the armature current. In other words, Equation 15.7 can be simplified to $T \propto I_T$. This can be derived by combining several of the preceding equations, as shown in Equation 15.9. First, note that the commutation angle is fixed at approximately 90° by the commutator, so in Equation 15.7, $\sin(\theta_E) \cong 1$. Next, recognize that the field flux (Φ_F)

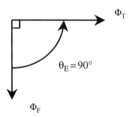

Figure 15-16. Vector diagram of field and armature flux with optimal $\theta_E = 90°$.

is fixed at a constant magnitude by the use of permanent magnets. At this point, we know $T \propto \Phi_T$. Finally, because the reluctance for the armature flux (Φ_T) is approximately constant, and because the MMF from armature current is proportional to armature current (I_T), Φ_T is approximately proportional to I_T:

$T_E \propto \Phi_T \times \Phi_F \times \sin(\theta_E)$	Start with Equation 15.7.
$T_E \propto \Phi_T \times \Phi_F$	$\sin(\theta_E) \cong 1$ because of commutator.
$T_E \propto \Phi_T$	Magnets make Φ_F approximately constant.
$T_E \propto \mathscr{F}_T/\mathfrak{R}_T$	$\Phi_T = \mathscr{F}_T/\mathfrak{R}_T$ (Equation 15.1).
$T_E \propto \mathscr{F}_T$	$\mathfrak{R}_T$ is approximately constant.
$T_E \propto n \times I_T$	Equation 15.6.
$T_E \propto I_T$	The number of turns (n) is constant (15.9)

The constant of proportionality in Equation 15.9 is called the motor torque constant, K_T:

$$T_E \cong K_T \times I_T \qquad (15.10)$$

The constant K_T is an approximation. The torque constant usually falls as I_T increases because of saturation of the steel in the path of the armature flux. This effect causes the average reluctance of the armature-flux path to increase, reducing the flux created by the electromagnetic MMF. The effective torque constant at peak current will be lower than that at low current, often by a factor of 20% or more. Be aware also that magnets usually weaken at high temperature, so the torque constant may fall further at peak operating temperature.

15.5.6 Motor Electrical Model

The electrical model describes the relationship between current, voltage, and motor speed. The model for an ideal permanent-magnet brush motor is based on the equality of the total applied voltage and the voltage drops in the motor. There are three voltage drops in the motor: the resistance drops, the inductance drops, and the back-EMF:

$$U = U_{\text{RESISTANCE}} + U_{\text{INDUCTANCE}} + U_{\text{BEMF}} \qquad (15.11)$$

The resistance drops are equal to the product of the current in the winding and the winding resistance: IR. The inductance drops equal the product of inductance and the change in current.

Back-EMF, or BEMF, is the phenomenon in which a PM motor generates a voltage U_{BEMF}, which is proportional to the rotor speed. BEMF is the basis of generators.

(Motors convert electrical energy to mechanical; generators convert mechanical energy to electrical.) The constant of proportionality for BEMF is defined as K_B:

$$U_{BEMF} = \text{Velocity} \times K_B \qquad (15.12)$$

K_B is proportional to K_T. Motors with large torque constants generate large BEMF. In fact, when using SI units for brush motors, $K_T = K_B$. In other words, a motor that generates 1 newton-meter of torque for 1 amp of current will generate 1 volt of BEMF when rotating 1 radian/second (9.55 RPM). BEMF is an undesirable although unavoidable effect in servomotors; it subtracts from the current-generating voltage applied from the controller. It limits the top speed of the motor for a given applied voltage because when the speed produces enough BEMF, the power stage no longer has sufficient voltage to force current into the armature. In fact, the theoretical maximum (unloaded, or "no-load") speed of a brush motor is the speed at which the BEMF is equal to the bus voltage.

So Equation 15.11 becomes

$$U = I \times R + L \times dI/dt + \text{Velocity} \times K_B \qquad (15.13)$$

Inductive losses reduce current-generating voltage when the current is changing rapidly. The inductive losses do not affect the steady-state torque because the current is not changing when steady-state torque is measured. However, if the motor is running at high speed and a rapid current change is commanded, the inductive losses together with the back-EMF losses can reduce the rate at which current can be generated. Because of this effect, it can take longer to generate torque in a motor rotating at high speed.

15.5.7 Control of PM Brush Motors

Brush motors are relatively easy to control because the commutation is mechanical. Referring to Figure 15-17, servo controllers create a torque command, either explicitly or implicitly, as the output of a velocity loop. The torque command is scaled, usually implicitly, by the estimated torque constant (K_T^*) to create an armature-current command, I_C. A current controller processes the difference of commanded and sensed

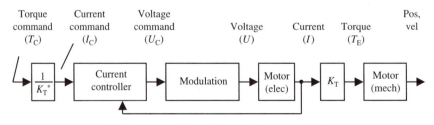

Figure 15-17. Overview of PM brush motor control.

current to generate commanded voltage, U_C. Commanded voltage is converted to an actual voltage (U), usually through pulse modulation. That is, the commanded voltage is converted to a series of on-and-off pulses to power transistors. The modulated voltage is applied to the motor to generate actual current (I), which, when scaled by K_T in the motor, generates torque that can be applied to the motor and load mechanical models.

15.5.7.1 Current Controller

Figure 15-17 is expanded to include the electrical model of a brush motor in Figure 15-18, starting with the current command. The current controller is PI, which is similar to the velocity PI controllers studied in the early chapters of this book except that the current loop, being inside the velocity loop, must have a much higher bandwidth. Figure 15-18 shows K_M, which is the approximate linear constant of the modulation process (see Section 12.4.9); more will be discussed later regarding modulation. The motor model is expanded to include the voltage drops from back-EMF (through K_B) and through electrical impedances (R and L).

Some lower-end servo controllers do not use a closed-loop current controller. Instead, they depend upon the resistance of the winding to limit the current. These controllers can have long delays in generating current because the electrical time constant of the motor, L/R, which limits the responsiveness, is usually equivalent to the delay of a 50- or 100-Hz loop, which is much slower than a current loop. Also, these controllers have difficulty protecting the motor and controller when the bus voltage is high (say, over 48 VDC) or during fast reversals. In a motor reversal, the applied voltage is quickly reversed, so the BEMF aids current production and large current spikes can be generated. Controllers that work without current loops usually rely on low bus voltages or high motor resistance to limit current.

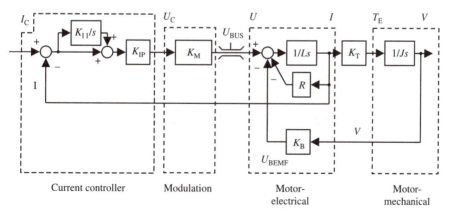

Figure 15-18. Example of PM brush motor control.

15.5.7.2 Voltage Modulation

Voltage modulation is the process of converting a voltage command to a series of on–off, high-voltage pulses. For example, if a current controller commands 75 volts and the bus voltage is 300 volts, the modulated output might be on for 25 μsec and off for 75 μsec. The average applied voltage would be 75 V because the modulated output was on for 25% and 25% × 300 V = 75 V. Modulation is used because power transistors are most efficient when they are fully on or fully off. When a transistor is fully off, there is no current flow, so no power is lost in the transistor. When the transistor is fully on, there is a small voltage drop (typically less than 2 V), so the power lost in the transistor ($V \times I$) is small even in the presence of high current.

Figure 15-19a shows a four-transistor "H-bridge," which is commonly used to power brush motors, driving current into a motor with a modulated output. The four transistors allow both positive and negative voltages to be applied to the winding.

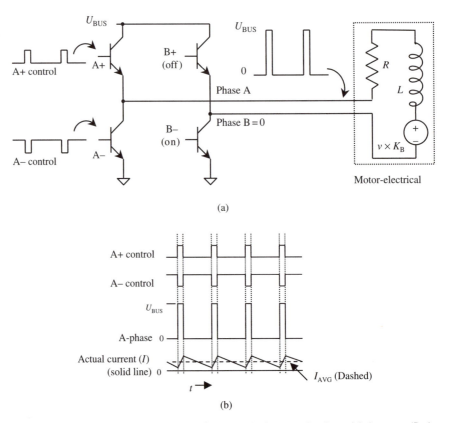

(a)

(b)

Figure 15-19. (a) Simple modulation stage. (b) Current and voltage vs. time in modulation stage (B-phase assumed = 0 volts).

The pulsed inputs to the power transistors connect the A-phase to either U_{BUS} (when A+ is on) or ground (when A− is on). This switching of A-phase drives current in and out of the winding, as shown in Figure 15-19b, which assumes that the B-phase is held at zero. The average current is equal to the current that would have been produced had the average (unmodulated) voltage been applied. The current ripple results because the current is pulsed. Modulation methods rely on motor inductance to smooth the current produced by pulsed voltages.

The most common modulation technique used is pulse-width modulation, or PWM. This method outputs voltage pulses at a fixed frequency and then varies the width of the pulse to increase or decrease the average applied voltage. The ripple in the current waveform is small if the modulation frequency is high relative to the inductance of the motor. For iron-core motors, typical inductance ranges from a few millihenrys to a few dozen millihenrys, and a PWM frequency of 8–16 kHz will usually work well. Some motors are wound without steel; they are sometimes referred to as *air-core* motors. For example, the disk motors discussed earlier in this chapter are normally air-core motors. The inductance of air-core motors can be very low — often just a few hundred microhenrys. In this case, a PWM frequency as high as 100 kHz may be required. If the PWM frequency is too low for motor inductance, the magnitude of ripple current will be excessive. This ripple generates heat without generating torque. Also, the ripple can vibrate the windings, causing audible noise. Sometimes it is necessary to add inductance in line with the motor; this is not normally desirable, because inductors are often large and generate significant heat.

Linear drives (for unmodulated voltage, not for linear motors) do not modulate; instead they output a continuous (nonpulsed) voltage to the motor. Linear drives are used in special applications or are applied at low power levels; they do not demonstrate the noisy characteristics so common with modulated-output amplifiers, but they have an order of magnitude greater power losses. These losses drive the cost and size of linear amplifiers out of the range of most servo applications.

15.5.8 Brush Motor Strengths and Weaknesses

The main strength of a brush motor is that commutation is mechanical, so the control of the motor is simple. Also, brush motors require only a single current sensor where other servomotor types usually require two. Brush motors require fewer power transistors, usually four instead of the six required by brushless motors. Another advantage is that brush motors can provide very smooth torque, in large measure because offsets in current sensors do not result in torque ripple; these offsets, which are common in current sensors, do cause ripple in brushless motors, as will be discussed.

Among the weaknesses of brush PM motors are that brushes wear, especially when exposed to contaminants such as silica, and must be replaced on a regular basis to avoid catastrophic failure, and brush wear results in carbon debris, which can cause contamination [21]. Electrical noise is a problem: When the commutator disconnects windings carrying heavy current, arcing results, which can generate substantial electrical noise. This noise can make applications more difficult to integrate, and brush motors are often avoided because of the nuisance of improving the wiring of a machine so the higher noise can be better tolerated.

Brush motors are usually larger than the equivalent brushless motor because of space taken by the commutator assembly; also, the armature, where most of the losses are generated, is on the rotor, which is usually inside the stator and thus more difficult to cool. The commutator is complex to manufacture. Brushes riding on the commutator generate audible noise at higher speeds. They also lose efficiency because of brush friction and because the voltage drops across the brush–commutator interface, both of which dissipate power. Because of mechanical commutation, the top speed of brush motors is limited.

The rotor of a brush motor is heavier than its brushless equivalent because, in most motors, the windings are wound around a steel core. Together, the copper wire and steel add up to a considerable inertia. In brushless motors, the magnets rotate; the brushless magnet assembly is light compared to a brush motor armature, especially when high-energy magnets such as NdBFe are used. Light inertia is often an advantage in servo applications where high acceleration is required. Reducing the motor inertia while providing the same torque often allows a smaller motor to do the same job. Brushless motors can provide as much as ten times the torque of a brush motor with the same rotor inertia [100].

Brush motors remain popular in some servo applications. The low cost of brush motor control makes brush motors appealing in cost-sensitive applications, especially in low-power applications where the cost of control is a larger factor in the overall system. Also, brush motors, when matched with a high-quality tachometer, can produce very smooth speed control.

It should also be noted that disk-style brush motors avoid many of the common problems of brush motors by winding the armature on fiberglass or other nonferrous material. This cuts rotor inertia considerably. Also, the lower inductance of disk motors reduces arcing and allows longer brush life. Thus, disk-style brush motors offer the simplicity and smooth torque of brush motors while enjoying comparatively light rotors and long brush life.

15.6 Brushless PM Motors

The weaknesses of the brush PM motor have caused the brushless DC motor (sometimes referred to as synchronous AC PM motor) to dominate many servo-motor markets. The brushless motor replaces the mechanical commutator with electronic commutation, eliminating the brushes and their problems. However, brushless motors are more difficult to control.

Brushless controllers must sense the electrical position of the motor with a feedback device, such as a resolver or encoder, or, in some cases, with coarse digital sensors called Hall-effect sensors. In many nonservo applications, the BEMF of the motor is used to measure position; this is called *sensorless* control. In all cases, the electrical position is used to calculate commanded phase currents, with the goal being to maintain the commutation angle (Θ_E in Equation 15.7) at or near the optimal 90°.

15.6.1 Windings of Brushless PM Motors

Windings of brushless PM motors are distributed about the stator in multiple phases. Usually there are three phases, each separated from the others by 120° (electrical). Brush motors can have many more phases, but a large number of phases in brushless PM motors is impractical because each phase must be individually controlled from the drive, implying a separate motor lead and set of power transistors for each phase. A simplified winding diagram of a three-phase motor is shown in Figure 15-20.

Brushless motors rely on electronic commutation. The drive monitors the rotor position and excites the appropriate winding to maintain a 90° commutation angle. Consider Figure 15-21, which shows a brushless rotor in a sequence of three positions as it rotates counterclockwise. The large arrows show the flux created by the windings. To simplify the drawing, the field flux is not shown, but recall that it points out of the north poles and into the south poles. Notice that the winding flux in each of the three motor positions is maintained in quadrature.

In brush motors, the commutation angle is maintained by mechanically switching phases in and out. Because the brush motor has many phases, each phase represents only a few electrical degrees of rotation and the torque from a brush motor is smooth. An equivalent technique is used on brushless motors in a commutation method called *six-step*, but it produces large torque perturbations at each transition because brushless motors usually have just three phases.

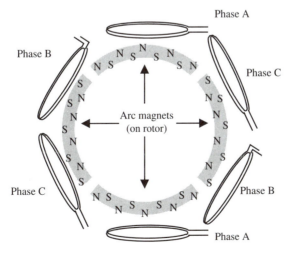

Figure 15-20. Simple winding set for a three-phase four-pole motor.

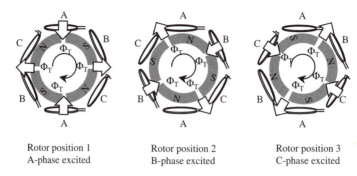

Rotor position 1
A-phase excited

Rotor position 2
B-phase excited

Rotor position 3
C-phase excited

Figure 15-21. Commutation sequence — maintaining the winding flux between the magnet poles.

15.6.2 Sinusoidal Commutation

Unlike a brush motor controller, a brushless motor controller controls current in multiple phases independently. This allows the controller to move the winding flux (Φ_T) angle in small increments. Figure 15-21 shows how flux created from the three windings interacts with flux from the rotor magnets. Were the position of the rotor in Figure 15-21 midway between positions 1 and 2, flux from the windings could be positioned properly by placing equal current in phase A and phase B. In general, quadrature can be maintained precisely by independently regulating the phase currents according to Equations 15.14–15.16:

$$I_A = I_S \times \sin(\theta_E) \tag{15.14}$$
$$I_B = I_S \times \sin(\theta_E - 120°) \tag{15.15}$$
$$I_C = I_S \times \sin(\theta_E - 240°) \tag{15.16}$$

where I_S is the magnitude of current in the motor and θ_E is the electrical position of the motor. This is called *sinusoidal commutation*.

Sinusoidal commutation provides smooth, efficient operation of the brushless motor. Torque is approximately proportional to I_S. In fact, brushless motors are usually given a torque constant based on I_S, so $T \approx K_T \times I_S$, assuming that commutation is performed correctly.

15.6.3 Phase Control of Brushless PM Motors

Phase control for brushless PM motors is shown in Figures 15-22 and 15-23. The concept is straightforward: Command each of the phase currents (I_{AC} commands I_A, and so on) to follow Equations 15.14–15.16, assuming $T = K_T \times I_S$. Phase control regulates each of the phase currents with independent current loops. Two current sensors are required; the third phase current is calculated from the other two because all three currents must sum to zero in a wye-connected three-phase motor, such as the

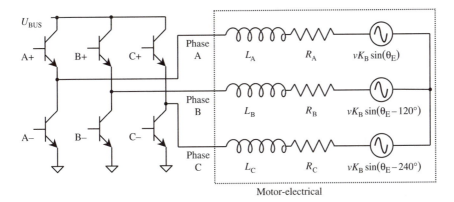

Figure 15-22. Three-phase modulator controlling a brushless motor.

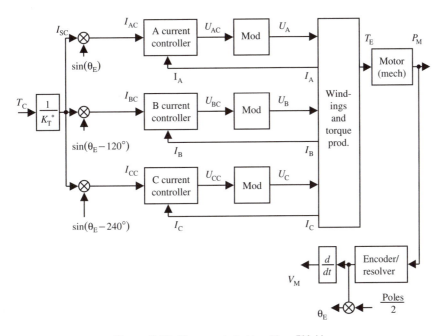

Figure 15-23. Phase-controlled brushless PM drive.

motor in Figure 15-22. The current commands for the three phases are calculated according to Equations 15.17–15.19:

$$I_{AC} = T_C / K_T^* \times \sin(\theta_E) \tag{15.17}$$

$$I_{BC} = T_C / K_T^* \times \sin(\theta_E - 120°) \tag{15.18}$$

$$I_{CC} = T_C / K_T^* \times \sin(\theta_E - 240°) \tag{15.19}$$

15.6.3.1 Modulation

In phase control, the modulation is equivalent to that of the brush motor, the biggest difference being that there are three phases to modulate rather than the two phases of Figure 15-19 [50,56,93]. The H-bridge is also nearly the same, except the brushless motor requires a third leg of the power stage, as shown in Figure 15-22.

The electrical model of the brushless motor is similar to that of the brush motor. Three copies of the electrical model of the brush motor are required, one for each phase. The main difference is that the BEMFs are sinusoidal when the motor is moving at a constant speed, whereas in a brush motor the BEMF is constant during constant speed. The commanded currents are sinusoidal at constant speed as well. Also, the inductive losses do affect steady-state torque in a brushless motor because phase currents are changing, even at constant speed and constant load. This is one factor that makes brushless motors more difficult than brush motors to control; the bandwidth of the current loop affects the torque–speed curve. Figure 15-23 shows a block diagram of a phase-controlled brushless PM drive.

Phase-controlled brushless motors produce smooth torque. However, there are torque perturbations, including those caused by the current sensors. Current sensors commonly have 1% or 2% current DC offset. In the brush motor, such an offset does not contribute to torque ripple; the brush motor will rotate smoothly, but the actual torque is offset from the command torque by a small amount. In the brushless motor, problems caused by current-sensor offset are more serious. DC offset in the current sensors causes ripple at the electrical frequency of the motor. To determine this frequency, multiply the motor speed in revolutions per second by poles over two. For example, if a six-pole motor were rotating at 300 RPM, offset in the current sensor would generate torque ripple at $300/60 \times 6/2 = 15\,\text{Hz}$. A 2% offset in a current sensor indicates that the current sensor may cause offset as much as 2% of the drive peak current. For a 10-A drive with a peak rating of 20 A, 2% would be 400 mA. Were the motor rotating with a small load (say, drawing just 1 A), the ripple caused by 400 mA of offset would be a problem for some applications. This is one reason it is important not to specify larger brushless drives than necessary; the offset increases with the drive rating, so oversized drives can cause unnecessary torque ripple. The area of three-phase modulation is well studied, including Refs. 41 and 51.

15.6.3.2 Angle Advance

The performance of brushless DC motors at higher speeds can be enhanced by advancing the commutation angle, that is, by adding an offset to θ_E in Equation 15.7. There are three reasons to advance the commutation angle. First, advancing the angle offsets the phase lag caused by the current loop. Second, the angle can be advanced to weaken the field flux. Third, some brushless motors can generate reluctance torque, and advancing the angle can optimize torque output. Each of these reasons is discussed in detail in the following.

15.6.3.3 Angle Advance for Current-Loop Phase Lag

Current loops, like all control loops, cannot produce an output that precisely mimics the command. As discussed throughout this book, control loops produce phase lag and attenuation at higher frequencies. Attenuation is of little concern for this discussion, but phase lag is important because it reduces torque output according to $\sin(\theta_E - \theta_{LAG})$.

For example, suppose a four-pole motor is rotating at 3000 RPM, creating an electrical frequency of 150 Hz. Suppose also that the current controller had a phase lag of 25° at 150 Hz. If the controller commanded current using Equations 15.15– 15.17, the resulting loss of torque would be $\sin(90°) - \sin(65°) = 10\%$.

Angle advance can cure phase loss by commanding a phase advance equal to the phase lag from the current loops. It can also anticipate the delay from sampling and correct for it as well. The higher the electrical frequency of the motor with respect to the current loop bandwidth, the more angle advance can be used to improve the commutation angle.

15.6.3.4 Field Weakening

The field flux Φ_F can be reduced by advancing the angle of the actual current (not just the commanded current as earlier) [42,53,84]. A sine wave that has been advanced can be considered to be the sum of two sine waves, one unadvanced and another advanced by 90°. For example, I_A from Equation 15.17 can be advanced 20°:

$$I_A = I_S \times \sin(\theta_E + 20°) \tag{15.20}$$

Now use the trigonometric identity

$$\sin(A + B) = \cos(A) \times \sin(B) + \sin(A) \times \cos(B)$$

to divide I_A into the two components:

$$\sin(\theta_E + 20°) = \cos(20°) \times \sin(\theta_E) + \sin(20°) \times \cos(\theta_E) \tag{15.21}$$

so that Equation 15.20 can be rewritten as

$$I_A = (0.94 \times I_S)\sin(\theta_E) + (0.35 \times I_S)\cos(\theta_E) \tag{15.22}$$

The sine term is 90° advanced from the field flux; the cosine term, which is another 90° advanced from the sine term, is thus 180° in front of the field flux. The sine term produces Φ_T in Equation 15.7; the cosine term generates flux in direct opposition to Φ_F. So with 20° of advance, 94% of the current magnitude produces torque and 35% of the current magnitude produces flux in opposition to the flux created by the magnet.

The flux in opposition to magnet flux reduces, or *weakens*, the field of the magnets. Reducing the flux from the magnets will reduce the BEMF constant of the motor and, since at any given bus voltage BEMF is the fundamental limit to motor top speed, reducing the BEMF allows higher-speed operation of the motor. Advancing the angle more will allow the motor to rotate at higher speeds. The angle should not be advanced more than is required to run at any given speed because excessive angle advance generates needless I^2R power losses.

Angle advance can be depicted graphically, as shown in Figure 15-24. The optimal angle between flux from the magnets (I_F) and from the winding in the absence of field weakening is 90° according to Equation 15.7. However, at high speed the angle can be advanced to weaken the field, as shown in Figure 15-24a. The components of the flux generated from the winding can be divided into Cartesian coordinates, with one component at the optimal 90° and the other 180° from the field, as shown in Figure 15-24b. The winding flux in opposition to the field ($\Phi_{F\text{-WINDING}}$) can be summed with the magnet flux ($\Phi_{F\text{-MAGNET}}$); the result is that the overall field flux is weakened, as shown in Figure 15-24c. The reduction in field flux results in a reduced BEMF constant (K_B) and a proportionally smaller motor torque constant (K_T).

Figure 15-25 shows the phase-current controller with angle advance. The commutation angle is enhanced with an advance angle, θ_A. A graph for angle advance in a typical brushless system is shown at the bottom of the figure (θ_A vs. V_M). The rest of the controller is identical to Figure 15-23.

15.6.3.5 Reluctance Torque

Reluctance torque is the torque generated because the motor is moving to a position where the reluctance seen by the armature flux is declining. A simple application of this principle is the refrigerator magnet, which is held in place by reluctance force. Because the reluctance along the path of the magnet flux is minimized when the magnet is as close as possible to (in contact with) the refrigerator, the magnet holds its position. Motors can be made to take advantage of this phenomenon by building the rotor to

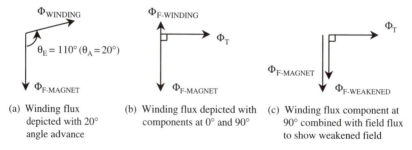

(a) Winding flux depicted with 20° angle advance

(b) Winding flux depicted with components at 0° and 90°

(c) Winding flux component at 90° combined with field flux to show weakened field

Figure 15-24. Angle advance depicted in three ways.

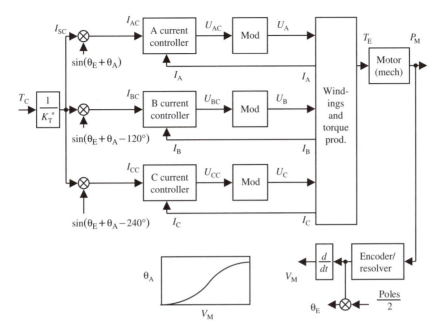

Figure 15-25. Brushless phase-current controller with angle advance.

have a lower reluctance to the winding flux than to the field flux, such as is the case for *interior permanent magnet*, or IPM, motors, as shown in Figure 15-26.

In the IPM motor of Figure 15-26, the reluctance seen by the field flux is much higher than the reluctance seen by the winding flux. The reason is that the only nonsteel material in the path of the winding flux is the motor air gap. However, the field flux must pass through the magnets as well as the air gap; the permeability of magnets is close to that of air, so the field flux endures two large-reluctance materials. The difference in reluctance is frequently 2:1 or more. Such motors are often called *hybrid PM/reluctance* motors. The optimum angle to apply current for the magnet component of torque is 90° (Equation 15.7); for Figure 15-26, the optimum angle of current (with respect to magnet flux) for the reluctance component of torque is 135°. The optimum angle for a hybrid motor at low speeds (that is, without field weakening) will be between 90° and 135°. Since the reluctance component is usually a fraction of the magnet torque in PM motors, the optimal angle is much closer to 90° than it is to 135°. Still, some angle advance (perhaps 10° or 15°) will often increase the torque of a motor as much as 15%. This is one example of how to take advantage of reluctance torque; there are other motor structures that use reluctance torque differently.

Figure 15-27 shows a PM rotor constructed with surface-mounted magnets. Here, the flux from the field sees about the same reluctance as the flux from the winding, because the permeability of magnet material for most modern brushless motors is so

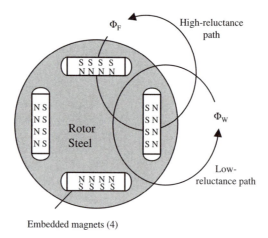

Embedded magnets (4)

Figure 15-26. Rotor of an interior permanent magnet (IPM) motor with high- and low-reluctance paths.

close to that of air. As a result, motors based on surface-mounted magnets do not produce reluctance torque.

15.6.4 DQ Control of Brushless PM Motors

A parallel method for phase control of brushless PM motors is direct-quadrature, or DQ, control [56]. DQ control rearranges the system by placing commutation inside the current loop. This improves some aspects of controller performance because the commutation frequency does not pass through the current controller. Thus, phase

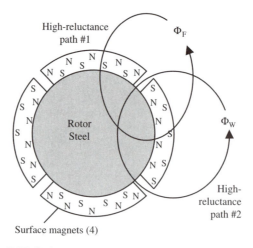

Surface magnets (4)

Figure 15-27. Surface magnet rotor with similar- (high-) reluctance paths.

lag caused by the current loop does not alter the commutation angle. Rather than regulating measured states (I_A, I_B, and I_C), DQ loops regulate calculated states, namely, the direct and quadrature currents, I_D and I_Q.

Direct and quadrature currents produce flux in relationship to the rotor, whereas phase currents produce flux in relationship to the stator. For example, if a motor were rotating at a constant speed and with a constant torque, the phase currents (I_A, I_B, and I_C) would be varying sinusoidally, but I_Q and I_D would be constant values. Phase currents are measured with respect to the stator frame of reference; a phase controller sees abundant activity in a motor spinning at high speed. DQ currents are measured with respect to the rotor frame; when DQ currents are measured on a spinning motor, there may be very little activity in the current controllers.

In DQ controllers, measured phase currents are combined to produce the state currents I_Q and I_D using trigonometry to translate from the stator frame to the rotor frame:

$$I_Q = I_A \sin(\theta_E) + I_B \sin(\theta_E - 120°) + I_C \sin(\theta_E - 240°) \qquad (15.23)$$
$$I_D = I_A \cos(\theta_E) + I_B \cos(\theta_E - 120°) + I_C \cos(\theta_E - 240°) \qquad (15.24)$$

The direct and quadrature currents are closely related to the winding and field fluxes that have been discussed throughout this chapter. In fact, I_Q generates Φ_T and I_D generates $\Phi_{F\text{-WINDING}}$. Figure 15-24b is redrawn in Figure 15-28 accordingly.

To better comprehend the operation of DQ control, consider the diagram in Figure 15-29, which shows a rotor in two positions, with winding flux (Φ_T) moved so that it is 90° ahead of the field flux (Φ_F) in both positions. Here the torque producing current is equal in both positions and there is no field weakening. Note that after 20° of rotation, the Φ_T has moved 20° with respect to the stator. However, the position of Φ_T is constant with respect to the magnet position; in both cases it is correctly aligned between the magnet poles. So while the phase currents would vary between these two positions, the quadrature current (generating Φ_T) and the direct current (generating $\Phi_{F\text{-WINDING}} = 0$) would not vary.

The DQ control system is shown in Figure 15-30. The torque command (T_C) is explicitly divided down by the estimated K_T (K_T^*), although this step is implicit in most drives. The quadrature current loop is closed before commutation. The direct current

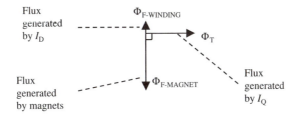

Figure 15-28. Flux generated by DQ currents shown in vector form.

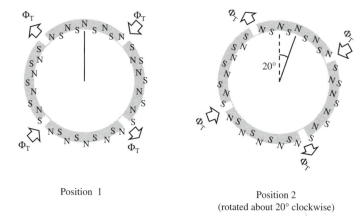

Position 1

Position 2
(rotated about 20° clockwise)

Figure 15-29. Two rotor positions, with the controlled flux vectors remaining constant.

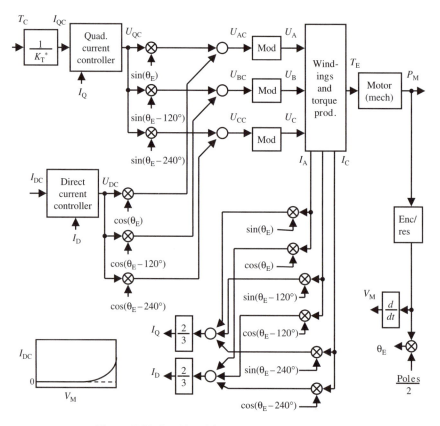

Figure 15-30. Brushless DQ controller with ID regulation.

loop is commanded to zero current at low speeds and increased for field weakening at higher speeds.

15.6.4.1 Modulation in DQ Control

The output voltage commands of the current controllers (U_{QC} and U_{DC}) are first commutated and then modulated. Two phase currents are measured (the third can be formed from the other two) and combined into I_D and I_Q and fed back to the current controllers.

15.6.4.2 Field Weakening DQ Control

Field weakening in DQ control it is a matter of regulating I_D as a function of speed. I_D is commanded to be zero at low speeds. When the speed increases so much that the BEMF is a large fraction of the applied bus voltage (U_{BUS}), the BEMF must be reduced. Increasing I_D reduces BEMF because the flux created by I_D ($\Phi_{F\text{-WINDING}}$) is in opposition to the field flux, just as it was for phase control (see Section 15.6.3.4).

15.6.5 Magnetic Equations for DQ

This section will provide a brief introduction to the magnetic equations of the DQ reference frame. For a more complete discussion on this topic, readers are referred to Ref. 56. One of the benefits of the DQ reference frame is that the magnetic equations of the motor are similar to those of the brush motor (Equation 15.13). Here there are two independent current paths, one for each of the direct and quadrature currents. As with Equation 15.13, applied voltage is balanced with voltage drops in the motor:

$$V_Q = I_Q \times R + L_Q \times dI_Q/dt + \text{Velocity} \times K_B - I_D \times L_D \times \text{Velocity} \times \text{Poles}/2 \quad (15.25)$$
$$V_D = I_D \times R + L_D \times dI_D/dt + I_Q \times \text{Velocity} \times L_Q \times \text{Poles}/2 \quad (15.26)$$

The parallels between Equations 15.13 and 15.25 are apparent: Both represent torque-producing current and both have resistive, inductive, and back-EMF losses. Equation 15.25 adds a fourth term, field weakening ($I_D \times L_D \times \text{Velocity}$). Here, I_D reduces the total flux, as shown in Figure 15-28.

Torque is produced as a combination of the direct and quadrature currents, as shown in Equation 15.27:

$$T_E = K_T \times I_Q + 3/2 \times \text{Poles} \times I_Q \times I_D \times (L_D - L_Q) \quad (15.27)$$

The first term on the right side is equivalent to torque in the brush motor (Equation 15.10). The second term represents reluctance torque (Section 15.6.3.5). Note that this term is nonzero only when the L_D and L_Q terms differ, as in the IPM motor (Figure 15-26).

15.6.6 Comparing DQ and Phase Control

There are many parallels between phase control and DQ control. Modulation is similar, and both control methods generally require two current sensors. Angle advance in phase control is equivalent to regulating I_D in DQ controls; angle advance and DQ both work to control the commutation angle across the speed range. Both phase-control and DQ-control current loops are usually PI loops. The key distinction in terms of the control system is that in DQ control, the commutation frequency does not pass through the current loops. As can be seen in the comparison of Figure 15-31, commutation comes before the current loops in phase control, so the commutation frequency must pass through the current loop; in DQ control, commutation is done inside the current loop, so the commutation frequency does not pass through the loop.

Moving commutation inside the current loop allows the DQ controller to get higher torque at high speeds. In the phase-control method, the current loops attenuate the current (and, thus, torque) because the current commands contain the commutation frequency; angle advance corrects for the phase lag in the current loop but does not correct for current-loop attenuation. As long as the current loops are not driven into saturation, attenuation does not occur in DQ control; when the motor has constant torque and speed, I_D and I_Q are constant and the current loops process only a DC signal.

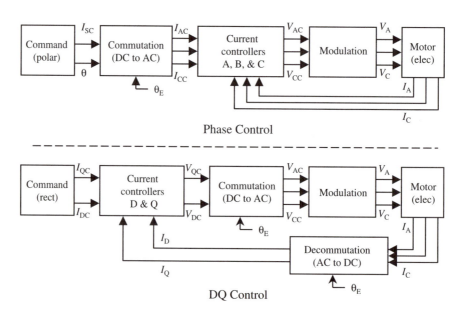

Figure 15-31. Schematic comparison of phase and DQ control.

The command format of the two methods differs because the phase controller takes the command in polar coordinates whereas the DQ controller takes the equivalent command in Cartesian coordinates. To convert from polar to rectangular, use

$$I_D = I_S \times \cos(\theta_A) \tag{15.28}$$

$$I_Q = I_S \times \sin(\theta_A) \tag{15.29}$$

To convert from rectangular coordinates to polar, use

$$I_S = \sqrt{I_Q^2 + I_D^2} \tag{15.30}$$

$$\theta_A = \text{Tan}^{-1}(I_Q/I_D) \tag{15.31}$$

15.7 Six-Step Control of Brushless PM Motor

An inexpensive alternative to phase control and DQ control is called six-step. Six-step control works with only one current path active at any time; for example, in one position, current may flow from phase A to phase B, but no current is allowed in phase C. There are six combinations of current flow in a wye-connected three-phase motor, such as the motor in Figure 15-22: A to B, A to C, B to C, B to A, C to A, and C to B. Six-step commutation requires a position sensor with only the coarse resolution of 60°, electrical. Often, an inexpensive magnet ring is fitted to the rotor and three magnetic sensors, called Hall-effect sensors, are positioned along its perimeter. Commutation is performed by modifying Equations 15.17–15.19 to accommodate the coarse resolution, as shown in Equations 15.29–15.31. This approximation is graphed for I_{AC} in Figure 15-32.

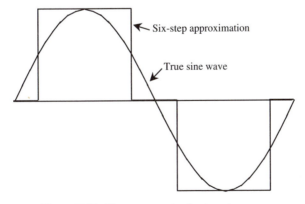

Figure 15-32. Six-step approximation to a sine wave.

$$I_{AC} = T_C/K_T^* \text{ for } 30° < \theta_E < 150°, -T_C/K_T^* \text{ for } 210° < \theta_E < 330°, \qquad (15.32)$$
$$0 \text{ elsewhere.}$$

$$I_{BC} = T_C/K_T^* \text{ for } -90° < \theta_E < 30° - T_C/K_T^* \text{ for } 90° < \theta_E < 210°, \qquad (15.33)$$
$$0 \text{ elsewhere.}$$

$$I_{CC} = T_C/K_T^* \text{ for } 150° < \theta_E < 270°, -T_C/K_T^* \text{ for } -30° < \theta_E < 90°, \qquad (15.34)$$
$$0 \text{ elsewhere.}$$

Six-step commutation is simpler than sinusoidal commutation for several reasons. As discussed, the commutation method requires only a coarse position sensor. Because current travels through only one winding path at a time, a single current loop can be used, although different phase currents must be switched in and out of current control at different positions. In fact, six-step commutation makes brushless motors almost as easy to control as brush motors. The chief problem with six-step commutation is torque ripple, For applications where torque ripple is of little consequence, six-step can be an appealing low-cost alternative to other commutation methods. Field weakening is used in six-step systems occasionally, especially in unidirectional motors, where the Hall sensors can be mechanically rotated slightly ahead.

15.7.1 Sensing Position for Commutation

Brushless motor drives must sense position to control the commutation angle. The most common feedback devices are encoders, resolvers, and Hall sensors. The use of encoders and resolvers for closing position and velocity loops was discussed in Chapter 14. Sensing position for those loops is much more demanding than sensing for commutation. However, commutation does have a few special requirements that bear discussion.

Commutation requires knowledge of rotor position with regard to the magnetic poles of the motor. For resolvers, which normally provide an absolute position within one revolution of the motor, this problem is most often solved by mechanically aligning the resolver to the magnetic poles of the motor. In that case, the electrical angle of the motor can be read from the resolver after multiplying by Poles/2 to convert the mechanical angle to electrical. Multispeed resolvers, resolvers with multiple electrical cycles per mechanical revolution, can be used for commutation, but the poles of the motor are usually an integer multiple (including "1") of the resolver pole pairs. For example, a three-speed (six-pole) resolver mounted on a six-pole motor reads electrical angle directly.

Most industrial encoders provide incremental A/B channels: These channel output pulses indicate that the motor has moved. However, upon power-up, the electrical position of the motor cannot be determined from the A and B channels. Some encoders provide coarse "Hall channels" and an index marker, which makes a transition one time for each revolution of the motor. The Hall channels allow six-step commutation upon power-up. After power-up, the drive is configured to monitor the index channel, which indicates a precise electrical position of the motor (assuming that the encoder is aligned to the magnetic poles of the motor, as it must be for the Hall tracks to work). After the index pulse is encountered, the encoder position is stored and the drive sums subsequent pulses from the A and B channels to measure the electrical position precisely. At this point, the electrical position can be determined well enough to support sinusoidal commutation.

Encoders with Hall channels work well, but there are two problems. First, many wires must be run to connect to the A, B, and index channels and to the Hall channels. Second, encoders with Hall channels are less common and frequently more expensive. This second problem can be addressed by having separate Hall sensors, but this requires a second mechanical assembly on the motor.

Another solution is to provide an initialization mode where the motor is excited by injecting current in many combinations (phase A to phase B, phase B to phase C, and so on) and then monitoring the direction in which the motor rotates to provide a crude electrical position for startup. This is sometimes called *wake and shake*. The drive is still configured to monitor the index channel; when the index pulse is encountered, it provides a much more accurate indication of electrical position. This is similar to having Hall sensors provide a coarse position for power-up, as discussed earlier, except the wake-and-shake method reduces wire count and does not require a special encoder. However, many applications cannot tolerate being moved about on power-up; for example, applications with vertical movement usually require full control anytime the holding brake is released.

Smart-format serial encoders, such as ENDAT encoders from Heidenhain, (www.heidenhain.com) and Hiperface encoders from Stegmann (www.stegmann.com), allow the drive to pole the motor for electrical position at power-up over a three- or four-wire communication network. These encoders reduce wire count and still support full sinusoidal commutation at power-up.

15.7.2 Comparison of Brush and Brushless Motors

Table 15-2 provides a brief comparison of brush, sinusoidally commutated brushless, and six-step brushless systems.

15.8 Induction and Reluctance Motors

There are alternative brushless technologies to those based on permanent magnets, the most popular being induction motors [62] and reluctance motors [49]. These motor types both avoid brushes while promising to reduce motor cost, the primary reason being the elimination of magnets, which are the most expensive material in PM motors. Induction and reluctance motors both offer wider speed ranges because field weakening is passive. Recall that for PM brushless motors, field weakening requires that flux be added to oppose the magnet flux. At very high speeds, the necessary field weakening may require

TABLE 15-2 COMPARISON OF BRUSH AND BRUSHLESS MOTORS

	Brush motors	Brushless PM, sine commutated	Brushless PM, six-step commutated
Maintenance	High due to brush wear	Low	Low
Electrical noise	High from arcing and PWM voltage transients	Medium due to PWM voltage transients	Medium due to PWM voltage transients
Motor friction	High due to brush contact	Low	Low
Speed	Commutator limits speed	Can rotate high speeds	Can rotate high speeds
Efficiency	Medium due to brush friction and voltage drop across commutator	High	High
Debris	Yes	No	No
Torque ripple	Can be very low depending on motor	Low (some from current sensor offset)	High due to commutation method
Motor size	Larger due to commutator and difficulty removing heat	Smaller	Smaller
Rotor size	Larger (winding assembly rotates)	Smaller (magnet assembly rotates)	Smaller (magnet assembly rotates)
Audible noise	High at high speeds because of brushes	Low	Low
Drive complexity	Simple	Complex	Simple
Number of current loops	1	3 (phase control) or 2 (DQ control)	1 in simple six-step drives
Position sensor required for commutation	None	High resolution (encoder, resolver)	Low resolution (Hall sensors)
Field weakening available?	No	Yes	Not normally

an impractical amount of current. For non-PM motors, the field must be created by current; thus, reducing the field requires reducing, not adding, flux, from the windings.

Induction and reluctance motors do have shortcomings. Induction motor rotors are complex to manufacture because the rotor contains an electrical circuit. Also, the rotor generates heat, which is difficult to eliminate, just as was the case for brush motors. The flux densities of high-energy rare-earth magnets commonly used in modern servo applications are not equaled by induction motors. Thus, induction motors are usually larger than PM brushless motors and have larger rotors, making them less suitable for high-acceleration applications. Reluctance motors have simple rotors. The most volumetrically efficient reluctance motors are called switched reluctance motors. These motors are simple and inexpensive to manufacture, but they are difficult to control and usually generate significant torque ripple.

15.9 Questions

1. What servo drive configuration is the least intelligent? the most intelligent?
2. Why do servomotors produce less torque at high speed with the same current?
3. What is the ideal commutation angle for a motor?
4. Why does the torque constant of a servo motor fall at high current?
5. What is the primary advantage of modulating voltage applied to a servomotor?
6. For a modulated power transistor on a 330-VDC bus, if the transistor is on 30% of the PWM cycle, what is the average output voltage?
7. List four advantages of brush motors (compared to brushless). List four advantages of brushless motors.
8. Which commutation method provides the smoothest output torque, sine-wave or six-step?
9. What is the primary goal of field weakening in brushless permanent magnet servomotors?
10. For brushless permanent-magnet motors, what is the primary advantage of DQ control over phase control?
11. What sensors are typically used to commutate six-step motors? Why are these sensors not used for sine-wave commutation?

Chapter 16

Compliance and Resonance

Mechanical resonance is one of the most pervasive problems in motion control [24, 25, 44, 88, 95]. Resonance is caused by compliance between two or more components in the mechanical transmission. Most often resonance is caused by compliance in transmission between motor and load. Resonance can also come from compliance between the motor and feedback, and sometimes it comes from compliance within the load, where the load can be thought of as multiple inertias connected together by compliant couplings. Also, resonance can be caused by a compliant motor mount so that the motor frame oscillates within the machine frame.

The most common cause of resonance is a compliant coupling between motor and load, as shown in the schematic of Figure 16-1; the block diagram for such a mechanism is shown in Figure 16-2. The commanded current (I_C) goes through the current controller to become feedback current (I_F), which is then multiplied by the motor torque constant (K_T) to produce electromagnetic torque (T_E). Electromagnetic torque directly drives the motor inertia, causing motor acceleration (A_M), which integrates to motor velocity (V_M) and then to motor position (P_M). Once the motor shaft begins to turn, the combined transmission compliance (here represented by the spring K_S) winds

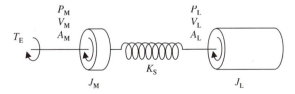

Figure 16-1. A two-mass model.

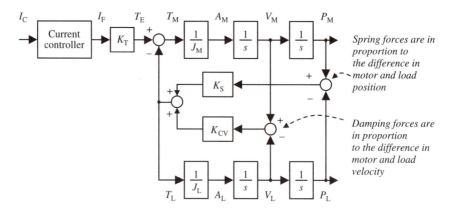

Figure 16-2. Block diagram for a resonant load.

up, producing torque to the load in proportion to the difference between motor and load positions; this is shown by the rightmost summing junction. The load torque accelerates the load inertia, which integrates to load velocity and then to load position. The same torque that the motor transmits to drive the load is transmitted back from the load to slow the motor. This can be seen in Figure 16-2, where the spring torque, $K_S \times (P_M - P_L)$, applies forward torque to the load and reverse torque to the motor. The damping torques, which are transmitted through the constant K_{CV}, are similar to spring torques, except they are based on the speed difference rather than the position difference.

The interaction of motor and load through a compliant coupling can be envisioned with a simple thought experiment. Imagine lifting a heavy clothes hanger with an outstretched rubber band. You cannot raise the hanger directly; instead, you can raise your hand to stretch the rubber band. This increases the force transmitted through the rubber band to the hanger, accelerating it upward. Your hand will feel the same force that is applied to the hanger, only opposite in direction.

Figure 16-2 includes a cross-coupled viscous damping term, K_{CV}. This term produces a torque proportional to the speed differences between the motor and load. Damping torques are proportional to the velocity difference; spring torques are proportional to the position difference. Damping helps stabilize the system considerably; unfortunately, the materials commonly used for transmissions, chiefly steel, provide little mechanical damping. Note that there are two other viscous damping terms in addition to K_{CV}: self-coupled damping of the motor to the frame and of the load to the frame. These terms provide viscous damping from the frame, such as damping from bearings connecting motor or load to the machine frame. In practice, these terms usually have small effect on resonance and so are neglected here. Note also that stiction and Coulomb friction have been neglected.

16.1 Equations of Resonance

Using the block diagram of Figure 16-2 and recognizing that $A_M \equiv s^2 \times P_M$, $V_M \equiv s \times P_M$, $A_L \equiv s^2 \times P_L$, and $V_L \equiv s \times P_L$, the following matrix equation can be derived:

$$\begin{bmatrix} T_E \\ 0 \end{bmatrix} = \begin{bmatrix} J_M s^2 + sK_{CV} + K_S & -sK_{CV} - K_S \\ -sK_{CV} - K_S & J_L s^2 + sK_{CV} + K_S \end{bmatrix} \begin{bmatrix} P_M \\ P_L \end{bmatrix} \tag{16.1}$$

Inverting the two-by-two matrix (see Appendix F) produces the transfer functions from electromagnetic torque (T_E) to motor position (P_M) in Equation 16.2. This equation is used for evaluating the response of motor–drive systems using a feedback sensor mounted to the motor:

$$\frac{P_M(s)}{T_E(s)} = \frac{1}{(J_M + J_L)s^2} \left\{ \frac{J_L s^2 + K_{CV}s + K_S}{\left(\frac{J_M J_L}{J_M + J_L}\right)s^2 + K_{CV}s + K_S} \right\} \tag{16.2}$$

The form of Equation 16.2 shows two effects. The term to the left of the curly brackets is equivalent to a noncompliant motor: a double integrator scaled by the total (load and motor) inertias. The term inside the brackets demonstrates the effect of the compliant coupling. At low frequencies, s is small and the term in the brackets approaches unity; in other words, there is little effect. At higher frequencies, the behavior is determined by the characteristics of the numerator and denominator. The cross-coupled damping (K_{CV}) is small in most machines, so the numerator and denominator act like filters with low damping ratios. An example of this transfer function is shown in Figure 16-3, where the load inertia (J_L) is 0.018 kg-m², the motor inertia (J_M) is 0.002 kg-m², the spring constant (K_s) is 2000 Nm/rad, and the cross-coupled viscous damping is 1 Nm-sec/rad.

As shown in Figure 16-3, the frequency where the gain is at the bottom of the trough is called the *antiresonant frequency* (F_{AR}). Here the numerator is at its minimum value. This occurs at

$$F_{AR} = \sqrt{\frac{K_s}{J_L}} \; rad/sec = \frac{1}{2\pi}\sqrt{\frac{K_s}{J_L}} \; Hz \tag{16.3}$$

Equation 16.3 can be derived by converting the numerator of Equation 16.2 to the form $J_L \times (s^2 + 2\zeta\omega_N s + \omega_N^2)$; here, it is easily seen that $K_S/J_L = \omega_N^2$.

The antiresonant frequency is the natural frequency of oscillation of the load and the spring — the motor inertia is not a factor. It's the same frequency at which the load would oscillate were it connected to ground through the compliant coupling. The motor is quite difficult to move at that frequency because all the energy fed into the motor flows immediately to the load. One misconception about antiresonance is that it

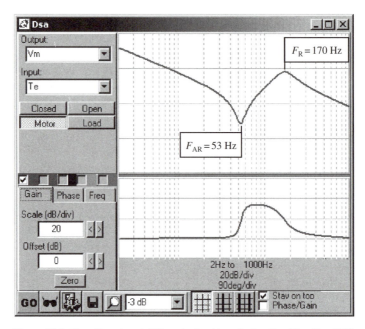

Figure 16-3. From Experiment 16B: motor-load transfer function (Equation 16.2).

can be found by searching for the frequency where the machine becomes quiet. While the motor may be nearly still at the antiresonant frequency, the same cannot be said for the load, which may be oscillating with great strength.

Also as shown in Figure 16-3, the *resonant* frequency (F_R) is the frequency where the gain is at a peak. At this frequency the denominator is minimized; this occurs at

$$F_R = \frac{1}{2\pi} \sqrt{\frac{K_S(J_L + J_M)}{(J_L \times J_M)}} \; \text{Hz} \qquad (16.4)$$

Similarly to Equation 16.3, Equation 16.4 can be derived by converting the denominator of Equation 16.2 to the form $(J_L \times J_M)/(J_L + J_M) \times (s^2 + 2\zeta\omega_N s + \omega_N^2)$. The motor–load combination puts up little resistance to motion at the resonant frequency. It's as if the total inertia became very small, causing the loop gain to become very large.

16.1.1 Resonance with Load Feedback

The inversion of Equation 16.1 can also be used to produce the transfer functions from electromagnetic torque (T_E) to load position (P_L), as shown in Equation 16.5. This equation is used for evaluating the response of motor–drive systems that rely wholly on load feedback.

$$\frac{P_{\mathrm{L}}(s)}{T_{\mathrm{E}}(s)} = \frac{1}{(J_{\mathrm{M}} + J_{\mathrm{L}})s^2} \left\{ \frac{K_{\mathrm{CV}}s + K_{\mathrm{S}}}{\left(\frac{J_{\mathrm{M}}J_{\mathrm{L}}}{J_{\mathrm{M}}+J_{\mathrm{L}}}\right)s^2 + K_{\mathrm{CV}}s + K_{\mathrm{S}}} \right\} \tag{16.5}$$

The load transfer function is similar to the motor transfer function, except there is no antiresonant frequency. A system that relies wholly on load feedback is usually more difficult to control. One reason is that Equation 16.5, compared to Equation 16.2, has 90° more phase lag at high frequency, owing to the loss of the numerator s^2 term. Most high-performance systems rely on motor feedback, at least to close the high-frequency zone control loops.[1] When load feedback is required to attain higher accuracy, many systems use a technique called *dual-loop* control (see Figure 17-22), which uses two feedback devices, one on the motor and another on the load.

16.2 Tuned Resonance vs. Inertial-Reduction Instability

Mechanical compliance can generate instability in two ways: tuned resonance and inertial-reduction instability.[2] These problems are similar, and both are often referred to as "*mechanical resonance*." While there are important distinctions between the two problems, both can be understood as resulting from the variation of effective inertia with frequency.

16.2.1 Tuned Resonance

With tuned resonance, the system becomes unstable at the resonant frequency of the combined motor and load (Equation 16.4); the motor and load oscillate at that frequency, moving in opposite directions as energy circulates between the two. At this frequency, the mechanism is easily excited, as if the total inertia were very small. Recalling that loop gain increases as inertia falls, this effective decrease of inertia increases the loop gain over a narrow band of frequencies. This increase in gain can greatly reduce gain margin.

Figure 16-4 shows a motor response (velocity vs. torque) for a mechanism that is likely to exhibit tuned resonance. The peaks at the resonant and antiresonant frequencies are sharp, moving 20–40 dB away from the baseline transfer function. The sharp peaks result from low cross-coupled viscous damping (K_{CV}). Referring to the

---

[1] The highest zone loops are the velocity for systems with velocity and position loops and the derivative gain of a PID position loop.

[2] In earlier writing, including the second edition of this book, the author referred to *tuned resonance* as "high-frequency resonance". *Inertial-reduction instability* was formerly referred to as "low-frequency resonance".

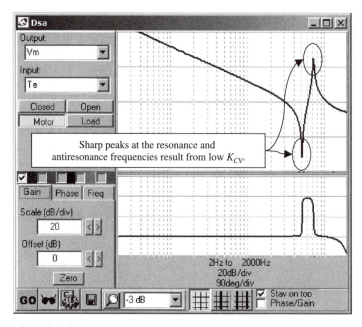

Figure 16-4. From Experiment 16A: Bode plot of motor response ($V_M(s)/T_E(s)$) for a mechanism likely to demonstrate tuned resonance.

numerator of transfer function (Equation 16.2), at the antiresonance frequency the $J_L s^2$ term cancels the K_S term, leaving only $K_{CV}s$. The smaller K_{CV}, the lower the magnitude of the numerator will be at the antiresonant frequency. Similarly, in the denominator, the s^2 and K_S terms cancel at the resonant frequency.

A common characteristic of tuned resonance is that the instability often occurs above the first phase crossover frequency. For example, the open-loop plot for a system with tuned resonance is shown in Figure 16-5. The first phase crossover occurs at 500 Hz; the resonant frequency is 700 Hz. In most control systems, gain declines as frequency increases. However, with tuned resonance, the loop gain can spike up over a narrow band of frequencies, eroding gain margin. Figure 16-5 compares the system with and without resonance. A dashed line in the gain shows the expected path of gain absent compliance, which declines at high frequencies due to integration in the motor. In the compliant system, the loop gain rises sharply at the resonant frequency, stealing about 30 dB of gain margin.

Machines that most often demonstrate tuned resonance are those with low cross-coupled viscous damping. Typically, these are sturdy machines, built with rigid frames and transmission components. Three examples are machine tools, electronically registered printing machines, and web-handling machines. In each of these cases, the needs for accuracy and mechanical rigidity demand the generous use of steel in frames and moving components, and the avoidance of flexible materials in the path of transmission.

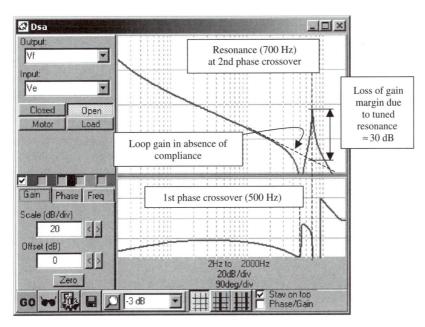

Figure 16-5. Open-loop Bode plot for a system with instability from tuned resonance.

Viscous damping dissipates energy each time a material deforms. In motion systems, that deformation often occurs in the torsion of the motor shaft and coupling. The load spins one way and the motor the other; motion stops and then both objects reverse direction. The shaft or coupling twists one way and then the other. When viscous damping is small, as it is in many rigid steel structures, the torsional bending between motor and load is nearly lossless.

Static friction (stiction) commonly has a significant effect on resonance. One explanation for this is that when the load is experiencing stiction, the load is virtually stationary, as if the load inertia were large without bound. The resonant frequency shifts down according to Equation 16.4, where the term under the radical reduces to K_S/J_M since $J_L >> J_M$. The shift in the resonant frequency can cause instability in an otherwise stable system. A commonly observed behavior is for the machine to be stable when moving but to break into oscillation when coming to rest.

Working on machines with tuned resonance can be exasperating. Much effort can be put into solving the problem when the machine is moving, only to have the problem return when bringing the system to rest. Because the behavior is so dependent on low damping, the oscillations may come and go with varying motor position because some positions will be subject to slightly more viscous damping than will others. Similarly, a machine may be stable when cold and then oscillate as transmission components warm up and viscous damping declines. In such cases, the designer would be well advised to run a Bode plot of the machine and ensure adequate margins of stability. Taking Bode plots in motion systems will be discussed in Chapter 17.

16.2.2 Inertial-Reduction Instability

With inertial-reduction instability, the system becomes unstable above the motor–load resonant frequency. Here, the flexible coupling essentially disconnects the load from the motor. Return to the example of holding a heavy clothes hanger by a rubber band. If your hand moves fast enough, the hanger remains almost stationary; at those frequencies the inertia of the "mechanism" (your hand, the rubber band, and the hanger) declines to that of your hand. Similarly, with inertial-reduction instability, the coupling disconnects the load so that the inertia at high frequency falls to that of the motor. Recalling that loop gain increases as inertia falls, the effective decrease of inertia increases loop gain in an amount approximately equal to $(J_M + J_L)/J_M$.

Figure 16-6 plots the transfer function of the compliant motor and load against dashed lines representing the rigid motor and load (below) and of the motor only (above). At low frequencies, the mechanism overlaps the rigid motor and load; at high frequencies, it overlaps the motor inertia. This behavior also can be seen in the transfer function (shown in Figure 16-6). At low frequencies, s is very small and the term in the brackets reduces to unity, leaving only $1/(J_M + J_L)s^2$. At high frequencies, where s is very large, the term in the brackets is dominated by the s^2 terms and reduces to $(J_M + J_L)/J_M$; a little algebra reduces the overall function to $1/J_M s^2$.

A key characteristic of inertial-reduction instability is that the frequency of oscillation occurs above, often well above, the resonant frequency of the motor and load. A closed-loop plot for a system controlling the motor and load of Figure 16-6 is shown in Figure 16-7. From Figure 16-6, the resonant frequency is about 165 Hz; however, the frequency of oscillation (that is, with maximum peaking) is about 465 Hz. Instability

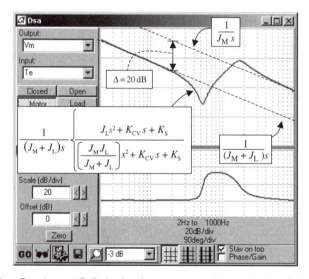

Figure 16-6. From Experiment 16B: Bode plot of motor response for a mechanism likely to demonstrate inertial-reduction instability.

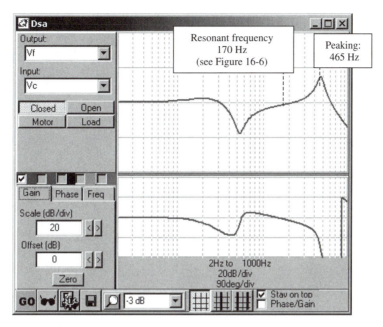

Figure 16-7. Closed-loop Bode plot of a system demonstrating inertial-reduction resonance.

in inertial-reduction instability comes from a gain increase that occurs over a wide frequency range, not from a narrow gain spike as with tuned resonance. Thus the frequency of oscillation can occur anywhere above the resonant frequency of the motor and load. Since the problem occurs over such a wide frequency range, referring to it as "resonance" may be a misnomer.

The systems with instability due to inertial reduction behave much as if the load inertia were removed, at least near the frequency of instability. This is demonstrated in Figure 16-8, which shows the command, V_C, and motor response, V_M, for the system of Figure 16-7. In Figure 16-8a, the step response is shown of the original system ($J_M = 0.002$ and $J_L = 0.02$ kg-m^2). In Figure 16-8b, the step response is shown of the same system, with the sole change being to set J_L to 0.0001 kg-m^2, a value so low that J_L has no significant effect. The behavior of the two systems at the frequency of instability is almost identical, as is demonstrated by similar frequencies, magnitude, and duration of high-frequency ringing.

The sound emitted by a machine often provides a clue as to whether the system suffers from tuned or reduced-inertia resonance. A machine with tuned resonance will usually emit a pure pitch, sounding much like a tuning fork. The frequency of oscillation is mechanically fixed, and changing controller gains may change the intensity of the oscillations but normally will not change the pitch. Inertial-reduction resonance is different; the frequency of oscillation is influenced by the control loop gain. Some variation of loop gain during operation is common, and each shift produces a corresponding shift in the frequency of instability. (For example, current-loop

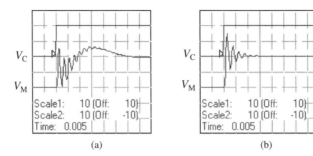

Figure 16-8. Command and motor response in a system (a) with inertial-reduction resonance and (b) without the load removed.

performance can vary because large current magnitude can reduce motor inductance.) Such variation causes the frequency of oscillation to vary, producing distorted tones. As a result, the sound is often a rough, grinding noise something between a foghorn and a garbage disposal.

Based on the author's experience, inertial-reduction instability is more common in industry, except for machine tool, web-handling, electronically registered printing, and a few other segments. The need in general industry for light, low-cost machines yields flexible machines with higher viscous damping. On the other hand, academic papers on resonance focus considerable attention on tuned resonance. This is probably because machine tool and web-handling industries have historically supported university research more than most industries. Also, be aware that some resonance problems are a combination of the two problems. In such cases, the frequency of oscillation will be above, but still near, the natural frequency of the motor and load.

16.2.3 Experiments 16A and 16B

The plots of this section were generated with *Visual ModelQ* Experiments 16A and 16B. These experiments are similar, the only differences being the mechanical parameters and control loop gains. Experiment 16A is configured to show tuned-resonance; Experiment 16B (Figure 16-9) demonstrates inertial-reduction instability.

16.3 Curing Resonance

There are numerous mechanical and electrical cures for resonance described in the technical literature; this section will discuss some of the better-known methods. The plots of this section are generated with Experiment 16C, which is shown in Figure 16-10. This is similar to Experiment 16B, except Experiment 16C adds lag and notch filters, two well-known cures for resonance. These filters are turned off by default but

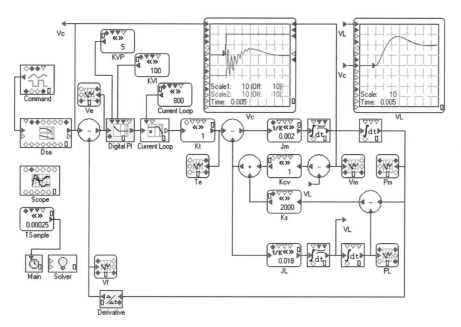

Figure 16-9. Experiment 16B: Inertial Reduction Instability.

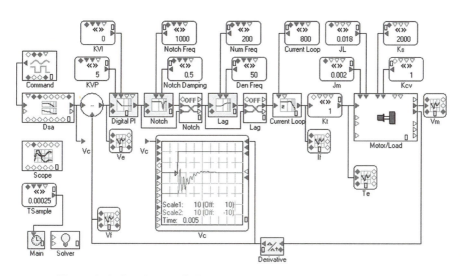

Figure 16-10. Experiment 16C: Experiment 16B with optional notch and lag filters.

can be enabled by double-clicking on the *Live Switches* "Notch" and "Lag." Also, the control law integrator has been turned off; most interesting issues of resonance occur in the highest servo frequencies, so a proportional-only controller makes side-by-side comparisons easier. Finally, the resonant load has been changed from the explicit four-integrator model of Experiment 16B to the *Visual ModelQ* built-in resonant load model. These two representations behave identically; Experiment 16B uses the explicit model for clarity, and Experiment 16C uses the *Visual ModelQ* block to save space in the model diagram.

16.3.1 Increase Motor Inertia/Load Inertia Ratio

Increasing the ratio of motor inertia to load inertia is one of the most reliable ways to improve resonance problems. This is because the smaller the ratio of load inertia to motor inertia, the less compliance will affect the system. Figure 16-6 shows the transfer function of a compliant mechanism, with dashed lines in the gain plot for the system with noncompliant inertias of motor only and motor plus load. The smaller the ratio of load inertia to motor inertia, the less variation in apparent inertia, as would be indicated by a shorter distance between the two parallel lines in Figure 16-6. This behavior can also be seen in the transfer function of Equation 16.4, where the term in brackets approaches unity when J_M becomes large with respect to J_L.

Reducing load inertia is the best way to reduce the ratio of load inertia to motor inertia. The load inertia can be reduced directly by reducing the mass of the load or by changing its dimensions. The load inertia felt by the motor (usually called the *reflected inertia*) can also be reduced indirectly by increasing the gear ratio. The inertia reflected from the load is reduced by N^2, where N is the gear ratio ($N > 1$ indicates speed reduction from motor to load). Even small increases in the gear ratio can significantly reduce the load inertia. Unfortunately, increasing the gear ratio can reduce the top speed of the application. Similar effects can be realized by changing lead screw pitch or pulley diameter ratios.

Any steps that can be taken to reduce load inertia will usually help the resonance problem; however, most machine designers work hard to minimize load inertia for nonservo reasons (cost, peak acceleration, weight, structural stress), so it is uncommon to be able to reduce the load inertia after the machine has been designed. The next alternative is to raise the motor inertia.

Increasing the motor inertia does help the resonance problem. As shown in Figure 16-11, raising the motor inertia by a factor of six (from $0.0002\,\text{kg-m}^2$ to $0.0012\,\text{kg-m}^2$) makes a significant improvement in resonance. (Note that because the total inertia is raised by 50% to $0.003\,\text{kg-m}^2$, K_{VP} must be increased by 50% to maintain the equivalent bandwidth.)

Unfortunately, raising motor inertia increases the total inertia, here by 50%. This reduces the total acceleration available by 50% or, equivalently, requires 50% more torque from the drive to maintain the acceleration. Increasing the motor inertia often increases the cost of both motor and drive. Even so, the technique is commonly used because it is so effective.

One example of this principle is that Kollmorgen servomotors are available in both low and medium inertia models. The low-inertia motors take advantage of high-energy

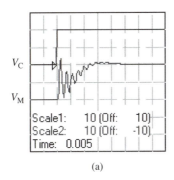
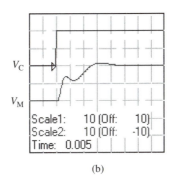

(a)   (b)

Figure 16-11. From Experiment 16C: Raising motor inertia improves resonant behavior: (a) 10:1 ratio and (b) 1.5:1 ratio.

rare-earth magnets to produce a motor with a very high torque-to-inertia ratio, which is ideal for providing high acceleration to light loads. The medium-inertia-motors are identical to the low-inertia motors magnetically and electrically; mechanically, they are similar, the only difference being that the medium-inertia motors have mass added to the shaft, so they have several times more inertia. These motors are commonly used to drive heavy loads, especially on compliant machines.

One common misconception about the load-inertia-to-motor inertia ratio is that it is optimized when the inertias are equal, or *matched*. This conclusion comes from an argument based on optimal power transfer. The argument is that, based on a fixed motor inertia and load inertia, the gear ratio that maximizes the power transferred from motor to load is the ratio that forces the reflected load inertia to be equal to the motor inertia. This is an interesting exercise but has little bearing on how motors and gear ratios are selected in practice because the assumption that the motor inertia is fixed is usually invalid; each time the gear ratio increases, the required torque from the motor decreases, allowing use of a smaller motor.

The primary reason that motor inertia and load inertia should be matched is to reduce resonance problems. Actually, to say the inertias should be matched is to oversimplify. Larger motor inertia improves resonance but increases cost. Experience shows that the more responsive the control system, the smaller the load-inertia-to-motor-inertia ratio. Ratios of 3–5 are common in typical servo applications, and pedestrian applications will work even with larger ratios. Nimble machines often require that the loads be about matched. The highest-bandwidth applications require that the load inertia be no larger than about 70% of the motor inertia. The load-inertia-to-motor-inertia ratio also depends on the compliance of the machine: Stiffer machines will forgive larger load inertias. In fact, direct-drive systems, where there is no transmission and the stiffness is so large as to eliminate resonance in most cases, allow load-inertia-to-motor-inertia ratios greater than 100 with high bandwidth.

16.3.2 Stiffen the Transmission

Stiffening the transmission often improves resonance problems. Figure 16-12, taken from Experiment 16D, shows improvement in reduced-inertia instability provided when mechanical stiffness is increased from 2000 Nm/rad to 10,000 Nm/rad. (K_{VP} was first raised to 1.7 to reduce margins of stability and make the improvement easier to observe.) The goal of stiffening the transmission is to raise the mechanism's resonant frequency, moving it out of the frequency range where it causes harm.

Some common ways to stiffen a transmission are:

- Widen belts, use reinforced belts, or use multiple belts in parallel.
- Shorten shafts; increase shaft diameter.
- Use stiffer gearboxes.
- Increase the diameter of lead screws and use stiffer ball nuts.
- Use idlers to support belts that run long distances.
- Reinforce the frame of the machine (to prevent the motor frame from oscillating in reaction to shaft torque).
- Oversize coupling components.

For the last suggestion, note that mechanical engineers usually select couplings based on catastrophic failure; this certainly sets the minimum size for the application. However, by using larger components, you can stiffen the transmission, especially when the coupling is the most compliant component in the transmission. Be cautious because this can add inertia, which may slow the peak acceleration.

When increasing the stiffness of a transmission, start with the loosest components in the transmission path. Spring constants combine as shown in Equation 16.6, so a single loose component can single-handedly reduce the overall spring constant, K_S, significantly. Note that the stiffness of most transmission components is provided by vendors, usually as catalog data.

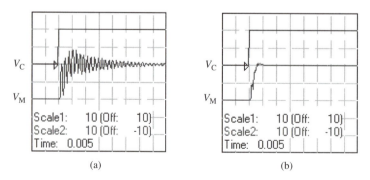

(a)   (b)

Figure 16-12. From Experiment 16D: Increasing stiffness can improve resonant behavior.

$$K_S = \cfrac{1}{\cfrac{1}{K_{Coupling}} + \cfrac{1}{K_{Gearbox}} + \cfrac{1}{K_{Leadscrew}}} \qquad (16.6)$$

Unfortunately, increasing transmission stiffness is not a reliable cure for resonance; in fact, it often worsens the problem. For example, while increasing stiffness improved stability margins in Experiment 16D, it does the opposite in Experiment 16C. This is demonstrated in Figure 16-13, which is taken from Experiment 16C. Here, the spring constant was increased from 2000 to 10,000. The open-loop Bode plot in Figure 16-14 shows how increasing the stiffness reduced the gain margin by about 5 dB.

Notice in Figure 16-14 that both open-loop plots have three gain crossover frequencies; most problems of resonance occur that the 3rd (highest) frequency. For Experiment 16C, increasing K_S did not reduce the gain, so the problems at the 3rd crossover remain. By contrast, in Experiment 16D (Figure 16-15), raising the stiffness eliminates the 2nd and 3rd gain crossover frequencies. Here, the resonant frequency moved far enough to the right that the peak of the open-loop gain no longer reached 0 dB. This can allow dramatically improved system performance. The main differences between Experiments 16C and 16D is that in Experiment 16D the motor inertia and load inertia were both reduced by a factor of 10; this moved the mechanism's resonant frequency higher and that dropped the gain at the resonant frequency about 6 dB.

Another problem that occurs with increasing stiffness is an unintended reduction of cross-coupled viscous damping (K_{CV}). This results because the couplings that have higher stiffness often use materials with low damping. For example, when changing from a "tire" coupling, such as a Dodge Paraflex™ coupling, which uses a polymer to transmit power, to a bellows coupling, which is all metal, viscous damping will decline sharply. This causes the narrow peaks of tuned resonance, as shown in Figure 16-4. Tuned resonance can be difficult to deal with, even when the resonant frequency is higher. As a result, changing to a stiffer (and often more expensive) coupling can produce systems with more resonance problems.

Of course, if increasing stiffness marginally often increases problems with resonance, then it is logical to assume that marginal reductions in stiffness could improve

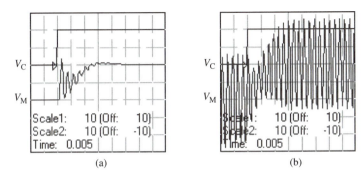

(a)   (b)

Figure 16-13. From Experiment 16C: Increasing stiffness can also worsen resonant behavior.

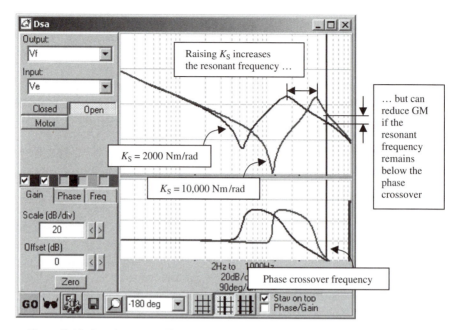

Figure 16-14. Open-loop gain in Experiment 16C shows how increasing K_s reduces GM by 5 dB.

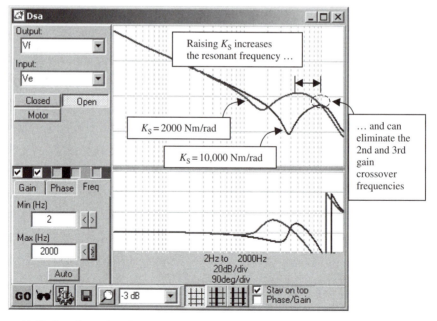

Figure 16-15. Open-loop gain in Experiment 16D shows how increasing K_s removes 2nd/3rd gain crossover.

system performance, especially when that reduction is accompanied by an increase in viscous damping. One difficulty with this approach is that the looser coupling reduces the control the motor has over the load; the motor may be stable, but disturbances on the load will flex a looser coupling more easily, which will harm performance for many machines. Aside from this, the evidence seems to support incremental reductions in stiffness to improve performance of resonant machines, although it must be said such an approach is used rarely in industry, and a person employing such an approach should do so with caution.

The problems associated with stiffening the transmission to improve resonant behavior demonstrate the need to model and measure machine performance. Measuring behavior on existing machines has become easier because more motion-control products have the ability to provide Bode plots. By taking open-loop Bode plots, effects of stiffening transmission components can be predicted with reasonable accuracy. This can guide mechanical engineers in selecting appropriate transmission components. Modeling machines before they are built presents a perplexing problem. Many mechanical parameters can be modeled with accuracy, especially load inertia and motor inertia and, often, transmission stiffness. However, viscous damping, an effect central to resonant behavior is often difficult to predict. Where these parameters can be estimated accurately, engineers would be well advised to model machines before building them; for other applications, machine builders will continue to rely on historical designs and trial and error.

16.3.3 Increase Damping

Another mechanical cure to resonance is to increase K_{CV}, the cross-coupled damping to the motor/load coupling. Almost any case of resonant behavior will respond positively to increasing K_{CV}, although the largest benefits are probably for tuned resonance, since this problem is directly due to low viscous damping. However, in practice it is difficult to add damping between motor and load. This is because the materials with large inherent viscous damping do not normally make good transmission components. There are a few well-tried techniques used to add damping. For example, the passive dampers used to reduce vibration on stepper motors, such as ferrofluidic dampers, can be mounted to the motor shaft to reduce resonance problems. And, faced with the challenging resonant behavior of systems using bellows couplings, more than one engineer has wrapped tape around the coupling. While the questionable long-term reliability of such a solution makes it difficult to recommend, it does beg the question of how manufacturers of bellows couplings could add a small amount of damping, perhaps though coating or filling the couplings.

Sometimes the unexpected loss of viscous damping can cause resonance problems in the field. When a machine is being assembled at the factory, it may have its largest viscous damping. The mechanical linkages are tight because the machine has not been operated. If the machine is tuned with aggressive gains, problems in the field are likely to occur. After the machine has been operated for a few weeks, the seals and joints will loosen, resulting in a net loss of damping. This can cause the machine to have resonance problems, so the machine may need to be retuned, often requiring a field service trip. Engineers responsible for products with a history of this problem would be well advised to purchase a DSA or motion-control equipment capable of generating Bode plots. This way margins of stability could be measured for each system, ensuring consistency and reducing the likelihood of field problems. Techniques for using a DSA to measure motion systems are discussed in Chapter 17.

16.3.4 Filters

The primary electrical cure for resonance is the low-pass filter; notch filters are also used. The filter is placed in the loop to compensate for the change in gain presented by the compliant load. The most common position for the filter is just before the current loop, like Notch and Lag in Figure 16-10. Note that this is the position that is commonly used to reduce resolution noise. Low-pass filters in this position can help either problem.

16.3.4.1 First-Order Filters

First-order filters, both low-pass and lag, work by reducing gain near and above the resonant frequency. They restore the gain margin, or a portion of it, that is taken by the increased gain of the motor/load mechanism at the resonant frequency and above. The cost of using a low-pass or lag filter is the phase lag it induces in the lower frequencies and the reduced phase margin that this implies.

Low-pass and lag filters are similar. Low-pass filters, with the transfer function $\omega/(s + \omega)$ or, equivalently, $1/(s/\omega + 1)$, attenuate at ever-increasing amounts as frequency increases above the break frequency, ω. Lag filters, with the transfer function $(s/\omega_1 + 1)/(s/\omega_2 + 1)$, where $\omega_2 < \omega_1$, have a maximum attenuation of ω_2/ω_1 at high frequency. The benefit of the lag filter is lower phase lag and thus less erosion of the phase margin. For that advantage, the lag filter is the focus of this section. The interested reader can replace the lag filter in Experiment 16C with a low-pass filter and repeat these steps; the results are similar. The benefit of the lag filter is shown in Figure 16-16, which shows the system of Experiment 16C with and without benefit of the lag filter. In Figure 16-16b, the lag filter reduces the tendency of the system to oscillate at the resonant frequency.

The Bode plot of the open loop, shown in Figure 16-17, also demonstrates the effects of the lag filter. The benefits occur at and around the resonant frequency. At the third crossover frequency, the lag filter increased gain margin by about 10 dB. However, the lag filter can inject phase lag at the gain crossover frequency, reducing phase margin.

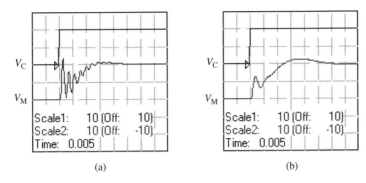

Figure 16-16. From Experiment 16C: Reduced-inertia resonance (a) without and (b) with a lag filter.

For tuned resonance (Figure 16-5), the effect of lag and low-pass filters is similar: to attenuate the open-loop gain at the resonant frequency. The issue here is that the peak may climb 20 or 30 dB, so a considerable attenuation may be required. The use of one or two low-pass filters to provide so much attenuation can require low break frequencies. It is not uncommon to see a system with tuned resonance at, say, 800 Hz, where multiple 100-Hz low-pass filters are required to provide sufficient attenuation; the bandwidth of the servo system with such severe filters may be limited to 20 or 30 Hz.

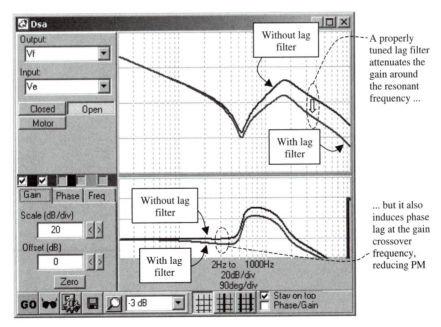

Figure 16-17. From Experiment 16C: the open-loop Bode plot of the reduced-inertia resonant system with and without a lag filter.

16.3.4.2 Second-Order Filters

Two types of second-order filters are also used to deal with resonance. Notch filters provide a great deal of attenuation over a narrow band of frequencies, and they do so without injecting significant phase lag at lower frequencies. Notch filters provide a narrow band of attenuation, which is effective for the narrow gain spike of tuned resonance. The interested reader can verify this by using the notch filter of Experiment 16C after copying in the mechanical parameters of the tuned resonant system in Experiment 16A; set the notch frequency to the natural frequency of the motor and load and enable the filter. Notch filters are not effective for reduced-inertia instability because this problem requires attenuation over a large frequency range.

While notch filters are effective for tuned resonance, the benefits are limited to a narrow frequency range. If the resonant frequency shifts even a small amount, the notch filter will become ineffective. Unfortunately, resonant frequencies often vary on practical machines. The load inertia may change, as was discussed in Section 12.4.4. The spring constant can also change. For example, the compliance of a lead screw will vary when the load is located in different positions, although this effect is not as dramatic as it might at first seem, since much of the compliance of a ball screw is in the ball-nut. Also, different copies of the same machine will often have significant variation of resonant frequency. Notch filters work best when the machine construction and operation allow little variation of the resonant frequency and when the notch can be individually tuned for each machine.

Two-pole low-pass filters are commonly used to deal with resonance. Often the damping ratio is set to between 0.5 and 0.7. In this range, the two-pole low-pass can provide attenuation while injecting less phase lag than the single-pole low-pass or lag filters. Bi-quad filters are also used to deal with resonance. The most common example is where bi-quad is configured to attenuate higher frequencies, that is, where the poles (denominator frequencies) are smaller than the zeros (numerator frequencies). In this case, the bi-quad operates like the lag filter, providing attenuation while injecting less phase lag than the low-pass filter.

16.4 Questions

1. Is the instability demonstrated by Experiment 16A caused by tuned resonance or inertial reduction? What evidence is there for your answer?
2. Repeat Question 1 for Experiment 16B.
3. Investigate the effects of increasing motor inertia on inertial-reduction instability using Experiment 16B. Tune the system setting the motor inertia to the following four values: 0.0018, 0.0045, 0.009, and 0.018 kg-m^2. Using the step response for tuning systems subject to inertial-reduction instability is difficult, so use closed-loop Bode plots. In each of the four cases, maximize K_{VP} with the gain above the antiresonant frequency remaining below 0 dB. (*Hint: For 0.0018 kg-m^2, this occurs at about 2.*) For simplicity, leave K_{VI} at zero for all

cases. Record the following: motor inertia, load/motor inertia ratio, K_{VP}, and bandwidth. Note that the bandwidth is the lowest frequency where the gain falls to -3 dB. What conclusions would you draw?

4. For question 3, what are the implications on system cost of raising motor inertia?

5. Using Experiment 16C, improve the performance of a system with reduced-inertia instability using low-pass and lag filters. Load and compile the model. Set $J_M = 0.004$ kg-m^2 and $K_{VI} = 0$.

   a. *Tuning without a filter*: Set K_{VP} to the maximum value that does not generate peaking while the lag filter is turned off. What is K_{VP}? What is the system bandwidth?

   b. *Tuning with a low-pass filter*: Turn on the lag filter, but set *Num Freq* to 1000 Hz (using such a high frequency for *Num Freq* makes the lag filter behave like a low-pass filter in the frequencies of interest). Adjust *Den Freq* and K_{VP} simultaneously so that K_{VP} is at its maximum value without peaking. What are *Den Freq* and K_{VP}? What is the system bandwidth?

   c. *Tuning with a lag filter*: Repeat Question 5b, but maintain *Num Freq* at four times *Den Freq*.

   d. What conclusions could you draw?

6. Continuing Question 5, turn off the lag filter and use a notch filter to improve the performance. Simultaneously adjust *Notch Freq* and K_{VP} to maximize K_{VP} without inducing peaking in the closed-loop Bode plot. (This induces tuned resonance.)

7. Using Experiment 16C, improve the performance of a system with tuned-resonance using low-pass and lag filters. Load and compile the model. Set $J_M = 0.004$ kg-m^2, $K_{CV} = 0.2$, $K_S = 20,000$, and $K_{VI} = 0$.

   a. *Tuning without a filter*: Set K_{VP} to the maximum value that does not generate peaking while the lag filter is turned off. What is K_{VP}? What is the system bandwidth?

   b. *Tuning with a low-pass filter*: Turn on the lag filter, but set *Num Freq* to 1000 Hz (using such a high frequency for *Num Freq* makes the lag filter behave like a low-pass filter in the frequencies of interest). Adjust *Den Freq* and K_{VP} simultaneously so that K_{VP} is at its maximum value without peaking. What are *Den Freq* and K_{VP}? What is the system bandwidth?

   c. *Tuning with a lag filter*: Repeat Question 5b, but maintain *Num Freq* at four times *Den Freq*.

   d. What conclusions could you draw?

8. Continuing Question 7, turn off the lag filter and use a notch filter to improve the performance. Simultaneously adjust *Notch Freq* and K_{VP} to maximize K_{VP} without inducing peaking in the closed-loop Bode plot. What is K_{VP}? What is the notch frequency? What is the bandwidth?

Chapter 17

Position-Control Loops

The majority of motion-control applications require precise velocity and position control. Most use position-control loops, which are control laws designed for the unique requirements of motion systems. These loops share much with those of Chapter 6. Both have proportional and integral gains, and both have derivative gains in many cases. The primary distinction for motion systems is that the plant has two stages of integration between excitation and feedback, one from current to velocity and the second from velocity to position. The most common plants (see Table 2-2) have one stage of integration from excitation to sensed feedback.

Several loop structures are commonly used in the motion-control industry. One technique is to place a velocity loop inside a position loop. Often, the velocity loop uses PI control and the position loop relies on P control; this will be called here P/PI control. In other cases, the integral is placed in the position loop, resulting in PI/P control. Another common structure is to rely solely on a position loop; this requires a derivative in the control law resulting in PID position control. The three methods, P/PI, PI/P, and PID, have important differences. The block diagrams look different and the tuning processes vary. However, the transfer functions of the three are similar, implying that the performance of these methods is similar when they are tuned appropriately.

The opening sections of this chapter discuss the three control laws. The block diagrams of each are presented and tuning methods are provided for each. Acceleration and velocity feed-forward paths are applied to each method. Software experiments are provided for each structure. The chapter concludes with several alternatives for measuring the dynamic response of machines.

17.1 P/PI Position Control

The P-position/PI-velocity loop structure is shown in Figure 17-1. A PI velocity loop is enclosed within a proportional position loop. The velocity loop is a single-integrating

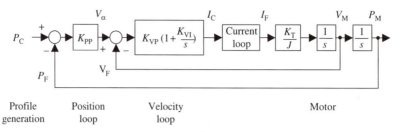

Figure 17-1. Block diagram of P/PI loops.

loop and thus can be controlled with any of the standard control loops from Chapter 6. The PI is chosen because it is simple to implement and simple to tune. On some occasions a PI+ loop is used. Normally, derivatives are avoided in the velocity loop because of noise in the velocity feedback signal, which is formed by differentiating a position sensor (see Section 14.4). The position loop is structured to provide a velocity command that is simply scaled position error. This is intuitive. If the position error is large, the system should move quickly; if there is no error, the commanded velocity should be zero.

The primary strength of P/PI control is simple tuning. The PI velocity loop is normally tuned for maximum performance, and then the position gain is increased to just below the value where overshoot appears. The primary shortcoming is that the method allows steady-state position error (following error) in the presence of a constant-velocity command. Following error is rarely desirable but is especially harmful in multiaxis applications, where it can cause significant errors in trajectory following. Fortunately, this can be corrected with velocity feed-forward, as will be discussed later. Another problem is that when the connection from position loop output to velocity loop input is analog, offset in the signal will generate proportionate position error when the system is at rest.

You may have noticed that the model of Figure 17-1 is simplified, omitting several high-frequency effects. For example, even though the velocity signal is normally formed by differentiating a sampled position signal, the imperfections of digital differentiation and sample/hold are ignored. Effects of feedback resolution and mechanical resonance also are disregarded, even though they are often important in motion-control applications. These effects are disregarded so we can better focus on the lower-frequency zones where the position-control methods are differentiated. High-frequency problems are affected almost solely by the high-frequency-zone gain (K_{VP} here), which is identical in all methods. In software experiments, we will set the high-frequency-zone gain the same for all methods and then investigate the differences in the lower frequencies. Of course, the higher value of the high-frequency-zone gain, the higher the level of performance; however, such an increase would be equally available to all three methods and would provide equal improvement.

K_{PP} has units of radians/second, producing velocity from position error (units of velocity/position are 1/sec, equivalent to rad/sec). This is a benefit because, assuming proper tuning, the bandwidth of the position loop will be approximately K_{PP}, independent of motor parameters. This allows you to estimate position loop performance knowing only the gain and its scaling. Sometimes K_{PP} is scaled in (m/sec)/mm (m/sec of velocity for each mm of position error) or (in/s)/mil. To convert rad/sec to (m/sec)/mm, simply multiply by 1000. Note that (m/sec)/m and rad/sec are identical.

One other detail to mention is the use of V_α, where V_C might have been expected. In motion-control applications, the commanded velocity, V_C, is normally the derivative of the position command, P_C, not the input to the velocity loop. The input to the velocity loop, V_α in Figure 17-1, is normally not an important signal in positioning applications.

17.1.1 P/PI Transfer Function

The transfer function of Figure 17-1 is derived using Mason's signal flow graphs and is shown as Equation 17.1. Notice the form is that of a low-pass filter: When s is low, the function is near unity; when s is high, the value declines to near zero. Notice that Equation 17.1 ignores the current loop; like sampling and resonance, current loops are high-frequency effects that often can be ignored when focusing on lower-frequency effects.

$$\frac{P_F(s)}{P_C(s)} = \frac{K_{VP}\,K_{PP}s + K_{VP}\,K_{VI}\,K_{PP}}{\frac{J}{K_T}s^3 + K_{VP}\,s^2 + K_{VP}(K_{VI} + K_{PP})s + K_{VP}\,K_{VI}\,K_{PP}} \qquad (17.1)$$

The benefit of the cascaded control loops can be investigated by replacing the inner loop (including the motor) with its transfer function. Using the $G/(1 + GH)$ rule, the velocity loop can be derived as Equation 17.2. Notice that this has the form of a low-pass filter; most importantly, the phase lag at low frequency (low values of s) is $0°$.

$$\frac{V_M(s)}{V_\alpha(s)} = \frac{K_{VP}s + K_{VP}\,K_{VI}}{\frac{J}{K_T}s^2 + K_{VP}s + K_{VP}\,K_{VI}} \qquad (17.2)$$

Figure 17-2 shows the standard P/PI control loop in an alternative representation: The velocity loop is replaced with Equation 17.2. In this position loop, the velocity loop occupies the place of the power converter, and the integrator that forms position

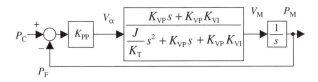

Figure 17-2. Alternative representation of Figure 17-1.

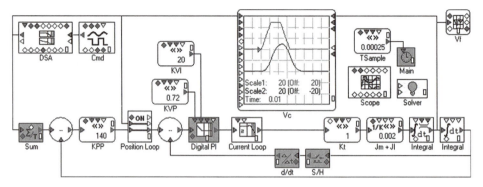

Figure 17-3. Experiment 17A: P/PI position control.

behaves like a single-integrating plant. From this diagram it should be clear that a proportional position loop can control the outer loop.

17.1.2 Tuning the P/PI Loop

This section will discuss a tuning procedure for the P/PI loop. The discussion starts with Experiment 17A, which is shown in Figure 17-3. A waveform generator, through a DSA, commands velocity, which is integrated in a sum block to form position command. The loop is the P/PI structure of Figure 17-1. A *Live Relay* is provided to change the configuration between position and velocity control to facilitate tuning. The sample rate is 4 kHz, and the current loop is an 800-Hz two-pole low-pass filter. Unlike the transfer function, the model does include the effects of sampling and digital differentiation, though their effects will be insignificant in these experiments.

17.1.2.1 Tuning the PI Velocity Loop

Begin tuning by configuring the system as a velocity controller. In Experiment 17A, double-click on the *Live Relay* "Position Loop" so that "OFF" is displayed. As with standard PI control, set the command as a square wave with amplitude small enough to avoid saturation; in this experiment you will need to change the waveform of the waveform generator "Cmd" to "Square." As discussed in Chapter 6, set the integral gain (K_{VI}) to zero and raise the proportional gain (K_{VP}) no higher than the high-frequency effects (current loop, sample rate, resonance, feedback resolution) allow without overshoot. Then raise the integral for about 5% overshoot.

In Experiment 17A, the value of K_{VP} has been set to 0.72. This produces a velocity loop bandwidth of about 75 Hz, which is typical for motion applications; this is shown in Figure 17-4. Many motion systems use higher gains, and velocity loop bandwidths of 200 Hz are common; in fact, K_{VP} in the system of Experiment 17A could have been

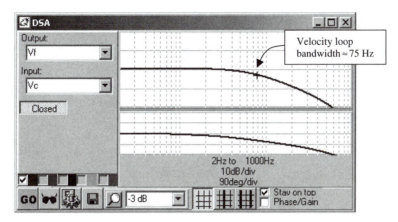

Figure 17-4. From Experiment 17A: Bode plot of velocity loop command response.

doubled without inducing overshoot. However, the value of 0.72 provides performance more common in motion systems, which is often limited by resolution and mechanical resonance. This value will be used as the starting point for all three methods, allowing the clearest comparisons. K_{VI} is set to 20, which induces about 4% overshoot.

The guidelines here are general, and there is considerable variation in the needs of different applications. Accordingly, be prepared to modify this procedure. For example, the value of integral is set low to minimize overshoot. The primary disadvantage of low integral gain is the loss of low-frequency stiffness. For applications where low-frequency disturbance response is important or where overshoot is less of a concern, the integral gain may be set to higher values. On the other hand, the integral may be zeroed in applications that must respond to aggressive commands without overshoot.

17.1.2.2 Tuning the P Position Loop

The position loop is tuned after the velocity loop. Start by configuring the system as a position loop. In Experiment 17A, double-click on the *Live Relay;* "Position Loop" should display "ON"; double-click on it if necessary. For point-to-point applications, apply the highest-acceleration command that will be used in the application. The waveform generator in Experiment 17A defaults to a trapezoidal command that reaches 50 RPM in about 10 msec (acceleration = 5000 RPM/sec). Always ensure that the system can follow the command. Commanding acceleration too great for the system peak torque will produce invalid results. As always in tuning, avoid saturating the current controller. The command in Experiment 17A requires about 1 A of current, well below the 15-A saturation level set in the PI controller.

The position loop gain, K_{PP}, is usually set to the largest value that will not generate overshoot. Overshoot is normally unacceptable because it implies that the load backs up at the end of the move. Normally, this causes a significant loss of accuracy, such as when backing up a load connected to a gear box results in lost motion. The result here is that K_{PP} is set to 140; this is shown in Figure 17-5a. For comparison, Figure 17-5b shows the overshoot that comes from setting K_{PP} to 200, about 40% too high.

Settling time is an important measure of trapezoidal response. It is commonly measured as the time between the end of the command to when the system comes to rest, as shown in Figure 17-5a. "Rest" implies that the position error is small enough for the system to be "in position," a measure that varies greatly between applications. It is common to minimize the total move time ($T_{MOVE} + T_{SETTLE}$ in Figure 17-5a) to maximize machine productivity. Typically, mechanical design and motor size fix T_{MOVE}; T_{SETTLE} is directly affected by tuning. Of course, higher loop gains reduce settling time, assuming adequate margins of stability.

The responsiveness of motion systems can also be measured with Bode plots, although this is less common in industry. The bandwidth is $P_F(s)/P_C(s)$, which is equal to $V_F(s)/V_C(s)$. This equality can be seen by multiplying $P_F(s)/P_C(s)$ by $s/s = 1$ and recalling that $s \times P(s) = V(s)$. Velocity signals are often used because many DSAs require an analog representation of input signals, something difficult to accomplish with position signals. The Bode plots in this chapter will rely on the ratio of velocities. Figure 17-6 shows the Bode plot for both position ("Position Loop" = ON) and velocity ("Position Loop" = OFF) operation. The bandwidth of the position loop is 33 Hz, a little less than half that of the velocity loop.

17.1.3 Feed-Forward in P/PI Loops

As discussed in Chapter 8, feed-forward can greatly increase command response without producing stability problems. Feed-forward in motion systems follows the principles of Chapter 8, but the implementation is unique because of the double-integrating plant. Velocity feed-forward is provided by most motion controllers and many drives. As shown in Figure 17-7, the velocity command from the profile generation is scaled and summed with the position loop output to form the velocity command.

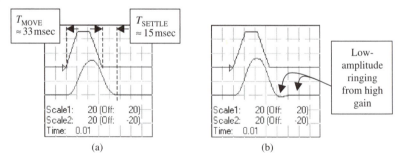

Figure 17-5. From Experiment 17A: response with (a) $K_{PP} = 140$ and (b) $K_{PP} = 200$.

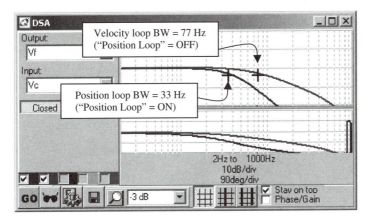

Figure 17-6. From Experiment 17A: Bode plots of position and velocity loop command response.

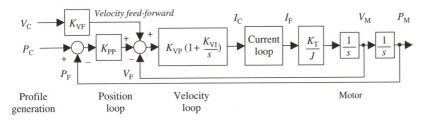

Figure 17-7. Block diagram of P/PI position control with velocity feed-forward.

The transfer function of the P/PI controller with velocity feed-forward is given in Equation 17.3. Note that V_C in Figure 17-7 is replaced by the derivative of P_C, $s \times P_C$. Figure 17-7 does not use this representation because velocity command is not calculated from position command in most modern motion controllers. In fact, the profile generator normally integrates commanded velocity to calculate position and, in doing so, avoids the problems of digital differentiation.

$$\frac{P_F(s)}{P_C(s)} = \frac{K_{VP} K_{VF}\, s^2 + K_{VP}(K_{VI}K_{VF} + K_{PP})s + K_{VP}\, K_{VI}\, K_{PP}}{\frac{J}{K_T}s^3 + K_{VP}\, s^2 + K_{VP}(K_{VI} + K_{PP})s + K_{VP}\, K_{VI}\, K_{PP}} \qquad (17.3)$$

Compared to Equation 17.1, Equation 17.3 adds an s^2 term to the numerator by the feed-forward path; this becomes the highest-frequency term in the numerator, improving the command response. The denominator is unchanged. As in Chapter 8, this implies that the command response improves, but neither the margins of stability nor the disturbance response is affected.

When K_{VF} is set to 100%, the velocity command from the profile generator, V_C, sends ahead the full command to the velocity loop. This eliminates steady-state following error, as can be seen first by noticing that the position loop guarantees that $V_C = V_F$ over the long term (otherwise, position error would grow to unbounded values during periods of constant-velocity command). Additionally, the integrator in the PI loop guarantees that $V_C \times K_{VF} + K_{PP} \times (P_C - P_F) - V_F = 0$ in steady state. Substituting $V_C = V_F$ from earlier and setting $K_{VF} = 1$ produces $K_{PP} \times (P_C - P_F) = 0$; thus, position error $(P_C - P_F)$ must be zero for any nonzero K_{PP} when $K_{VF} = 1$.

Unfortunately, $K_{VF} = 100\%$ is often impractical in point-to-point applications, as well as in many contouring applications, because it generates too much overshoot. Normally, K_{VF} is set below 70%. Even $K_{VF} = 70\%$ generates so much overshoot that K_{PP} normally must be reduced by 30% or more. The higher the feed-forward gain, the more K_{PP} normally must be reduced, further compromising disturbance response in the frequency range dominated by K_{PP}.

17.1.4 Tuning P/PI Loops with Velocity Feed-Forward

Experiment 17B, shown in Figure 17-8, is Experiment 17A with several modifications. First, velocity feed-forward was added, as was acceleration feed-forward, which will be discussed in the following section. Position error is displayed to make it easier to observe the effects of various gains on this signal. The *Live Relay* "Position Loop" has been removed; it will be unnecessary to configure this experiment as a velocity

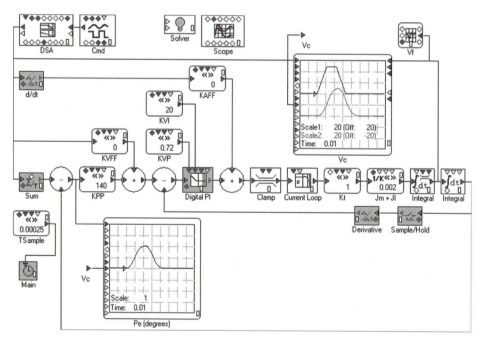

Figure 17-8. Experiment 17B: P/PI position control with acceleration and velocity feed-forward.

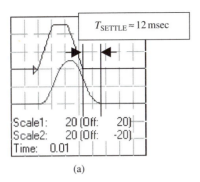 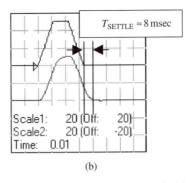

(a)   (b)

Figure 17-9. From Experiment 17A: trapezoidal velocity command (above) and response (a) without feed-forward and (b) with 60% velocity feed-forward.

controller because the velocity loop tuning values from Experiment 17A will be used. A clamp was added before the current loop to ensure that the acceleration feed-forward path cannot increase the current beyond the system maximum. This will not be an issue if the system is kept out of saturation, as should always be the case.

The results of tuning velocity feed-forward for point-to-point operation are shown in Figure 17-9. Figure 17-9a shows the system without feed-forward; Figure 17-9b shows K_{VF} set to 0.6, a common value. K_{PP} is reduced from 140 to 100 to reduce overshoot. Feed-forward improves the settling time by 33%, but it will worsen disturbance response marginally, owing to the lower value of K_{PP}.

17.1.5 Acceleration Feed-Forward in P/PI Loops

Acceleration feed-forward is a natural extension of velocity feed-forward. As shown in Figure 17-10, a current proportional to the acceleration command is added to the velocity loop output. The predominant benefit of acceleration feed-forward is that it removes the overshoot induced by velocity feed-forward without lowering the position

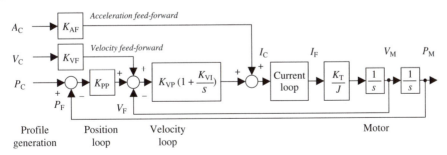

Figure 17-10. Block diagram of P/PI loops with acceleration and velocity feed-forward.

loop gain. In combination with velocity feed-forward, this supports much faster response than does velocity feed-forward alone. Acceleration feed-forward is implemented less often than velocity feed-forward in industrial equipment, but its popularity is growing.

Note that the K_{AF} converts acceleration to amps, implying that the term must include an implicit conversion factor, J/K_T (scaling by J converts acceleration to torque; dividing by K_T converts torque to amps). Accordingly, K_{AF} must be set as a fraction of J/K_T, and it must be adjusted whenever the load inertia changes. This problem is not present in velocity feed-forward because K_{VF} is unitless, providing one velocity signal as a fraction of another.

The transfer function for full (acceleration and velocity) feed-forward is shown in Equation 17.4. Note that the acceleration term adds an s^3 term to the numerator, further improving response without affecting the structure of the denominator. In the ideal case ($K_{AF} = J/K_T$ and $K_{VF} = 1$), the response becomes unity, indicating that the motor follows the command perfectly. Such a result is impractical in systems with aggressive commands because the current-loop response is limited. As discussed in Chapter 8, imperfections in the power converter induce overshoot when 100% feed-forward is applied. However, both gains can often be set to more than 80% of their ideal values, considerably enhancing command response.

$$\frac{P_F(s)}{P_C(s)} = \frac{K_{AF}\,s^3 + K_{VP}\,K_{VF}\,s^2 + K_{VP}(K_{VI}\,K_{VF} + K_{PP})s + K_{VP}\,K_{VI}\,K_{PP}}{\frac{J}{K_T}s^3 + K_{VP}\,s^2 + K_{VP}(K_{VI} + K_{PP})s + K_{VP}\,K_{VI}\,K_{PP}} \qquad (17.4)$$

17.1.6 Tuning P/PI Loops with Acc/Vel Feed-Forward

Tuning the P/PI control loop with feed-forward begins by tuning the system without any feed-forward, as discussed in Section 17.1.2. Then set the velocity feed-forward to a high value, typically between 0.7 and 0.9, with higher values yielding faster command response. Next, use acceleration feed-forward to eliminate the overshoot caused by velocity feed-forward (there is usually no need to lower K_{PP} as discussed in Section 17.1.2). The result when this process is applied to Experiment 17B is shown in Figure 17-11; here, $K_{VF} = 0.91$ and $K_{AF} = 0.00146$, about 73% of J/K_T. Notice that the settling time, which was 8 msec with velocity feed-forward only (Figure 17-9b), is so small it cannot be measured here, yet overshoot is not a problem.

Feed-forward improves the bandwidth of the position controller, as is demonstrated by the Bode plot of Figure 17-12. Three plots are shown:

 a. No feed-forward, with a bandwidth of 35 Hz (compare to Figure 17-5a)
 b. With velocity feed-forward, with a bandwidth of 53 Hz (compare to Figure 17-9b)
 c. With velocity/acceleration feed forward, with a bandwidth of 570 Hz (compare to Figure 17-11)

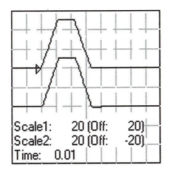

Figure 17-11. From Experiment 17B: trapezoidal velocity command (above) and response with velocity and acceleration feed-forward.

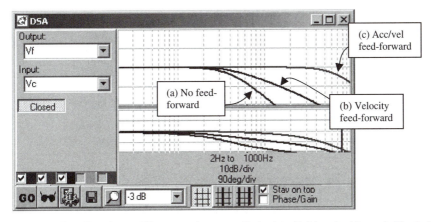

Figure 17-12. From Experiment 17B: command response Bode plot with (a) no feed-forward, (b) velocity feed-forward, and (c) acceleration and velocity feed-forward.

Feed-forward allows the command response of an outer loop to approach that of an inner loop. Here, velocity feed-forward allows the position controller bandwidth to approach the velocity loop bandwidth. Acceleration and velocity feed-forward together allow the position-controller bandwidth to approach the current-loop bandwidth.

It should be pointed out that the process used here is conservative. You can often adjust the position loop gain to higher values and adjust the feed-forward gains (K_{VF} down and K_{AF} up) to eliminate the overshoot. This will normally have little effect on the command response, but disturbance response will improve in the lower-frequency range, where K_{PP} is most effective. On the other hand, this process may be too aggressive if the system inertia changes during normal operation; larger feed-forward gains normally cause system performance to vary more with changing inertia.

The tuning process here is most appropriate for point-to-point applications. When contouring, such as machine tool and robotic applications, trajectories are often less severe, so overshoot is less likely. As a result, feed-forward gains can often be set somewhat higher. In fact, feed-forward gains are often set to 100%, eliminating steady-state following error during constant-speed and constant-acceleration motion. In multiaxis contouring applications, the elimination of steady-state error is important because it results in more accurate path following.

17.2 PI/P Position Control

An alternative to P/PI control is cascading a PI position loop with a proportional velocity loop. This is shown in Figure 17-13. One advantage of PI/P control over P/PI control is that during constant speed, steady-state following error is eliminated by the position loop integrator. A problem is that, in the absence of acceleration feed-forward, any amount of integral gain causes overshoot. Because acceleration feed-forward is often unavailable, the integral gain is commonly zeroed, and the advantage of zero following error does not apply.

The transfer function of PI/P can be derived using Mason's signal flow graphs on Figure 17-13; this is shown in Equation 17.5. This is identical to that of P/PI (Equation 17.4), except for the s^1 terms.

$$\frac{P_F(s)}{P_C(s)} = \frac{K_{AF}\, s^3 + K_{VP}\, K_{VF}\, s^2 + K_{VP} K_{PP} s + K_{VP} K_{PI} K_{PP}}{\frac{J}{K_T} s^3 + K_{VP}\, s^2 + K_{VP}\, K_{PP}\, s + K_{VP}\, K_{PI}\, K_{PP}} \tag{17.5}$$

17.2.1 Tuning PI/P Loops

Tuning PI/P loops is similar to tuning P/PI loops. This section will discuss tuning using Experiment 17C, which, though not shown, is identical to Experiment 17B except that the PI and P controllers are exchanged. The velocity loop proportional gain, K_{VP}, is tuned to the same gain as in P/PI. The position loop proportional gain is then

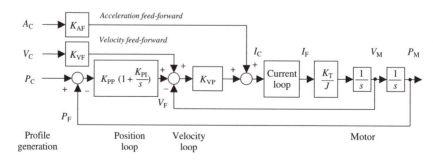

Figure 17-13. Block diagram of PI/P loops with feed-forward.

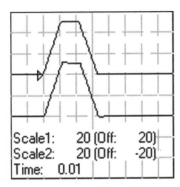

Figure 17-14. From Experiment 17C: step response of PI/P with full feed-forward.

increased to just below the value where overshoot occurs. Any nonzero value of the position loop integral gain will produce overshoot in the absence of acceleration feed-forward. As result, this process assumes that acceleration feed-forward is available (if not and if overshoot is not acceptable, K_{PI} will either be zeroed or set to a small enough value that the overshoot is unnoticeable).

The process from this point varies from that for P/PI. First, set the velocity feed-forward gain according to the needs of the application. Typically, values of 0.8–0.95 are the largest values that can be used without generating overshoot. Then adjust acceleration feed-forward to eliminate overshoot. Finally, add in K_{PI} until just before overshoot begins. When this process is applied to Experiment 17C, the result is as shown in Figure 17-14. Here, $K_{VP} = 0.72$, $K_{PP} = 170$, $K_{PI} = 20$, $K_{VF} = 0.91$, and $K_{AF} = 0.00146$. The result is nearly identical to the results for P/PI (see Figure 17-11). The bandwidth is 590 Hz.

The process here is conservative. Often, you can increase servo gains and adjust feed-forward gains to eliminate overshoot. However, use caution when tuning servo gains in the presence of high feed-forward gains because high feed-forward gains can mask stability problems.

17.3 PID Position Control

PID position control provides a single loop to control position; there is no velocity loop. The PID position loop deals with the 180° phase lag of the double-integrating plant by adding a derivative term in series with it. The structure is shown in Figure 17-15.

The transfer function for the PID controller is derived from Figure 17-15 and is shown as Equation 17.6:

$$\frac{P_F(s)}{P_C(s)} = \frac{K_{AF}\, s^3 + (K_{PD} + K_{VF})s^2 + K_{PP}\, s + K_{PI}}{\frac{J}{K_T}s^3 + K_{PD}\, s^2 + K_{PP}\, s + K_{PI}} \tag{17.6}$$

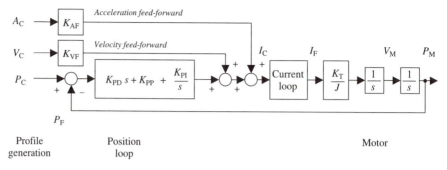

Figure 17-15. Block diagram of PID loops with feed-forward.

This structure is fundamentally different from PID control for single-integrating plants as presented in Chapter 6. For example, K_{PD} here behaves like K_P in Chapter 6 and K_{PP} here behaves like K_I in Chapter 6. This is because the second integrator in the plant adds one level of integration to the loop path, shifting the effect of each constant in the control law up one level of integration compared to systems with single-integrating plants.

17.3.1 Tuning the PID Position Controller

The principles of tuning a PID position controller will be demonstrated using Experiment 17D, which is shown in Figure 17-16. This model is similar to those of Experiments 17B and 17C, with two exceptions. First, there is a section that turns off the integral gain when motion is commanded; these blocks are in the lower right of the diagram. Second, there is a disturbance DSA, because disturbance response will be a topic of particular interest with the PID position controller. Both functions will be discussed in later sections.

Tuning the PID position controller follows the standard zone-based tuning process presented in Chapter 3. K_{PD} is the highest zone, K_{PP} is the middle zone, and K_{PI} is the lowest zone. (Note that this is different from Chapter 6, where the K_P gain was tuned first and K_D augmented it; this difference results because of the double-integrating plant.) To begin, zero the two lower-zone gains and increase K_{PD} until overshoot appears. As with the other loops, the value of 0.72 was selected as typical of the gains that would be seen in industrial applications. The value here can be increased, but this gain has identical behavior to K_{VP} in the cascaded loops, so using the same value provides a more direct comparison of the loop structures. Continuing, K_{PP} was set to 14 and K_{PI} was set to 500. The integral reset is disabled. The results are shown in Figure 17-17.

17.3.1.1 Selective Zeroing of the PID Integral Term

One of the difficulties with the PID position loop is that it is prone to overshoot; this forces the integral gain to lower values than might be desirable. A common cure for

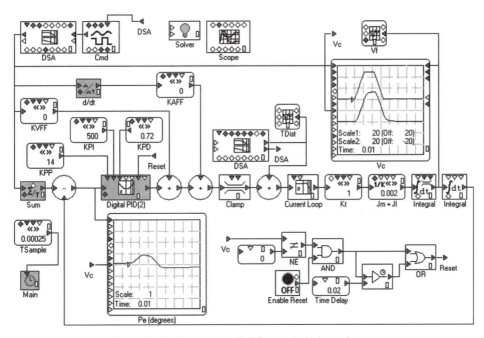

Figure 17-16. Experiment 17D: PID control with integral reset.

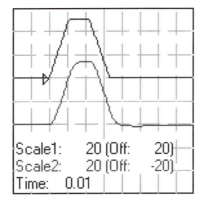

Figure 17-17. From Experiment 17D: trapezoidal velocity command response of a PID position loop.

this problem is to selectively enable the integral, zeroing it any time motion is commanded. This is the purpose of the seven blocks in the lower right of Figure 17-16. The not-equal comparison is high whenever commanded velocity is not zero. The inverted timer delays the comparison output 20 msec. The OR-gate is true during the move and for 20 msec afterwards. The *Live Button* "Enable Reset" allows you to disable this feature.

The zeroing of the integrator during motion allows substantial increase in integral gains, as is demonstrated in Figure 17-18. For this figure, the integral gain is increased from 500 to 2000. The resulting overshoot in Figure 17-18a is removed when the nonzero motion integral reset is enabled, as shown in Figure 17-18b.

17.3.2 Velocity Feed-Forward and the PID Position Controller

A fundamental difference between PID position loop and the cascaded loops is the role of velocity feed-forward. Compare the s^2 term in the numerators of the three transfer functions (Equations 17.4–17.6); in the cascaded forms, that term is $K_{VF} \times K_{VP}$, whereas it is $K_{VF} + K_{PD}$ in PID control. The most useful values of feed-forward in the cascaded form were between 0.6 and 0.9, yielding an s^2 term in the numerator of $0.6K_{VP}$ to $0.9K_{VP}$. Recognizing that K_{VP} and K_{PD} are, in fact, one and the same, the only way to get an equivalent velocity feed-forward in PID position loops is for K_{VF} to be between $-0.4K_{PD}$ and $-0.1K_{PD}$. Unfortunately, few position controllers allow negative values of K_{VF}. As a result, the implicit velocity feed-forward is normally higher than desired, causing the method to be more susceptible to over-shoot than the cascaded loops; this in turn forces K_{PP} low to avoid overshoot. This is especially true when acceleration feed-forward is unavailable.

If negative velocity feed-forward is available, set K_{VF} to -40% to -10% of K_{PD}; this provides the equivalent of 60% to 90% feed-forward in the cascaded loops. The result is shown in Figure 17-19a, where $K_{VF} = -0.228$. K_{PI} was set to zero and K_{PP} was retuned, yielding $K_{PP} = 85$. K_{PI} was then retuned to 500. This provides performance equivalent to the cascaded loops with velocity feed-forward only. Notice that K_{PP} is much higher here than it was with $K_{VF} = 0$, as predicted.

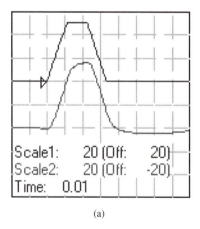
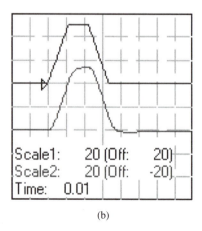

(a)   (b)

Figure 17-18. From Experiment 17D: Resetting integral during command motion removes overshoot. Command response with nonzero-motion integral reset (a) disabled and (b) enabled.

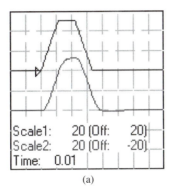 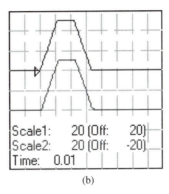

Scale1:	20 (Off:	20)
Scale2:	20 (Off:	-20)
Time:	0.01	

(a)

Scale1:	20 (Off:	20)
Scale2:	20 (Off:	-20)
Time:	0.01	

(b)

Figure 17-19. Trapezoidal velocity command response of PID with (a) velocity feed-forward and (b) velocity and acceleration feed-forward.

17.3.3 Acceleration Feed-Forward and the PID Position Controller

Tuning acceleration feed-forward in PID position loops is similar to the cascaded loops; it removes the overshoot added by velocity feed-forward (or by the lack of sufficient negative velocity feed-forward in PID controllers). Ideally, you set $K_{VF} = -K_{VP}$ and tune the PID gains. Then set K_{VF} to between -10% and -40% of K_{VP} (equivalent to 60%–90% in the cascaded loops). Adjust K_{AF} to remove overshoot. This process yielded $K_{VP} = 0.72$, $K_{PP} = 100$, $K_{PI} = 2000$, $K_{VF} = -0.082$, and $K_{AF} = 0.00146$; integral reset was disabled; enabling it allows higher values of K_{VI}. The result is shown in Figure 17-19b. Note that the response is nearly identical to the cascaded loops with full feed-forward (Figures 17–11 and 17–14).

For position controllers that do not support negative velocity feed-forward, setting K_{VF} to zero usually provides reasonable results. This is especially true when substantial acceleration feed-forward is applied so that the optimal value of K_{VF} is very near zero. These guidelines apply to tuning for aggressive point-to-point moves. Contouring applications are often tuned with full feed-forward ($K_{VF} = 0$ and $K_{AF} = J/K_T$) to eliminate steady-state following error.

17.3.4 Command and Disturbance Response for PID Position Loops

The command responses of the three cases discussed earlier are shown in Figure 17-20. The systems with negative velocity feed-forward (Figure 17-19a) and with velocity/acceleration feed-forward (Figure 17-19b) had command response bandwidths of 53 Hz and 580 Hz, respectively. These systems were tuned to be equivalent to the cascaded loops and provide similar performance (see Figure 17-12). The system with

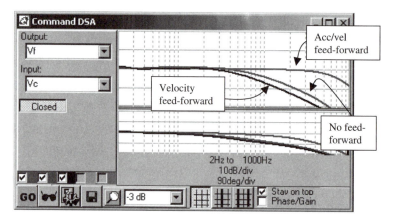

Figure 17-20. From Experiment 17D: command response Bode plot with no feed-forward, velocity-only feed-forward, and acceleration/velocity feed-forward.

no feed-forward is a different case. It has 50% higher bandwidth than the system with feed-forward. This results, of course, from the characteristic of PID position loops that zero velocity feed-forward behaves very much like cascaded systems with 100% velocity feed-forward. The bandwidth is high (80 Hz), but the proportional gain is low, 25, or about one-fourth of the values when negative velocity feed-forward was used.

The penalty of low proportional gain is the poorer disturbance response. This is demonstrated using the DSA "Disturbance DSA" in Experiment 17D (Figure 17-21). The disturbance responses of the two systems with feed-forward (Figure 17-19a and b) are identical because the loop gains are the same (feed-forward gains do not affect disturbance response). This is compared to the disturbance response of the non-feed-forward system (Figure 17-17). In the middle frequency ranges, where K_{PP} has the dominant effect, the disturbance response is clearly poorer for the zero-feed-forward system. At its worst, around 5 Hz, the response to a disturbance is about 10 dB, or about three times worse; that is, a torque disturbance at 5 Hz will generate three times more velocity and three times more position error in the non-feed-forward system. This results from the value of K_{PP}, which is about one-fourth the magnitude in non-feed-forward systems. The lesson should be clear: Excessive values of feed-forward, whether explicit gains or implicit in a control method, induce overshoot, forcing loop gains lower and degrading disturbance response.

17.4 Comparison of Position Loops

The three position loops presented here have many similarities. The transfer function and command/disturbance responses of all are about the same when tuned the same way. All the methods provide good performance. This stands to reason; control system users demand high-quality performance to remain competitive. Were one of the three methods significantly inferior to the others, it would quickly fall into disuse.

The equivalence of the methods can be seen in the transfer functions. Table 17-1 compares each of the coefficients in the numerator and denominator of the transfer function. Performance will be equivalent if the terms of the numerator and denominator are the same or nearly the same.

The gain sets from the three examples in this chapter that use acceleration and velocity feed-forward are shown for reference in Table 17-2. These gains are used to calculate the transfer function coefficients for each of those examples; the coefficients are shown in Table 17-3. Although the values do not correspond exactly, they are quite close. This is the expected case. If two systems respond in substantially the same way (as these examples did), it is likely they have similar transfer functions.

17.4.1 Positioning, Velocity, and Current Drive Configurations

The functions of the position loops described earlier are implemented jointly by the motion controller and the drive. For example, if the drive is configured as a current drive (that is, the position controller sends a current command to the drive), then the drive closes current loops and commutates the motor. At the other end of the spectrum is the positioning drive, in which case all loops are closed by the loop. Positioning drives are

TABLE 17-1 TRANSFER FUNCTION COEFFICIENTS FOR THREE POSITION CONTROL LAWS

Numerator	P/PI (Equation 17.3)	PI/P (Equation 17.4)	PID (Equation 17.5)
s^3 term	K_{AF}	K_{AF}	K_{AF}
s^2 term	$K_{VP} \times K_{VF}$	$K_{VP} \times K_{VF}$	$K_{PD} + K_{VF}$
s^1 term	$K_{VP}(K_{VI} \times K_{VF} + K_{PP})$	$K_{VP} \times K_{PP}$	K_{PP}
s^0 term	$K_{VP} \times K_{VI} \times K_{PP}$	$K_{VP} \times K_{PP} \times K_{PI}$	K_{PI}

Denominator	P/PI	PI/P	PID
s^3 term	J/K_T	J/K_T	J/K_T
s^2 term	K_{VP}	K_{VP}	K_{PD}
s^1 term	$K_{VP}(K_{VI} + K_{PP})$	$K_{VP} \times K_{PP}$	K_{PP}
s^0 term	$K_{VP} \times K_{VI} \times K_{PP}$	$K_{VP} \times K_{PP} \times K_{PI}$	K_{PI}

TABLE 17-2 TUNED GAINS FOR THREE EXAMPLE SYSTEMS

Loop type	Reference figure	High-frequency zone	Middle-frequency zone	Low-frequency zone	Velocity feed-forward	Acceleration feed-forward
P/PI	Figure 17-11	$K_{VP} = 0.72$	$K_{VI} = 20$	$K_{PP} = 140$	$K_{VF} = 0.91$	$K_{AF} = 0.00146$
PI/P	Figure 17-14	$K_{VP} = 0.72$	$K_{PP} = 170$	$K_{PI} = 20$	$K_{VF} = 0.91$	$K_{AF} = 0.00146$
PID	Figure 17-19b	$K_{PD} = 0.72$	$K_{PP} = 100$	$K_{PI} = 2000$	$K_{VF} = -0.082$	$K_{AF} = 0.00146$

TABLE 17-3 EVALUATION OF TABLE 17-1 FOR THE GAIN SETS OF TABLE 17-2

	P/PI	PI/P	PID
Numerator s^3 term	0.00146	0.00146	0.00146
Numerator s^2 term	0.655	0.655	0.638
Numerator s^1 term	114	122.4	100
Numerator s^0 term	2016	2448	2000
Denominator s^3 term	0.002	0.002	0.002
Denominator s^2 term	0.72	0.72	0.72
Denominator s^1 term	115	122.4	100
Denominator s^0 term	2016	2448	2000

sometimes stand-alone "single-axis controllers" that control simple machines. Other times, positioning drives are connected to a central controller through a digital communication link called a *field bus*; examples of such buses are DeviceNet, Profibus, and SERCOS. All loops discussed in this chapter can easily adapt to current or position drives.

A third drive configuration is called a *velocity drive*. In this case, the position loop is executed by a central positioner, producing a velocity command for the drive. Only the cascaded loops support this architecture. The velocity drive does bring benefits to some applications. With it, the drive can offload the positioner. Also, velocity drives typically have less phase lag than current drives, at least with microprocessor based drives. These drives convert an analog command to digital, typically every 100–200 μsec. If this delay is placed in the velocity command, as it would be for a system using a velocity drive, it becomes phase lag in the position loop. This short delay causes few problems for that loop. However, that same delay in a current drive becomes phase lag in the velocity loop, where it is more harmful.

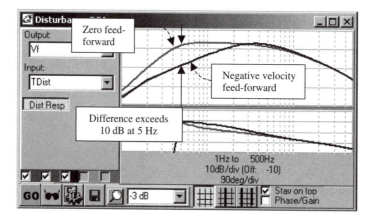

Figure 17-21. Disturbance response of PID position with and without feed-forward.

Velocity drives do have shortcomings. The servo gains, which are application dependent, are split; velocity loop gains are in the drive, and feed-forward and position loop gains are in the controller. This means that during commissioning, the human-interface software may need to communicate with two products (controller and drive) that may be different. Second, if the velocity command from the controller is analog, it will be sensitive to electrical noise; the command to a current drive is less sensitive to this noise.

17.4.2 Comparison Table

Perhaps the biggest difference is that in systems with a PID position loop that provides neither negative velocity feed-forward nor acceleration feed-forward, the tendency to overshoot can force K_{PP} lower, reducing disturbance response. There are several other differences that have been discussed in this chapter; they are shown in Table 17-4.

17.4.3 Dual-Loop Position Control

In most applications, the motor feedback device is used to calculate the load position. This calculation requires scaling to account for the transmission components, such as gear trains, belt/pulley sets, and lead screws. However, this practice suffers from inaccuracy because errors in transmission components distort the relationship between motor feedback and load position. Neither gear teeth nor threads on lead screws can

TABLE 17-4 COMPARISON OF POSITION LOOPS

P/PI	PI/P	PID
Strengths		
Integral gain does not imply overshoot, even when acceleration feed-forward is not used	There is no steady-state following error when K_{PI} is greater than zero	There is no steady-state following error when K_{PI} is greater than zero
Supports positioning, velocity, and current drive configurations	Supports positioning, velocity, and current drive configurations	Higher integral gains are possible when the integral is reset during moves; this function is commonly available
Weaknesses		
K_{VF} must $= 1$ to eliminate steady-state following error	Integral gain usually implies overshoot when acceleration feed-forward is not used	High tendency to overshoot when negative K_{VF} is not available, forcing K_{PP} lower than the equivalent gain would be with cascaded loops
		Does not support velocity drive configuration

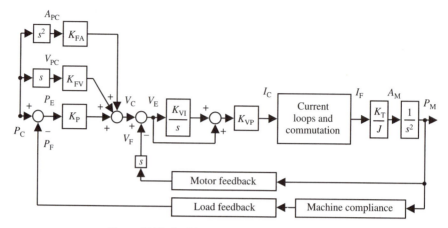

Figure 17-22. Dual-loop control applied to a P/PI loop.

be placed perfectly. In applications that require higher accuracy, a sensor can be placed directly on the load to be used in conjunction with the motor sensor. (See Section 12.4.7 for an alternative solution.)

Dual-loop position control uses two feedback devices: one from the motor and a second from the load. The motor feedback device is fast and thus ideal for closing the high-frequency zone (K_{VD} or K_{PD}). A load sensor measures load directly; inaccuracy in the transmission components does not affect its accuracy. However, because the load is connected to the motor through the compliant transmission, the load sensor does not react immediately to changes in motor position; the time lag between motor rotation and load movement makes the load sensor too slow to close a fast loop. The solution to this problem is to connect the load sensor to the position loop, where accuracy is most important, and to connect the motor sensor to the velocity loop, where concern for speed dominates; this is called *dual-loop control* and is shown in Figure 17-22. Note that dual-loop control can be implemented in PID position loops by using the motor feedback for the K_{PD} term and the load feedback for the K_{PP} and K_{PI} terms.

Beyond correcting for accuracy of transmission components, dual-loop control has other advantages. It makes the control system much less sensitive to backlash. Because the position loop is closed on actual load position, backlash (lost motion) between the motor and load is corrected by the position loop. Dual-loop control also provides the control laws access to the load position and, by calculation, the load velocity.

17.5 Bode Plots for Positioning Systems

As has been discussed throughout this book, measuring the frequency-domain response of the machine provides numerous benefits. Some manufacturers of motion-control equipment include an internal DSA as part of their products; the number of such products is small but growing. Even if the drive or positioner you are using does not provide a DSA, you may be able to make these measurements with a laboratory DSA. However, since most DSAs rely on analog signals, this alternative is most practical when using an analog velocity or current drive; this section will discuss how to generate Bode plots in that case. It begins with a technique for systems with velocity drives; after that, a technique for systems with current drives will be discussed.

Frequency-domain measurements should usually be made when the machine is moving. If the machine is at rest, stiction (Section 12.4.5) may be the primary physical effect and the measurements will sense the Dahl effect rather than the effects of load inertia. Inject a small offset velocity during measurements to avoid this problem. Of course, you should avoid saturation and other non-LTI behaviors.

17.5.1 Bode Plots for Systems Using Velocity Drives

If you are using a velocity drive with an internal DSA, obtaining the velocity Bode plot is usually easy, since the drive manufacturer provides this as built-in function. For example, Danaher Motion provides several such drives. If not and if you are using an analog velocity drive, a laboratory DSA can be connected to control the drive. In either case, this approach will measure the Bode plot of the velocity loop. Normally, this is sufficient because most performance issues are related to the velocity loop. For example, mechanical resonance and slow current loops directly affect the velocity loop. Figure 17-23 shows a DSA controlling a velocity drive.

To estimate gain and phase margin, the open-loop velocity plot can be derived from the closed loop. The closed-loop transfer function ($T_{VCL}(s)$) in Figure 17-23 is

$$V_C(s)/V_F(s) = T_{VCL}(s) = G(s)/(1 + G(s) \times H(s)) \qquad (17.7)$$

Equation 17.7 can be reformulated to provide the open-loop Bode plot. If $H(s)$ is assumed to be unity, Equation 17.7 becomes

$$T_{VCL}(s) = G(s)/(1 + G(s)) \qquad (17.8)$$

And, with a little algebra, $G(s)$, the loop gain when $H(s) = 1$, is seen to be

$$G(s) = T_{VCL}(s)/(1 - T_{VCL}(s)) \qquad (17.9)$$

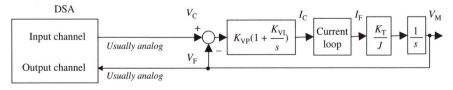

Figure 17-23. DSA measures closed-loop response of PI velocity controller.

To improve accuracy, factor in sample delay, which is typically about 1.5 T. Set $H(s) = e^{-1.5sT}$ in Equation 17.7. Also, note that the analog conversion of V_F may add one or two samples; contact the drive manufacturer to determine this. If so, be sure to advance the measured closed-loop function, for example, by e^{+sT} for one delay, to eliminate this lag; it should be eliminated because it is part of the DSA measurement but does not affect drive operation.

If you need the Bode plot of the position loop after you have the velocity loop plot, there are at least two alternatives. The first is to calculate it based on the measured velocity loop performance and the known position loop gains. Similar to Figure 17-2, the velocity controller, current loop, motor, and load can all be combined into a single block, $T_{VCL}(s)$. The remaining terms are the position loop and feed-forward gains, which can be determined from the controller. An alternative is to measure the open-loop gain of the position controller and to calculate the closed-loop response. This will be discussed in the next section.

17.5.2 Bode Plots for Systems Using Current Drives

When using current drives, you can measure the open-loop response of the position loop and then calculate the closed-loop response. Often, you will need to add a summing junction in the loop, as shown in Figure 17-24. The summing junction can be created with an op-amp circuit. Note that for systems with analog velocity drive, you can use a similar technique; however, the method of Section 17.5.1 is preferred because it is usually easier and it works well.

Notice in Figure 17-24 that the DSA has a separate generator signal; most DSAs have separate input and generator channels, but they are often connected, as is implicit in Figure 17-23. Here the generator signal must be separate so that the DSA can add an excitation signal to the current command and measure the loop gain, which is formed by

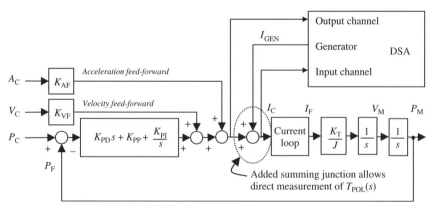

Figure 17-24. Adding an analog summing junction in the position loop allows measurement of open-loop gain.

the current loop, the motor, and the control law. Now the open-loop transfer function of the system can be measured, including the current controller, the plant, and all the loop gains. This does not, of course, include the effects of feed-forward gains.

If no feed-forward is used in the system, the closed-loop command response can be calculated from the open-loop, $T_{POL}(s)$:

$$T_{PCL}(s) = T_{POL}(s)/(1 + T_{POL}(s)) \qquad (17.10)$$

If feed-forward is used, the closed-loop response can be calculated by adding terms to Equation 17.10.

$$T_{PCL-FF}(s) = (K_{AF} \times s^3 + K_{VF} \times s^2 + T_{POL}(s))/(1 + T_{POL}(s)) \qquad (17.11)$$

Incidentally, torque-disturbance response can be measured from the system of Figure 17-24 almost directly:

$$T_{DIST}(s) = V_M(s)/(K_T \times I_{GEN}(s)) \qquad (17.12)$$

where K_T is required to convert the excitation current to a torque signal.

17.6 Questions

1. Using Experiment 17A, practice tuning by first setting K_{VP} as high as possible with the velocity loop responding to a square wave without overshooting. Retune K_{VI} and K_{PP} as discussed in Section 17.1.2. Be sure to turn "Position Loop" off when tuning or measuring the velocity loop.
   a. What are the values of K_{VP}, K_{VI}, and K_{PP}?
   b. What is the velocity loop bandwidth?
   c. What is the position loop bandwidth?
2. Transfer the values from Question 1a into Experiment 17B. Set the feed-forward gains to the maximum values without generating overshoot in response to the default trapezoidal command in the experiment.
   a. What are the values of K_{VF} and K_{AF}?
   b. What is the position loop bandwidth with feed-forward?
3. Transfer the values from Question 1a into Experiment 17B. Set the velocity feed-forward gain to 0.5 and the acceleration feed-forward gain to zero. Reduce position loop gain to eliminate the overshoot caused by velocity feed-forward.
   a. What are the values of K_{VF} and K_{PP}?
   b. What is the position loop bandwidth with feed-forward?
4. Use Experiment 17B to investigate the sensitivity of the P/PI loop with respect to changing inertia.

      a. Tune the system without feed-forward, as in Question 1. What is the apparent effect on performance when inertia falls from 0.002 kg-m^2 to 0.001 kg-m^2?

      b. Now use the values with velocity and acceleration feed-forward, as in Question 2. What is the apparent effect on performance when inertia falls from 0.002 kg-m^2 to 0.001 kg-m^2? How much peaking is induced at high frequency?

      c. Now use the values with just velocity feed-forward, as in Question 3. What is the apparent effect on performance when inertia falls from 0.002 kg-m^2 to 0.001 kg-m^2?

      d. What conclusion would you draw?

5. Table 17-3 shows that the gains resulting from the tuning procedures of Chapter 17 turned out to provide nearly equivalent transfer functions for P/PI, PI/P, and PID position control. This could also have been done analytically.

      a. Using the relationships shown in Table 17-1, convert the three loop gains (K_{VP}, K_{VI}, and K_{PP}) from P/PI compensation to make exact equivalence in the *denominator* of PI/P.

      b. Repeat for PID control.

      c. Convert the two feed-forward gains from P/PI compensation to make exact equivalence in two of the *numerator* terms of PI/P.

      d. Repeat for PID control.

      e. Compare the numerator terms. Which, if any, are different, and by how much?

Chapter 18

Using the Luenberger Observer in Motion Control

This chapter will discuss several applications of the Luenberger observer in motion-control systems. First, applications that are likely to benefit from observer technology are presented. The chapter then presents the use of the observer to improve two areas of performance: command response and disturbance response. Command and disturbance response are improved because observed feedback, with its lower phase lag, allows higher loop gains. Disturbance response is further improved through the use acceleration feedback. As with earlier chapters, key points will be demonstrated with software experiments.

18.1 Applications Likely to Benefit from Observers

Here are five characteristics of applications that will benefit from the Luenberger observer:

- The need for high performance
- The availability of computational resources in the controller
- The ability of the average user to install and configure the system
- The availability of a highly resolved position feedback signal
- The presence of phase lag in the position or velocity feedback signals

The first two guidelines are critical — without the need for an observer or a practical way to execute observer algorithms, the observer would not be chosen. The remaining guidelines are important. The more of these guidelines an application meets, the more likely the observer can substantially improve system performance.

18.1.1 Performance Requirements

The first area to consider is the performance requirements of the application. Machines that demand rapid response to command changes, stiff response to disturbances, or both will likely benefit from an observer. The observer can reduce phase lag in the servo loop, allowing higher gains, which improve command and disturbance response. For machines where responsiveness is not an issue, there may be little reason to use an observer.

18.1.2 Available Computational Resources

The second factor to consider is the availability of computational resources to implement the observer. Observers almost universally rely on digital control. If the control system is to be executed on a high-speed processor such as a DSP, where computational resources sufficient to execute the observer are likely to be available, an observer can be added without significant cost burden. In addition, if digital control techniques are already employed, the additional design effort to implement an observer is relatively small. However, if the system uses a simple analog controller, putting in place a hardware structure that can support an observer will require a large effort.

18.1.3 Controls Expertise in the User Base

Another factor to consider is the user base — the engineers and technicians who purchase, install, and maintain the equipment. Observers require some level of controls expertise for installation and configuration. The user base must be capable of understanding the features of an observer if it is to provide benefit.

18.1.4 Sensor Noise

Luenberger observers are most effective when the position sensor produces limited noise. Sensor noise is often a problem in motion-control systems. Noise in servo systems comes from two major sources: EMI generated by power converters and transmitted to the control section of the servo system, and resolution limitations in sensors, especially in the feedback sensor. EMI can be reduced through appropriate wiring practices [67] and through the selection of components that limit noise generation.

Resolution noise from sensors is difficult to deal with. Luenberger observers often exacerbate sensor-noise problems. The availability of high-resolution feedback sensors raises the likelihood that an observer will substantially improve system performance.

18.1.5 Phase Lag in Motion-Control Sensors

The two predominant sensors in motion-control systems are incremental encoders and resolvers. Incremental encoders respond to position change without substantial phase lag. As discussed in Chapter 14, resolver signals are commonly processed with a tracking loop, which generates substantial phase lag in the position signal. Because resolvers produce more phase lag, their presence makes it more likely that an observer will substantially improve system performance. The sine encoder is often processed in a manner similar to the way resolver signals are processed. As was the case with the resolver, the tracking loop can inject substantial phase lag in the sine encoder signals.

Independent of the feedback sensor, most motion-control systems generate phase lag in the control loop when they derive velocity from position. Velocity is commonly derived from position using simple differences. It is well known to inject a phase lag of half the sample time. This phase lag also provides an opportunity for the Luenberger observer to improve system performance.

18.2 Observing Velocity to Reduce Phase Lag

The remainder of this chapter will cover ways to use Luenberger observers in motion systems. This section discusses the use of observers to reduce phase lag within the servo loop. Removing phase lag allows higher control law gains, improving command and disturbance response. A common source of phase lag in digital velocity controllers is the use of simple differences to derive velocity from position. In a resolver-based system, additional phase lag often comes from the method employed to retrieve position information from the resolver.

18.2.1 Eliminate Phase Lag from Simple Differences

The use of simple differences to derive velocity from position is common in digital motor controllers that rely on a position sensor for feedback. The difference between the ideal 90° and simple differences at any given frequency is equivalent to half the sample time, as was discussed in Section 5.6.2.

18.2.1.1 Form of Observer

The observer structure that eliminates the phase lag generated by simple differences is shown in Figure 18-1. The feedback current, I_F, is scaled by K_T to produce *electromagnetic torque*, T_E, in the physical system and by K_{TEst} to produce estimated electromagnetic torque, T_{EEst}. The term *electromagnetic torque* describes the torque generated from the windings of the motor. In the actual system, the electromagnetic torque is summed with the disturbance torque, $T_D(s)$, to form the total torque. Total

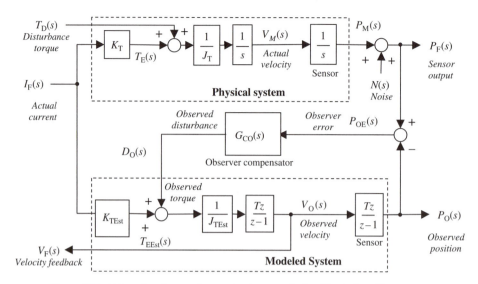

Figure 18-1. Using a Luenberger observer to observe velocity with a fast position sensor.

torque is divided by inertia to produce motor acceleration; acceleration is integrated twice to produce velocity and position. The model system is similar, except integration, $1/s$, is replaced by its digital equivalent, $Tz/(z-1)$, and the observed disturbance is used in the calculation of observed torque.

The second integration of both the model and the actual system is considered part of the sensor rather than part of the motor. Because the state of concern is velocity, the assumption in this observer is that the velocity is the plant output and that the integration stage that creates position is an artifact of the measuring system. This observer assumes that the position sensor provides feedback without substantial phase lag. This is a reasonable assumption for an incremental encoder. As was discussed in Chapter 10, phase lag incurred when measuring the actual state can be reduced using an observer. In this case, the goal of the observer is to remove the phase lag of simple differences, which normally corrupts the velocity measurement.

18.2.1.2 Experiment 18A: Removal of Phase Lag from Simple Differences

The benefits of the removal of phase lag from simple differences are demonstrated in Experiment 18A. The block diagram of that model is shown in Figure 18-2.[1] This model will be reviewed in detail. The velocity command (V_C) comes from a command generator (*Command*) through a dynamic signal analyzer (*Command Dsa*). The

---

[1]Occasionally, Experiment 18A and some of the other models in this chapter will become unstable. This usually occurs when changing a control loop gain. In some cases, restoring the model to its original settings will not restore stability. If this happens, click *File, Zero All Stored Outputs*.

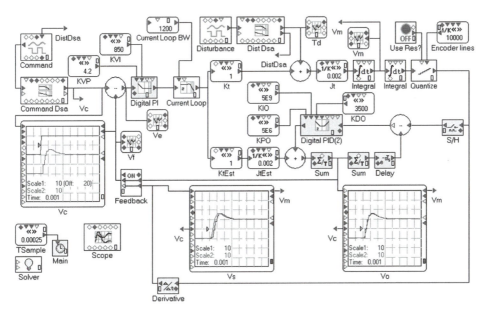

Figure 18-2. Experiment 18A: investigating the effects of reducing phase lag from simple differences.

command velocity is compared to the feedback velocity and fed to a PI controller (*Digital PI*), which is configured with two *Live Constants*, K_{VP} and K_{VI}. These parameters are set high ($K_{VP} = 4.2$, $K_{VI} = 850$), yielding a bandwidth of about 660 Hz and a 25% overshoot to a step command; this is responsive performance for the modest 4-kHz sample rate used in this model. As will be shown, the observer is necessary to attain such a high performance level.

The output of the velocity controller is a current command, which feeds the current controller (*Current Loop*). The current controller is modeled as a two-pole low-pass filter with a bandwidth of 1200 Hz set by a constant (*Current Loop BW*), and a fixed damping of 0.7. This is consistent with current-loop controllers used in the motion-control industry.

The output of the current loop is torque-producing current. It feeds the motor through the torque constant (K_T) to create electromagnetic torque. Simultaneously, a disturbance generator (*Disturbance*) feeds a second DSA (*Dist Dsa*) to create disturbance torque (T_D). The two torque signals are summed to create total torque. The total torque is divided by the total inertia (J_T) to produce motor acceleration. (Note that the J_T block is an *Inverse Live Constant*, indicating the output is the input divided by the constant value.) This signal is integrated once to create motor velocity (V_M) and again to create motor position.

There is an optional resolution, which is unused by default. To enable resolution, double-click on the *Live Button* "Use Res?". This model assumes an instantaneous feedback signal from the position sensor, which is consistent with the operation of encoder systems. Accordingly, the units of resolution are encoder lines, where each line translates into four counts via ×4 *quadrature*, a common method used with optical encoders (Section 14.2). The scaling to convert encoder lines to radians (SI units of angle) is implemented with the *Mult* node of the *Live Constant* named *Encoder Lines* being set to $4/2\pi$ or 0.6366, where the 4 accounts for quadrature and the 2π converts revolutions to radians.

The position signal feeds the observer loop, which uses a PID control law (*Digital PID (2)*) configured with the *Live Constants* K_{IO}, K_{PO}, and K_{DO}. In the ideal case, the output of the PID controller is equal to acceleration caused by the disturbance torque. The PID output is summed with the power-converter path of the observer, which is the output current scaled by the estimated torque constant (K_{TEst}) and divided by the estimated total inertia (J_{TEst}). The result is summed twice, approximating the two integrals of the physical motor. The observer produces the observed velocity output, V_O, which is displayed on the rightmost *Live Scope*. The delay of one step is added to allow the observer to be constructed, since the loop must have some starting point.

One difference between the block diagram of Figure 18-1 and the model of Figure 18-2 is that the observer has been reconfigured to remove the effects of inertia from the observer loop. Notice that J_{TEst} is directly after K_{TEst} and outside the observer loop. This implies that the units of the observer PID output are acceleration, not torque as in Figure 18-1. This will be convenient because changing estimated inertia will not change the tuning of the observer loop; without this change, each time estimated inertia is varied, the observer must be retuned. This observer structure will be used in experiments throughout this chapter.

The model for Experiment 18A includes three *Live Scopes*. At left is a display of the step response: the command (above) is plotted against the actual motor velocity. Along the bottom are two scopes:

> Center: the actual motor velocity, V_M, vs. the sensed velocity, V_S
> Right: V_M vs. the observed velocity, V_O

The sensed velocity, V_S, lags the motor velocity about 250 μsec because of the phase lag injected by simple differences. As expected, the observed velocity shows no signs of phase lag.

Experiment 18A allows the selection of either of the velocity signals as the feedback signal for the control loop. Double-click on the *Live Switch* Feedback to alternate between observed (default) and sensed velocity. This feature will be used to show the benefit of using the observer to measure velocity.

The key variables used in this and other experiments in this chapter are detailed in Table 18-1. All velocity variables display in RPM.

The results of Experiment 18A are shown in Figure 18-3. There are two plots, one for the system configured using each of V_S and V_O as velocity loop feedback. The sensed feedback signal has substantial phase lag, which produces ringing in the step response. The observed velocity has conservative margins of stability. The difference in stability between sensed and observed feedback is due wholly to the phase lag induced by simple differences. Note that the velocity loop gains ($K_{VP} = 4.2$, $K_{VI} = 850$) were

TABLE 18-1 KEY VARIABLES OF EXPERIMENTS 8A–8F

Variable	Description
Current Loop BW	Bandwidth of current loop, which is modeled as a two-pole low-pass filter with a damping ratio of 0.7
J_T, J_{TEst}	Actual and estimated total inertia of motor and load. This model assumes a rigid coupling between motor and load
K_{PO}, K_{IO}, K_{DO}	Proportional, integral, and derivative gains of observer
K_T, K_{TEst}	Actual and estimated torque constant of motor. The torque output of the motor windings is the actual current (I_F) multiplied by the torque constant.
K_{VP}, K_{VI}	Proportional and integral gains of velocity loop
T_D	Disturbance torque
V_C	Command velocity
V_E	Velocity error, $V_C - V_F$
V_F	Feedback velocity, the velocity signal used to close the velocity loop, which can be set to V_S or V_O at any time through the *Live Switch* "Feedback"
V_M	Actual motor velocity
V_O	Observed motor velocity
V_S	Sensed velocity, the velocity derived from the feedback position using simple differences

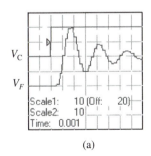 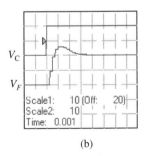

(a)   (b)

Figure 18-3. Results of Experiment 18A: Square wave command (above) and response with feedback as (a) sensed velocity (V_S) and (b) (V_O) Luenberger observer output. Gains are $K_{VP} = 4.2$ and $K_{VI} = 850$ in both cases.

adjusted to maximize their values when observed velocity is used for feedback; the same values are used in both cases of Figure 18-3.

The improvement in stability margins can also be seen in Bode plots from Experiment 18A, as in Figure 18-4. Here, the use of the observed velocity reduces peaking by more than 10 dB, allowing much higher command response than can be supported by sensed feedback. Note that these Bode plots are generated using the *Command Dsa.*

18.2.1.3 Experiment 18B: Tuning the Observer

Experiment 18B isolates the observer to focus on tuning the observer loop. The model is shown in Figure 18-5. The motor and the velocity controller have been removed. The velocity command is summed (i.e., digitally integrated) to produce the position command. That command is fed into the observer loop. The observer feed-forward path has been removed so that only the observer loop remains. The model includes a DSA that can show the frequency-domain response of the observer loop. A *Live Scope* shows the step velocity response of the observed velocity. Finally, the variables V_{OE} and V_{O2} are used to display the open-loop response of the observer; these variables will be discussed later in this section.

The *Live Scope* in Figure 18-5 shows slight ringing for the selected tuning values ($K_{PO} = 5 \times 10^6$, $K_{IO} = 5 \times 10^9$, $K_{DO} = 3500$). These values were determined experi-

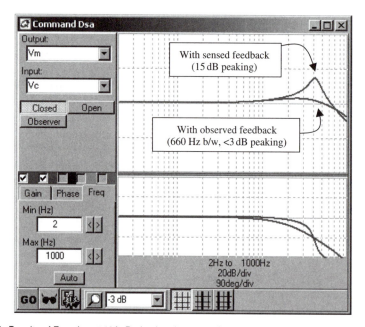

Figure 18-4. Results of Experiment 18A: Bode plot of command response with sensed feedback and observed (V_O) feedback. Gains are $K_{VP} = 4.2$ and $K_{VI} = 850$ in both cases.

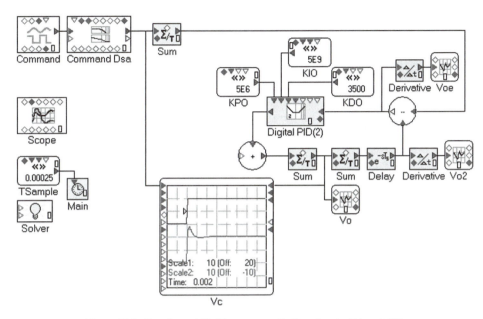

Figure 18-5. Experiment 18B: Observer used in Experiments 18A and 18E.

mentally, tuning one observer gain at a time. The step response of the observer (V_O) for each of the three gains is shown in Figure 18-6. K_{DO} is set just below where overshoot occurs. K_{PO} is set for about 25% overshoot; K_{IO} is set for a small amount of ringing.

The frequency response of the standard observer velocity (V_O) is shown in the Bode plot of Figure 18-7. The bandwidth of the observer is greater than 1000 Hz. In fact, the bandwidth is too great to be measured by the DSA, which samples at 4 kHz. There is 4 dB of peaking, verifying that the observer tuning gains are moderately aggressive.

Plotting the open-loop response of the observer loop brings up some interesting modeling issues. First, the loop command, as generated by the waveform generator and DSA, is summed to create the position command to the observer. A detail of this

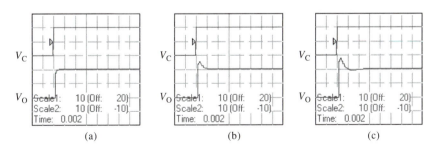

Figure 18-6. Step response of V_O in the tuning of the observer of Experiment 18B and viewing the standard observer output. (a) $K_{DO} = 3500$; (b) add $K_{PO} = 5 \times 10^6$; (c) add $K_{IO} = 5 \times 10^9$.

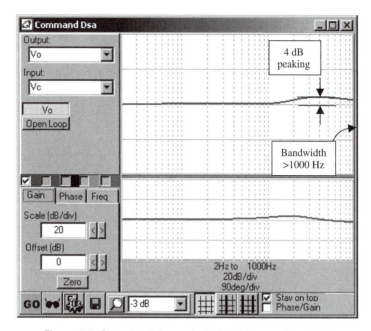

Figure 18-7. Bode plot of observed velocity (V_O) in experiment 18B.

feature, found in Figure 18-5, is shown in Figure 18-8. The implication is that the command is a commanded velocity, which is then integrated to create the commanded position. This is required to keep the position command continuous over time. Large step changes in position create enormous swings in velocity and thus can cause saturation in the observer. This structure avoids this problem.

A second feature of this model is the use of velocity signals (as opposed to position signals) to provide a Bode plot of the open-loop transfer function of the observer. This is done because the method used to create Bode plots in *Visual ModelQ*, the fast Fourier transform, or FFT, requires the start and end values of each signal being measured to be identical. The position generated from integration of the DSA random number generator does not necessarily meet this requirement. However, the model is configured to start and stop the Bode plot excitation at zero speed, so the velocities, which are the derivatives of the position signals, always start and stop at zero. Velocity signals can be used in place of position signals because the transfer function of two position signals is the same as the transfer function of their respective velocities. For example,

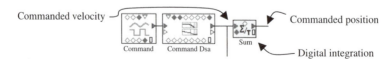

Figure 18-8. Detail of observer command for Experiment 18B (taken from Figure 18-5).

$P_M(s)/P_C(s) = V_M(s)/V_C(s)$. In the s-domain this would be equivalent to multiplying the numerator and denominator by s (s being differentiation), which would produce no net effect. So the open-loop transfer function of the observer, shown in a dashed line in Figure 18-9, is observer position error to observed position; however, that is identical to the derivative of both of those signals (shown by the blocks "Derivative"), which produce V_{OE} and V_{O2}. The DSA is preset to show the open loop as V_{O2}/V_{OE}.

The open-loop Bode plot of the observer is shown in Figure 18-10. The gain crossover is at about 700 Hz, and the phase margin is about 42°. There are two phase

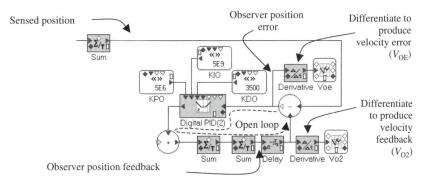

Figure 18-9. Detail of signals for Bode plot from Experiment 18B (taken from Figure 18-5).

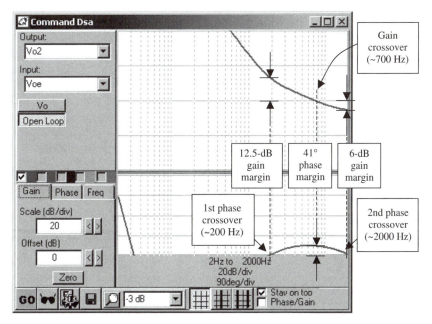

Figure 18-10. Bode plot of observer open-loop gain and phase showing margins of stability.

crossover frequencies, one at about 200 Hz and the other at about 2000 Hz. The gain margin is 12 dB at 200 Hz but only 6 dB at 2000 Hz. The crossover at 200 Hz generates the peaking in Figures 18-7 and the slight ringing in Figure 18-6c.

The crossover at 200 Hz results because the low-frequency open-loop phase is $-270°$, $-90°$ from each of the two integrators in the observer loop (marked *Sum*) and another $-90°$ from the integral gain in the PID controller, which dominates that block at low frequency. The gain margin at 200 Hz is set almost entirely by the integral gain, K_{IO}. The interested reader can verify this using Experiment 18B (see Question 18-1).

18.2.2 Eliminate Phase Lag from Conversion

The observer of Figure 18-1 is designed with the assumption that there is no significant phase lag induced in the measurement of position. This is valid for incremental (A-quad-B) encoders, but not for most resolver and sine-encoder conversion methods. Unlike incremental encoders, the conversion of resolver and sine-encoder inputs normally injects significant phase lag. For these sensors, the model needs to be augmented to include the phase lag induced by the conversion of feedback signals. Such an observer is shown in Figure 18-11. Here, the model system sensor includes the R-D converter transfer function of Equation 14.1 as $G_{RDEst}(z)$. Of course, the model of the physical system has also been modified to show the effects of the actual R-D converter on the actual position, $P_M(s)$, using $G_{RD}(s)$.

The observer of Figure 18-11 assumes the presence of a hardware R-D converter. Here, the process of creating the measured position takes place outside of the digital

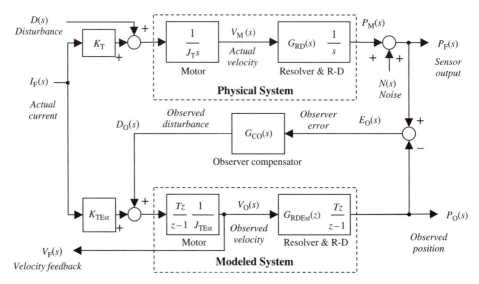

Figure 18-11. Using a Luenberger observer to observe velocity with resolver feedback.

control system. When the process of R-D conversion is done in the digital controller [52], the observer can be incorporated into the R-D converter, as is discussed in Ref. 26.

18.2.2.1 Experiment 18C: Verifying the Reduction of Conversion Delay

The effect of conversion delay is evaluated using Experiment 18C, as shown in Figure 18-12. This experiment uses a two-pole filter as a model of R-D conversion both in the physical system (the block *R-D* at the upper right of the figure) and in the observer model (the block *R-D Est* at the center right of the figure). The filter used to simulate the R-D converter is a low-pass filter with the form

$$R-D(s) = \frac{2\zeta\omega_N + \omega_N^2}{s^2 + 2\zeta\omega_N + \omega_N^2} \tag{18.1}$$

This is an *LPF2A* block in *Visual ModelQ*. This filter has the same form as Equation 18.1. The filter used in the observer, *R-D Est(z)*, is the *z*-domain equivalent of *R-D(s)*.

The observer compensator must change from PID to PID-D^2, where the D^2 indicates a second-derivative term. This second-derivative term is necessary because

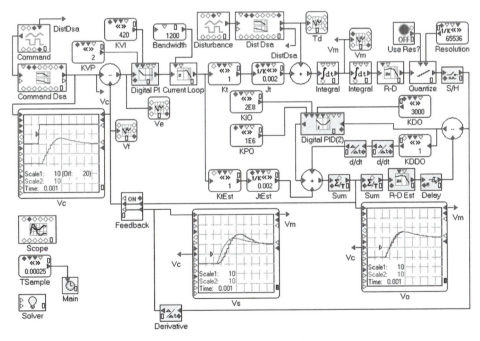

Figure 18-12. Experiment 18C: reducing phase lag from the R-D converter.

the *R-D Est* block, being a low-pass filter, adds substantial phase lag. The phase lead added by the derivative term improves stability margins, allowing higher observer gains. Without the second derivative, the observer gains are limited to low values, as can be verified using Experiment 18D, which will be discussed shortly. In Figure 18-12, the scaling of the second derivative is named K_{DDO}; this term is added to the observer PID controller output.

As with Experiment 18A (Figure 18-2), Experiment 18C includes three *Live Scopes*. At center left is a display of the step response: The command (above) is plotted against the actual motor velocity. Along the bottom are two scopes: the sensed velocity (V_S) and the observed velocity (V_O); each are plotted against motor velocity (V_M).

The sensed velocity, V_S, lags the motor velocity by about two-thirds of a division, much more than the lag seen in Experiment 18A. The phase lag here comes from two sources: simple differences, which Experiment 18A also displayed, and that from R-D conversion, which is the dominant source of phase lag in this model. As with Experiment 18A, Experiment 18C allows the selection of either velocity signal as feedback. Double-click on *Feedback* any time to change the feedback signal.

The benefits of using observed velocity feedback are readily seen by changing the control loop feedback source via the *Live Switch* named *Feedback*. By default, the source is the observed velocity, V_O. By double-clicking on *Feedback*, the feedback source will change to the sensed velocity, V_S, which is the simple difference of the output of the R-D converter. The step response of the system with these feedback signals is shown in Figure 18-13; in both cases, the control loop PI gains are the same: $K_{VP} = 2$, $K_{VI} = 420$. These values are similar to those used in Experiment 18A except that K_{VI} was reduced to increase stability margins. The results are that the system using V_S is unstable but that the observed signal produces reasonable margins of stability.

The Bode plot of command response for the gains $K_{VP} = 2$ and $K_{VI} = 420$ is shown in Figure 18-14. It confirms the results of Figure 18-13. The system based on sensed velocity feedback has over 20 dB of peaking. In fact, the value K_{VP} had to be reduced slightly (from 2.0 to 1.85) to provide sufficient stability margins so that a Bode plot could be calculated as Bode plots cannot be generated from unstable systems. The case using observed feedback has a modest 2–3 dB of peaking. The dramatic difference in

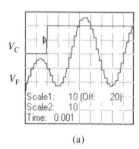
(a)

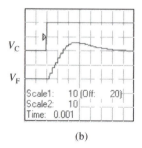
(b)

Figure 18-13. From Experiment 18C: Step response with two feedback signals. For both cases, $K_{VP} = 2$ and $K_{VI} = 420$. (a) Sensed feedback ($V_F = V_S$); (b) observed feedback ($V_F = V_O$).

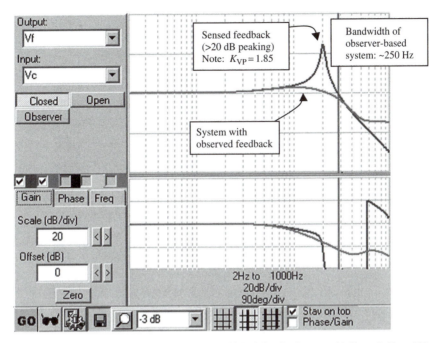

Figure 18-14. From Experiment 18C: command response with both feedback types with $K_{VP} = 2$, $K_{VI} = 420$, (K_{VP} was reduced to 1.85 for sensed feedback for stability).

these plots is caused by the ability of the observer to provide a feedback signal that does not include the delays of R-D conversion. In fact, for the observed feedback, the value of K_{VP} can be increased well above 2 while maintaining reasonable margins of stability.

For reference, the R-D converter filters are set for about a 400-Hz bandwidth, which is a typical bandwidth of R-D conversion in industrial systems. Notice that, for the systems based on observed feedback, the closed-loop bandwidth is about 250 Hz. Attaining such high bandwidth with a 400-Hz R-D conversion is very difficult without using observer techniques.

18.2.2.2 Experiment 18D: Tuning the Observer in the R-D–Based System

The R-D converter for Experiment 18C is tuned to about 400 Hz. In industry, R-D converters are commonly tuned to between 300 and 1000 Hz. The lower the bandwidth, the less susceptible the converter will be to noise; on the other hand, higher-bandwidth-tuning gains induce less phase lag in the velocity signal. The bandwidth of 400 Hz was chosen as being representative of conversion bandwidths in industry. The response of the model R-D converter is shown in Figure 18-15. The configuration parameters of the filters *R-D* and *R-D Est* were determined by experimentation to achieve 400-Hz bandwidth with modest peaking: *Frequency* = 190 and *Damping* = 0.7.

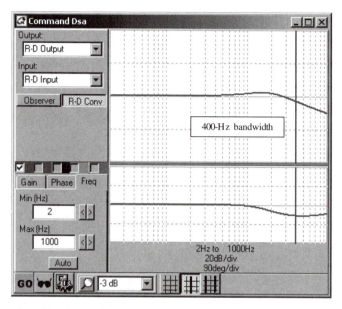

Figure 18-15. From Experiment 18D: Bode plot of R-D converter response showing 400-Hz bandwidth.

The process to tune the observer is similar to the process used in Experiment 18B, with the exception that a second-derivative term, K_{DDO}, is added. Experiment 18D, shown in Figure 18-16, isolates the observer from Experiment 18C, much as was done in Experiment 18B for the system of Experiment 18A.

The process to tune this observer is similar to that used to tune other observers. First, zero all the observer gains except the highest-frequency gain, K_{DDO}. Raise that gain until

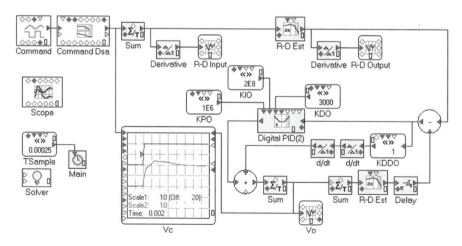

Figure 18-16. Experiment 18D: tuning the R-D converter.

a small amount of overshoot to a square wave command is visible. In this case, K_{DDO} is raised to 1, and the step response that results has a small overshoot, as shown in Figure 18-17a. Now, raise the next-highest-frequency gain, K_{DO}, until signs of low stability appear. In this case, K_{DO} was raised a bit higher than 3000 and then backed down to 3000 to remove overshoot. The step response is shown in Figure 18-17b. Next, K_{PO} is raised to 1×10^6; the step response, shown in Figure 18-17c, displays some overshoot. Finally, K_{IO} is raised to 2×10^8, which generates a slight amount of ringing, as shown in the *Live Scope* of Figure 18-16. The Bode plot of the response of the observer gains is shown in Figure 18-18. The bandwidth of the observer is approximately 880 Hz.

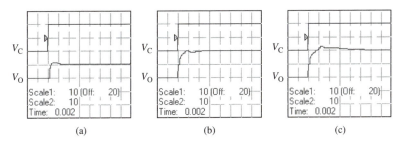

Figure 18-17. From Experiment 18D: step response in the tuning of the observer. (a) $K_{DDO} = 1$; (b) add $K_{DO} = 3000$; (c) add $K_{PO} = 1 \times 10^6$.

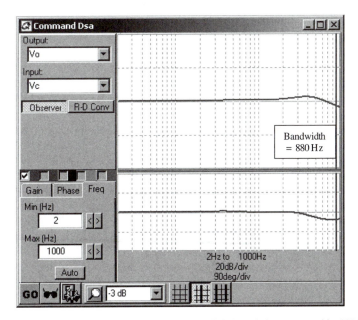

Figure 18-18. From Experiment 18D: Bode plot of response of R-D–based observer tuned for 880-Hz bandwidth ($K_{DDO} = 1$, $K_{DO} = 3000$, $K_{PO} = 1 \times 10^6$, $K_{IO} = 2 \times 10^8$).

18.3 Acceleration Feedback

Acceleration feedback can be used to improve disturbance rejection in motion control systems. Acceleration feedback works by slowing the motor in response to measured acceleration [24–26,48,55,57,59,70,87]. The acceleration of the motor is measured, scaled by K_{AFB}, and then used to reduce the acceleration (current) command. The larger the actual acceleration, the more the current command is reduced. K_{AFB} has a similar effect to increasing inertia; this is why acceleration feedback is sometimes called *electronic inertia* or *electronic flywheel* [58,59]. The idealized structure is shown in Figure 18-19 [24].

The effect of acceleration feedback is easily seen by calculating the transfer function of Figure 18-19. Start by assuming current loop dynamics are ideal: $G_{PC}(s) = 1$. Applying the $G/(1 + GH)$ rule to Figure 18-19 produces the transfer function

$$\frac{P_M(s)}{I_C(s)} = \frac{K_T/J_T}{1 + (K_T/J_T) \times (J_T/K_T)K_{AFB}} \times \frac{1}{s^2} \quad (18.2)$$

which reduces to

$$\frac{P_M(s)}{I_C(s)} = \frac{K_T/J_T}{1 + K_{AFB}} \times \frac{1}{s^2} \quad (18.3)$$

It is clear upon inspection of Equation 18.3 that any value of $K_{AFB} > 0$ will have the same effect as increasing the total inertia, J_T, by the factor of $1 + K_{AFB}$. Hence, K_{AFB} can be thought of as electronic inertia.

The primary effect of feeding back acceleration is to increase the effective inertia. However, this alone produces little benefit. The increase in effective inertia actually reduces loop gain, reducing system response rates. The benefits of acceleration feedback are realized when control loop gains are scaled up by the amount that the inertia increases, that is, by the ratio of $1 + K_{AFB}$. This is shown in Figure 18-20. Here, as K_{AFB} increases, the effective inertia increases, and the loop gain is fixed so that the stability margins and command response are unchanged.

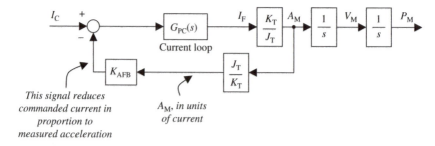

Figure 18-19. Idealized acceleration feedback.

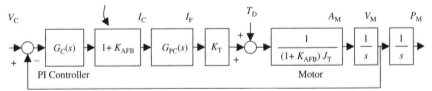

Scale the PI output up by the same amount effective inertia increases

Figure 18-20. Velocity controller based on idealized acceleration feedback.

Using the $G/(1 + GH)$ rule and allowing the $1 + K_{AFB}$ terms to cancel, the command response for the system of Figure 18-20 is

$$\frac{V_M(s)}{V_C(s)} = \frac{G_C(s) \times G_{PC}(s) \times K_T/(J_T s^2)}{1 + G_C(s) \times G_{PC}(s) \times K_T/(J_T s^2)} \tag{18.4}$$

or

$$\frac{V_M(s)}{V_C(s)} = \frac{G_C(s) \times G_{PC}(s) \times K_T/J_T}{s^2 + G_C(s) \times G_{PC}(s) \times K_T/J_T} \tag{18.5}$$

Notice that the command response is unaffected by the value of K_{AFB}. This is because the loop gain increases in proportion to the inertia, producing no net effect.

The disturbance response of Figure 18-20 is improved by acceleration feedback. Again, using the $G/(1 + GH)$ rule, the disturbance response is

$$\frac{V_M(s)}{T_D(s)} = \frac{K_T/[(1 + K_{AFB})J_T]}{s^2 + G_C(s) \times G_{PC}(s) \times K_T/J_T} \tag{18.6}$$

For the idealized case of Equation 18.6, the disturbance response is improved through the entire frequency range in proportion to the term $1 + K_{AFB}$. For example, if K_{AFB} is set to 10, the disturbance response improves by a factor of 11 across the frequency range. Unfortunately, such a result is impractical. First, the improvement cannot be realized significantly above the bandwidth of the power converter (current loop). This is clear upon inspection because the acceleration feedback signal cannot improve the system at frequencies where the current loop cannot inject current. The second limitation on acceleration feedback is the difficulty in measuring acceleration. Although there are acceleration sensors that are used in industry, few applications can afford either the increase in cost or the reduction in reliability brought by an addition sensor and its associated wiring. One solution to this problem is to use observed acceleration rather than measured. Of course, the acceleration can be observed only within the capabilities of the observer configuration; this limits the frequency range over which the ideal results of Equation 18.6 can be realized.

18.3.1 Using Observed Acceleration

Observed acceleration is a suitable alternative for acceleration feedback in many systems where using a separate acceleration sensor is impractical. Such a system is shown in Figure 18-21. The observed acceleration, A_O, is scaled to current units and deducted from the current command. The term $1 + K_{AFB}$ scales the control law output, as it did in Figure 18-20.

18.3.2 Experiment 18E: Using Observed Acceleration Feedback

Experiment 18E models the acceleration-feedback system of Figure 18-21 (see Figure 18-22). The velocity loop uses the observed velocity feedback to close the loop. A single term, K_{TEst}/J_{TEst}, is formed at the bottom center of the model to convert the current-loop output to acceleration units; to convert the observed acceleration to current units, the term $1 + K_{AFB}$ is formed via a summing block (at bottom center) and used to scale the control law output, consistent with Figure 18-20. An explicit clamp is used to ensure that the maximum commanded current is always within the ability of the power stage. The *Live Scope* shows the system response to a torque disturbance (the command generator defaults to zero).

The improvement of acceleration feedback is evident in the step-disturbance response, as shown in Figure 18-23. Recall that without acceleration feedback, the control law gains were set as high as was practical; the non-acceleration-feedback

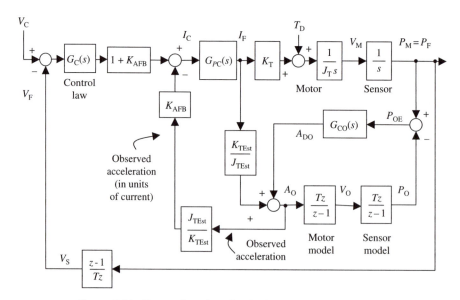

Figure 18-21. Observer-based acceleration feedback in motion systems.

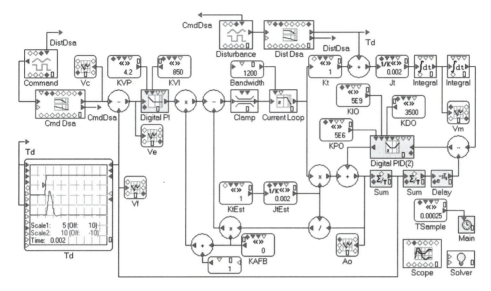

Figure 18-22. Experiment 18E: using observed acceleration feedback.

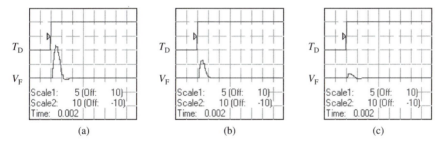

Figure 18-23. Response to a 5-Nm step disturbance without and with acceleration feedback: (a) without acceleration feedback ($K_{AFB} = 0.0$); (b) with minimal acceleration feedback ($K_{AFB} = 1.0$); (c) with more acceleration feedback ($K_{AFB} = 10.0$).

system took full advantage of the reduced phase lag of the observed velocity signal. Still, acceleration feedback produces a substantial benefit.

The Bode plot of the system with and without acceleration feedback is shown in Figure 18-24. The acceleration feedback system provides benefits at all frequencies below about 400 Hz. Figure 18-24 shows two levels of acceleration feedback: $K_{AFB} = 1.0$ and $K_{AFB} = 10.0$. This demonstrates that as the K_{AFB} increases, disturbance response improves, especially in the lower frequencies, where the idealized model of Figure 18-19 is accurate. Note that acceleration feedback tests the stability limits of the observer. Using the observer as configured in Experiment 18E, K_{AFB} cannot be much above 1.0 without generating instability in the observer. The problem

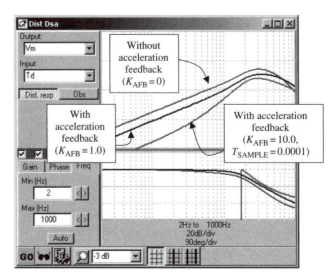

Figure 18-24. Bode plot of velocity loop disturbance response without ($K_{AFB} = 0$) and with ($K_{AFB} = 1.0$, 10.0) acceleration feedback.

is cured by reducing the sample time of the observer through changing the *Live Constant* named T_{SAMPLE} to 0.0001. This allows K_{AFB} to be raised to about 15 without generating instability. It should be pointed out that changing the sample time of a model is easy but may be quite difficult in a working system. See Question 18-5 for more discussion of this topic.

18.4 Questions

1. Verify that at the first phase crossover (200 Hz), the gain margin of the observer in Experiment 18B is set almost entirely by the value of the integral gain, K_{IO}. (*Hint: Take the open-loop plot gain with $K_{IO} = 0$ and compare to open-loop plot with default K_{IO} at and around 200 Hz.*)

2. Use Experiment 18A to compare the disturbance response of observer- and non-observer-based systems with identical control law gains. Compile and run the model. Change the input from command to disturbance as follows. Double-click on *Command* Wave Gen and set the amplitude to 0.0001 (this is set low enough to be insignificant but high enough to continue to trigger the *Live Scopes*). Similarly, set the amplitude of the *Disturbance* Wave Gen to 2.0. The *Live Scopes* are now showing the velocity signals in response to a disturbance.

   a. Compare the two velocity signals. Notice that the two *Live Scopes* at the bottom compare the two velocity signals to the actual motor velocity.

b. Using *Dist DSA*, measure and compare the disturbance response based on both of the signals using the default control-loop gains.

3. Using the *Live Scope* displays at the bottom of Experiment 18A, compare the noise sensitivity of the two velocities' signals. Enable resolution by double-clicking on the *Live Button* "Use Res?" so it displays "ON".

4. Use Experiments 18A and 18C to evaluate the robustness of an observer-based system in the presence of fluctuations of total inertia (J_T).

   a. Find the PM of the encoder system of Experiment 18A with nominal parameters, using V_O as feedback.

   b. Repeat with the total inertia reduced from 0.002 to 0.0014.

   c. Repeat part a for the resolver-based system of Experiment 18C.

   d. Repeat part b for the resolver-based system of Experiment 18C.

5. Use Experiment 18E to study the relationship between acceleration feedback and system sample time. Use the *Live Scope* display to see instability caused by excessive values of K_{AFB}.

   a. Make a table that shows the relationship between the maximum allowable K_{AFB} with the system-observer sample time (T_{SAMPLE}) set at 0.00025, 0.0002, and 0.00015.

   b. Compare the improvement in low-frequency disturbance response (say, 10 Hz) for each of the three settings to the baseline system ($K_{AFB} = 0$). (Set *TSample* to 0.0001s for this part; this improves the readings from the DSA without requiring you to change the FFT sample time.)

   c. Compare the results in part b to the ideal improvement gained by using acceleration feedback (see Equation 18.3).

Appendix A
Active Analog Implementation of Controller Elements

This appendix will discuss the implementation of many of the operations from Table 2-1 based on op-amps.

Integrator

Figure A-1 shows the standard op-amp integrator. The ideal transfer function of the circuit is

$$T(s) = \frac{1}{R_1 C_1 s} \tag{A.1}$$

Notice that Figure A-1 shows the output as −*output* since this is an inverting integrator. The zener diodes D_1 and D_2 clamp the op-amp output voltage; without these diodes, the output will be limited by the power supply voltages, or *rails*. Zener diodes *leak*, an undesirable characteristic where a small amount of current flows through the diode even though the applied voltage is well below the diode's zener ratings. This degrades the integrator accuracy when the input voltage is small. Capacitors also leak; usually low-leakage capacitors, such as monolithic ceramic capacitors, are used to reduce this behavior. Low-leakage op-amps are often selected for the same reason.

If the input has high-frequency content such as noise, consider a low-pass R-C as an input prefilter. High-frequency noise fed directly into an op-amp summing junction can generate offset on the output.

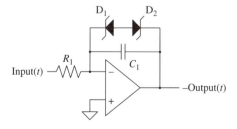

Figure A-1. Op-amp–based integration.

Differentiator

An op-amp–based differentiator is shown in Figure A-2; the transfer function is

$$T(s) = \frac{R_1}{\frac{1}{sC_1} + R_2} = \frac{1}{C_1 R_2 s + 1} sC_1 R_1 \tag{A.2}$$

This circuit includes a low-pass filter to reduce the noise that will inevitably occur in differentiation; the break frequency is at $1/(C_1 R_2)$ rad/sec.

The implementation of differentiators can require empirical optimization, where the break frequency of the filter is set as low as possible (i.e., raise the value of resistor R_2 as high as possible). When the break frequency is reduced, noise is reduced, but the circuit acts less like a differentiator near or above the break frequency because the low-pass filter adds phase lag. This reduces the phase lead of the transfer function further below the ideal 90° as the excitation frequency approaches the low-pass break frequency. For example, when the excitation frequency equals the low-pass break frequency, the phase lead of this network is just 45° because of the 45° phase lag from the low-pass filter.

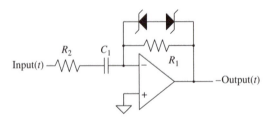

Figure A-2. Op-amp–based differentiator.

Lag Compensator

Figure A-3 shows an op-amp lag circuit. The transfer function of the circuit is

$$T(s) = K\left(\frac{\tau_Z s + 1}{\tau_P s + 1}\right) \tag{A.3}$$

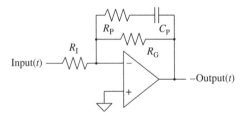

Figure A-3. Op-amp–based lag compensator.

where

$$K = R_G/R_I$$
$$\tau_Z = R_P C_P$$
$$\tau_P = (R_P + R_G)C_P$$

Lag compensators are commonly used with infinite DC gain (remove R_G), in which case they become PI compensators. The transfer function becomes

$$T(s) = K_P\left(1 + \frac{K_I}{s}\right) \tag{A.4}$$

where

$$K_P = R_P/R_I$$
$$K_1 = 1/(R_P C_P)$$

Lead Compensator

Figure A-4 shows an op-amp lead circuit. The ideal transfer function of the circuit is

$$T(s) = K\frac{\tau_Z s + 1}{\tau_P s + 1} \tag{A.5}$$

where

$$K = R_G/R_F$$
$$\tau_Z = (R_Z + R_F)C_Z$$
$$\tau_P = R_Z C_Z$$

Lead circuits are commonly used in series with feedback to advance the phase. The maximum phase advance from a lead network occurs at the geometric mean of $1/\tau_Z$

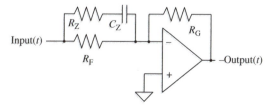

Figure A-4. Op-amp–based lead compensator.

and $1/\tau_P(\sqrt{1/\tau_Z \times 1/\tau_P})$; when using lead networks, you should place this frequency at the gain crossover frequency to maximize phase margin.

The ratio of τ_Z to τ_P is often called α. The larger α is, the more phase is added, but also the more high-frequency noise the lead circuit will pass. For analog circuits, α is usually limited to values of 4 to 10 because of noise considerations.

Lead-Lag Compensator

The lead and lag compensators of Figures A-3 and A-4 can be combined into a single lead-lag compensator, as shown in Figure A-5. The lead network operates only on the feedback; the lag network operates on both the command and the feedback. This network allows many degrees of freedom for tuning but requires only one op-amp. This method is discussed in Section 6.4.1.5.

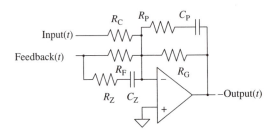

Figure A-5. Op-amp–based lead-lag compensator.

Sallen-and-Key Low-Pass Filter

Among the numerous op-amp-based two-pole low-pass filters available [16,34,72], the most popular may be the Sallen-and-Key filter shown in Figure A-6. The transfer function of this circuit is

$$T_{SK}(s) = \frac{\omega_N^2}{s^2 + 2\zeta\omega_N s + \omega_N^2} \tag{A.6}$$

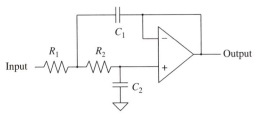

Figure A-6. Sallen-and-Key low-pass filter.

where

$$\omega_N^2 = \frac{1}{R_1 C_1 R_2 C_2}$$

$$2\zeta\omega_N = \frac{R_1 + R_2}{R_1 R_2 C_1}$$

Adjustable Notch Filter

There are numerous active analog notch filters. One practical example is shown in Figure A-7. This filter allows the adjustment of the notch frequency with a trimpot (R_4). The transfer function is:

$$T(s) = \frac{s^2 + \omega_N^2}{s^2 + 2\zeta\omega_N s + \omega_N^2} \tag{A.7}$$

Set the filter components as follows:

1. Assume a setting for the trimpot R_4 below 50% ($x \approx 0.25$).
2. Set $R_1 = R_2$.
3. Set $R_3 = R_4$.
4. Select the desired ζ of the filter (usually between 0.2 and 0.5).
5. Select the desired notch frequency, F_C.
6. Select $C_1 = \zeta/(R_4 \times 2\pi \times F_C)$.
7. Select $C_2 = 1/\{\zeta R_4 \times 2\pi \times F_C x(1 - x)\}$.
8. Adjust F_C, the actual center frequency, using R_4.

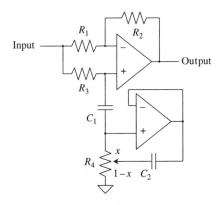

Figure A-7. Analog notch filter.

Appendix B
European Symbols for Block Diagrams

This appendix lists the symbols for the most common function blocks in formats typically used in North America and in Europe. Block diagram symbols in most North American papers, articles, and product documentation rely on text. Most linear functions are described by their s-domain or z-domain transfer functions; non-linear functions are described by their names (*friction* and *sin*). On the other hand, block diagram symbols in European literature are generally based on graphical symbols. Linear functions are represented by the step response. There are exceptions in both cases. In North American documentation, saturation and hysteresis are typically shown symbolically, whereas European literature uses text for transcendental functions.

Part I. Linear Functions

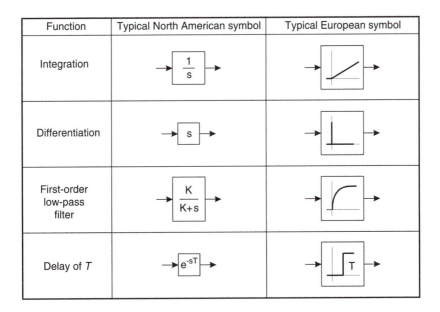

Function	Typical North American symbol	Typical European symbol
Integration	$\dfrac{1}{s}$	
Differentiation	s	
First-order low-pass filter	$\dfrac{K}{K+s}$	
Delay of T	e^{-sT}	T

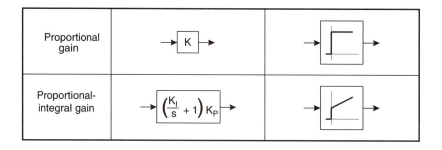

Proportional gain	→ K →	
Proportional-integral gain	→ $\left(\dfrac{K_I}{s} + 1\right) K_P$ →	

Part II.   Nonlinear Functions

Function	Typical North American symbol	Typical European symbol
Rectification		
3-Phase rectification		
AC inverter		
D-to-A and A-to-D		
Resolver-to-digital converter		
Hysteresis		

Saturation (clamping)	*Clamp*	
Saturation (clamping) with synchronization (anti wind-up)	*Anti wind-up* $\left(\dfrac{K_I}{s} + 1\right) K_P$	
Inductance with saturation	v — $\dfrac{1}{(L_0 + L_2 i^2)s}$ — i	
Transcendental functions (e.g., sin)	sin	sin
Voltage-controlled oscillator (VCO)	VCO	VCO
Friction	Friction	

Appendix C
The Runge–Kutta Method

The Runge–Kutta method is a differential equation solver [82]. It provides the value of the states of the system based on the value of the state derivatives (see Section 11.3). It is a popular method because it is relatively simple and it converges well. One popular version of the method is referred to as *fourth-order*. This algorithm is based on four vectors, each of length n, where n is the order of the system:

$\mathbf{x}(t)$	The value of the system states after the algorithm has run
$\mathbf{x}_0$	The value of the system states before the algorithm has run
$\mathbf{x}(t)$	The value of the system state derivatives input to the algorithm
$\mathbf{x}_{\text{SUM}}(t)$	TA vector for storing intermediate values

The method is based on a fixed time step, h. The algorithm calls the procedure model four times during each time step; each call to model provides the value of the derivatives of the states under slightly different conditions.

The Runge–Kutta Algorithm

The Runge–Kutta algorithm is

$$\mathbf{x}_0 = \mathbf{x}(t)$$
$$\mathbf{x}'(t) = \text{model}(\mathbf{x}(t),\ t)$$
$$\mathbf{x}_{\text{SUM}}(t) = \mathbf{x}'(t)$$
$$\mathbf{x}(t) = \mathbf{x}_0 + \mathbf{x}'(t) \times h/2$$
$$t = t + h/2$$
$$\mathbf{x}'(t) = \text{model}(\mathbf{x}(t),\ t)$$
$$\mathbf{x}_{\text{SUM}}(t) = \mathbf{x}_{\text{SUM}}(t) + 2 \times \mathbf{x}'(t)$$
$$\mathbf{x}(t) = \mathbf{x}_0 + \mathbf{x}'(t) \times h/2$$
$$\mathbf{x}'(t) = \text{model}(\mathbf{x}(t),\ t)$$
$$\mathbf{x}_{\text{SUM}}(t) = \mathbf{x}_{\text{SUM}}(t) + 2 \times \mathbf{x}'(t)$$
$$\mathbf{x}(t) = \mathbf{x}_0 + \mathbf{x}'(t) \times h$$
$$t = t + h/2$$
$$\mathbf{x}'(t) = \text{model}(\mathbf{x}(t),\ t)$$
$$\mathbf{x}_{\text{SUM}}(t) = \mathbf{x}_{\text{SUM}}(t) + \mathbf{x}'(t)$$
$$\mathbf{x}(t) = \mathbf{x}_0 + \mathbf{x}_{\text{SUM}}(t) \times h/6$$

Basic Version of the Runge–Kutta Algorithm

The fourth-order Runge–Kutta algorithm is implemented in the following BASIC program. The algorithm is implemented starting at line 9000. The earlier lines provide an example with initialization (lines 10–230) and a model (lines 2000–2090).

```
10      REM RUNGE–KUTTA Simulation Program with example
20      REM
30      REM A program to simulate the time response of the fourth-
40      REM order Bessel filter. It uses fourth-
50      REM order Runge–Kutta to solve differential equations.
60      DIM X(20), X0(20), DX(20), SUM(20)   :REM Workspace
80      A = 4437   :REM These constants are for Bessel filter
90      B = 8868000!
100     C = 9.17E + 09
110     D = 4.068E + 12
120     TMAX = .01   :REM Run model for 10 msec
130     MAXPOINTS = 100   :REM Number of time steps for model
140     PRINTPOINTS = INT(MAXPOINTS/20)   :REM Print 20 points of data
150     H = TMAX/MAXPOINTS   :REM Runge-Kutta time increment
160     N = 4   :REM 4th-order filter
170     PRINT "Fourth-Order Bessel Filter Response To Step Input"
180     FOR J = 0 TO MAXPOINTS
190     REM
200     IF (INT(J/PRINTPOINTS) = J/PRINTPOINTS) THEN PRINT USING
        "t = ##.#### Vo = ###.#####"; T,X(1)
210     GOSUB 9000   :REM CALL RUNGE-KUTTA
220     NEXT J
230     END
2000    REM Place model at line 2000.
2010    REM
2020    REM This is a time-based model for a 4th-order Bessel filter
2040    VI = 1   :REM Input is a step of 1 volt
2050    DX(1) = X(2)   :REM X(1) is Vo(t), X(2) is Vo'(t)
2060    DX(2) = X(3)   :REM X(3) is Vo''(t)
2070    DX(3) = X(4)   :REM X(4) is Vo'''(t)
2080    DX(4) = −A * X(4) − B * X(3) − C * X(2) − D * X(1) + D * VI
2090    RETURN   :REM end model with return
9000    REM This is a fourth-order RUNGE-KUTTA differential equation
9010    REM solver. The equations may be time varying and/or nonlinear.
9020    REM
9030    REM n order of equation
9040    REM x vector of length n containing the initial values
```

```
9050   REM of the variable on entry to Runge and containing
9060   REM the final values (value at t + h) at return.
9070   REM t time step in seconds (set to t + h at return).
9080   REM h step size in seconds
9090   REM dx, xo, n length vectors for workspace.
9100   REM sum
9110   REM The model is a user-supplied subroutine beginning at
9120   REM line 2000.
9130   REM
9140   REM IMPORTANT!! RUNGECTR and SUM are internal to this routine.
9150   REM Do not use them elsewhere in your program.
9160   REM
9170   REM
9180   FOR RUNGECTR = 1 TO N
9190   X(RUNGECTR) = X0(RUNGECTR)
9200   NEXT RUNGECTR
9210   GOSUB 2000
9220   FOR RUNGECTR = 1 TO N
9230   SUM (RUNGECTR) = DX(RUNGECTR)
9240   X(RUNGECTR) = XO(RUNGECTR) + H * DX(RUNGECTR)/2
9250   NEXT RUNGECTR
9260   T = T + H/2
9270   GOSUB 2000
9280   FOR RUNGECTR = 1 TO N
9290   SUM(RUNGECTR) = SUM(RUNGECTR) + 2 * DX(RUNGECTR)
9300   X(RUNGECTR) = X0(RUNGECTR) + H * DX(RUNGECTR)/2
9310   NEXT RUNGECTR
9320   GOSUB 2000
9330   FOR RUNGECTR = 1 TO N
9340   SUM(RUNGECTR) = SUM(RUNGECTR) + 2 * DX(RUNGECTR)
9350   X(RUNGECTR) = X0(RUNGECTR) + H * DX(RUNGECTR)
9360   NEXT RUNGECTR
9370   T = T + H/2
9380   GOSUB 2000
9390   FOR RUNGECTR = 1 TO N
9400   SUM(RUNGECTR) = SUM(RUNGECTR) + DX(RUNGECTR)
9410   X(RUNGECTR) = X0(RUNGECTR) + SUM(RUNGECTR) * H/6
9420   X0 (RUNGECTR) − X(RUNGECTR)
9430   NEXT RUNGECTR
9440   RETURN
```

C Programming Language Version of the Runge–Kutta Algorithm

The fourth-order Runge–Kutta algorithm is shown here.

```
/*      runge( )
 *
 *      runge ( ) is a double-precision fourth-order RUNGE-KUTTA
 *      differential equation solver. The equations may be time
 *      varying and/or nonlinear.
 *
 *      n order of equation
 *      x vector of length n containing the initial values
 *      of the variable on entry to runge and containing
 *      the final values (value at t + h) on exit.
 *      t time step in seconds must be incremented in the model
 *      h step size in seconds
 *      model the user-supplied external routine that will supply
 *      runge ( ) with the values of the derivatives (dx) when
 *      supplied with the values of the functions (x) and time
 *      (t).
 *      dx, xo, sum n length vectors for workspace.
 *
 *      This routine is based on the development given by Gupta,
 *      Bayless, and Peikari in "Circuit Analysis with
 *      Computer Applications to Problem Solving," Intext Educational
 *      Publishers, 1972, pages 236 and following.
 *
 */
void    model (int, double *, double, double *);
void
runge (int n, double *x, double t, double h,
        int (*model)(int, double *, double, double *),
        double *dx, double *x0, double *sum)
  {
  int i;
  for (i = 0; i < n; x0[i] = x[i],i + +);      /*assign x[ ] to x0[ ] */
  (*model)(n,x,t,dx);                          /*execute the model*/
  for (i = 0; i < n; i + +)
     {
     sum [i] = dx[i];                          /*sum dx */
     x[i] = x0[i] + .5 * h* dx[i];             /*load x[ ]*/
     }
  (*model)(n,x,t +.5 * h, dx);                 /*execute the model again at a
                             *different time */
  for (i = 0; i < n; i + +)
```

```
    {
    sum [i]+ = 2 * dx[i];                    /*sum dx */
    x[i] = x0[i] + .5 * h* dx[i];            /*load x[ ] */
    }
(*model) (n,x,t + .5 * h,dx);                /*execute the model a third time */
for (i = 0; i < n; i + +)
    {
    sum[i]+ = 2 * dx[i];                     /*sum dx */
    x[i] = x0[i] + h * dx[i];                /*load x[ ] */
    }
(*model)(n,x,t + h,dx);                      /*execute the model the fourth and last
                         *time at time t + h */
for (i = 0; i < n; i + +)
    {
    sum [i]+ = dx[i];                        /*sum dx last time */
    x[i] = x0[i] + sum[i] * h/6.;            /*finally, calc x[ ] */
    }
}
```

H-File for C Programming Language Version

This is the h-file that accompanies the preceding C programming language version:

```
/*runge.h*/
void runge(int,double *, double,double,int *( ), double *, double *, double *);
```

Appendix D
Development of the Bilinear Transformation

Bilinear Transformation

The bilinear transformation is so named because it approximates z with a ratio of two linear functions in s. Begin with the definition of z:

$$z \equiv e^{sT} = \frac{e^{sT/2}}{e^{-sT/2}} \qquad \text{(D.1)}$$

The Taylor series for e^{sT} is

$$z = 1 + sT + \frac{(sT)}{2!} + \frac{(sT)^3}{3!} + \cdots \qquad \text{(D.2)}$$

Using the first two terms of the Taylor series for both the numerator and the denominator of Equation D.1 produces

$$z \cong \frac{1 + sT/2}{1 - sT/2} \qquad \text{(D.3)}$$

Some algebra rearranges the equation to

$$s \cong \frac{2}{T}\left(\frac{z-1}{z+1}\right) \qquad \text{(D.4)}$$

As an alternative to Table 5-1, Equation D.4 can be used to provide a transfer function in z that approximates any function of s.

Prewarping

Prewarping the bilinear transformation causes the phase and gain of the s-domain and z-domain functions to be identical at the prewarping frequency. This is useful where exact equivalence is desired at a particular frequency, such as when using a notch filter. Prewarping modifies the approximation of Equation D.4 to

$$s \approx \frac{\omega_0}{\tan(\omega_0 T/2)} \left(\frac{z-1}{z+1} \right) \tag{D.5}$$

where ω_0 is the prewarping frequency, the frequency at which exact equivalence is desired. Recalling Euler's formulas for sine and cosine:

$$\cos(x) = \frac{e^{jx} + e^{-jx}}{2}, \qquad \sin(x) = \frac{e^{jx} - e^{-jx}}{2j} \tag{D.6}$$

and recalling that $\tan(x) = \sin(x)/\cos(x)$, Equation D.5 can be rewritten as

$$\begin{aligned}
s &= \omega_0 \cdot \frac{2j}{e^{j\omega_0 T/2} - e^{-j\omega_0 T/2}} \cdot \frac{e^{j\omega_0 T/2} + e^{-j\omega_0 T/2}}{2} \left(\frac{z-1}{z+1} \right) \\
&= j\omega_0 \cdot \frac{e^{j\omega_0 T/2} + e^{-j\omega_0 T/2}}{e^{j\omega_0 T/2} - e^{-j\omega_0 T/2}} \left(\frac{z-1}{z+1} \right) \\
&= j\omega_0 \left(\frac{e^{j\omega_0 T/2} + e^{-j\omega_0 T/2}}{e^{j\omega_0 T/2} - e^{-j\omega_0 T/2}} \right) \left(\frac{e^{sT} - 1}{e^{sT} + 1} \right)
\end{aligned} \tag{D.7}$$

Our interest here is in steady-state response, so $s = j\omega$:

$$s = j\omega_0 \left(\frac{e^{j\omega_0 T/2} + e^{-j\omega_0 T/2}}{e^{j\omega_0 T/2} - e^{-j\omega_0 T/2}} \right) \left(\frac{e^{j\omega T} - 1}{e^{j\omega T} + 1} \right) \tag{D.8}$$

Now, if $e^{j\omega T/2}$ is divided out of both the numerator and the denominator (on the right side), the result is

$$s = j\omega_0 \left(\frac{e^{j\omega_0 T/2} + e^{-j\omega_0 T/2}}{e^{j\omega_0 T/2} - e^{-j\omega_0 T/2}} \right) \left(\frac{e^{j\omega T/2} - e^{-j\omega T/2}}{e^{j\omega T/2} + e^{-j\omega T/2}} \right) \tag{D.9}$$

So when $\omega = \omega_0$, most of the factors cancel out, leaving the exact value for s:

$$s = j\omega$$

which means that when the transfer function is evaluated at the prewarping frequency, the approximation is exactly correct.

Factoring Polynomials

Most methods of approximating functions of s with functions of z require that the polynomials, at least the denominator, be factored. The bilinear transformation does not have this requirement, though the factored form usually requires less algebra, as this example shows. Compare this function factored (Equation D.10) and unfactored (Equation D.11):

$$T(s) = \frac{1}{(s+1)^4} \tag{D.10}$$

$$T(s) = \frac{1}{s^4 + 4s^3 + 6s^2 + 4s + 1} \tag{D.11}$$

The factored form can be converted to z almost directly:

$$\frac{[T \times (z+1)]^4}{[2(z-1) + T(z+1)]^4} = \frac{[T/(T+2)]^4(z+1)^4}{[z + (T-2)/(T+2)]^4} \tag{D.12}$$

However, the unfactored form would require a considerable amount of algebra to convert.

Phase Advancing

The approximation $z + 1 \approx 2\angle\omega T/2$ can be used to advance the phase of the z function when the s function has fewer zeros than poles.

To begin:

$$z + 1 = e^{j\omega T} + 1 \tag{D.13}$$

Dividing $e^{j\omega T/2}$ out of the right side yields

$$z + 1 = e^{j\omega T/2}(e^{j\omega T/2} + e^{j\omega T/2}) \tag{D.14}$$

Recalling Euler's formula, $2\cos(x) = e^{jx} + e^{-jx}$, produces

$$z + 1 = 2e^{j\omega T/2}\cos\left(\frac{\omega T}{2}\right) \tag{D.15}$$

Finally, when $\omega T/2$ is small, $\cos(\omega T/2) \approx 1$, so

$$z + 1 \approx 2e^{j\omega T/2} = 2\angle\left(\frac{\omega T}{2}\right) \tag{D.16}$$

This approximation is accurate enough for most applications since it is usually not important that the gains of the s and z functions match at high frequencies.

Appendix E
The Parallel Form of Digital Algorithms

As an alternative to the "controllable form" of Section 5.5, z-domain transfer functions can be implemented with the parallel form. A transfer function of Equation 5.38 will be an example of the five-step process to implement the parallel form:

$$T(z) = 1.4 \left(\frac{z - 0.8}{z - 0.6} \right) \left(\frac{z}{z - 1} \right) \tag{E.1}$$

1. Write the transfer function as the ratio of two (unfactored) polynomials. The example from Equation E.1 is rewritten as

$$T(z) = 1.4 \left(\frac{z^2 - 0.8z}{z^2 - 1.6z + 0.6} \right) \tag{E.2}$$

2. Convert the function to the sum of a constant and a ratio of two polynomials, where the order of the denominator is greater than the order of the numerator. The preceding example becomes

$$T(z) = 1.4 + \frac{1.12z - 0.84}{z^2 - 1.6z + 0.6} \tag{E.3}$$

3. Factor the denominator:

$$T(z) = 1.4 + \frac{1.12z - 0.84}{(z - 0.6)(z - 1)} \tag{E.4}$$

4. Divide the ratio of polynomials into the sum of the different factors using partial fraction expansion. The preceding example is rewritten as a sum:

$$T(z) = 1.4 + \frac{0.42}{z - 0.6} + \frac{0.7}{z - 1} \tag{E.5}$$

5. Write the algorithm for this form. For our example, the algorithm is in three steps:

   a. $D_n = 0.6D_{n-1} + 0.42R_{n-1}$ \hfill (E.6)

$$\text{b. } E_n = E_{n-1} + 0.7R_{n-1} \tag{E.7}$$

$$\text{c. } C_n = 1.4R_n + D_n + E_n \tag{E.8}$$

This method can develop numerical difficulties when poles are close together. For example, consider this transfer function:

$$T(z) = \left(\frac{z}{z - 0.5}\right)\left(\frac{z}{z - 0.501}\right) \tag{E.9}$$

$$T(z) = \frac{z^2}{z^2 - 1.001z + 0.2505} \tag{E.10}$$

The controllable form of the algorithm is

$$C_n = R_n + 1.001C_{n-1} - 0.2505C_{n-2} \tag{E.11}$$

However, the parallel form of this equation is

$$T(z) = 1 + \frac{1.001z - 0.2505}{(z - 0.5)(z - 0.501)} \tag{E.12}$$

$$T(z) = 1 - \frac{250}{z - 0.5} + \frac{251.001}{z - 0.501} \tag{E.13}$$

The parallel algorithm is

$$\text{a. } D_n = 0.500D_{n-1} - 250R_{n-1} \tag{E.14}$$

$$\text{b. } E_n = 0.501E_{n-1} + 251.001R_{n-1} \tag{E.15}$$

$$\text{c. } C_n = R_n + D_n + E_n \tag{E.16}$$

The first few steps will demonstrate the difficulties with the parallel form. Assume that C_0, D_0, and E_0 are zero and that the input is a unit step: $R_n = 1$. The first few steps of the standard (nonparallel) form are

$$C_0 = 1$$
$$C_1 = 1 + 1.001 \cdot 1 = 2.001$$
$$C_2 = 1 + 1.001 \cdot 2.001 - 0.2505 \cdot 1 = 2.752501$$

The first few steps of the parallel form are

$$D_0 = 0$$
$$E_0 = 0$$
$$C_0 = 1 + 0 + 0 = 1$$
$$D_1 = -250$$
$$E_1 = +251.001$$
$$C_1 = 1 - 250 + 251.001 = 2.001$$
$$D_2 = 0.500(-250) - 250 \cdot 1 = -375$$
$$E_2 = 0.501 \cdot 251.001 + 251.001 = +376.752501$$
$$C_2 = 1 - 375 + 376.752501 = 2.752501$$

The parallel form requires two extra places of accuracy (for $D_n + E_n$) in calculations as compared to the controllable form because the two poles are so close to each other.

One advantage of this form is that fewer old values need be stored. All calculations are based on information from the last cycle (or two if there are complex poles). However, the same space for storage is required (D_n and E_n compared to C_{n-1} and C_{n-2} in the standard form).

Appendix F
Basic Matrix Math

An $n \times m$ matrix (or array) has n rows and m columns. The element in row i and column j of matrix $\mathbf{A}$ is $\mathbf{A}(i,j)$, or a_{ij}. For example, a two-by-two matrix is

$$\mathbf{A} = \begin{bmatrix} a_{11} & a_{12} \\ a_{21} & a_{22} \end{bmatrix} \tag{F.1}$$

A vector is a matrix with one row ($n \times 1$).

Matrix Summation

If two matrices have the same number of columns and the same number of rows, they can be added together. Each element in the resulting matrix is the sum of the corresponding elements in each of the original matrices. For example,

$$\mathbf{C} = \mathbf{A} + \mathbf{B} = \begin{bmatrix} a_{11} & a_{12} \\ a_{21} & a_{22} \end{bmatrix} + \begin{bmatrix} b_{11} & b_{12} \\ b_{21} & b_{22} \end{bmatrix} = \begin{bmatrix} a_{11} + b_{11} & a_{12} + b_{12} \\ a_{21} + b_{21} & a_{22} + b_{22} \end{bmatrix} \tag{F.2}$$

Matrix Multiplication

Two matrices can be multiplied together if the second matrix has the same number of rows as the first matrix has columns. The multiplication for the element in column i and row j is carried out according to Equation F.3:

$$C_{ij} = \sum_{k=1}^{m} (a_{ik} \times a_{kj}) \tag{F.3}$$

where m is the number of columns in the first matrix or, equivalently, the number of rows in the second matrix. For example,

$$\begin{aligned} \mathbf{C} = \mathbf{A} \times \mathbf{B} &= \begin{bmatrix} a_{11} & a_{12} \\ a_{21} & a_{22} \end{bmatrix} \times \begin{bmatrix} b_{11} & b_{12} \\ b_{21} & b_{22} \end{bmatrix} \\ &= \begin{bmatrix} a_{11} \times b_{11} + a_{12} \times b_{21} & a_{11} \times b_{12} + a_{12} \times b_{22} \\ a_{21} \times b_{11} + a_{22} \times b_{21} & a_{21} \times b_{12} + a_{22} \times b_{22} \end{bmatrix} \end{aligned} \tag{F.4}$$

Matrix Scaling

Any matrix can be scaled by a constant by scaling each element of the matrix by that constant. For example,

$$\mathbf{A}/k = \begin{bmatrix} a_{11}/k & a_{12}/k \\ a_{21}/k & a_{22}/k \end{bmatrix} \tag{F.5}$$

Matrix Inversion

A square matrix (i.e., one where the number of rows and the number of columns are equal) can be inverted. The matrix $\mathbf{B}$ is the inversion of $\mathbf{A}$ if

$$\mathbf{A} \times \mathbf{B} = \mathbf{I} \tag{F.6}$$

where $\mathbf{I}$ is the *identity* matrix. The identity matrix has 1 in all the diagonal positions (a_{11}, a_{22}, and so on) and zero in all other positions. An example of a two-by-two identity matrix is

$$\mathbf{I} = \begin{bmatrix} 1 & 0 \\ 0 & 1 \end{bmatrix} \tag{F.7}$$

In the general case, inverting matrices is complicated, requiring sophisticated numerical algorithms [82]. However, for the two-by-two matrix, $\mathbf{B}$ is the inverse of $\mathbf{A}$ if

$$\mathbf{B} = \frac{\begin{bmatrix} a_{22} & -a_{12} \\ -a_{21} & a_{11} \end{bmatrix}}{a_{11} \times a_{22} - a_{12} \times a_{21}} \tag{F.8}$$

This was used to derive Equation 16.2 from Equation 16.1.

Appendix G
Answers to End–of–Chapter Questions

Chapter 2

1. a. $-1.4\,\text{dB}\angle-32°$
    b. $-5.1\,\text{dB}\angle40°$
    c. $5.1\,\text{dB}\angle-40°$
    d. $-1.7\,\text{dB}\angle21°$
2. a. 1.0 (0 dB)
    b. 0.2 (-14 dB)
    c. 5.0 (14 dB)
    d. 0.6667 (-3.5 dB)
3. a. $\frac{C(s)}{R(s)} = \frac{\omega}{s+\omega}$, evaluated at $s = j\omega$: $-3\,\text{dB}\angle-45°$
    b. same as part a, evaluated at $s = 0.1j\omega$: $-0.04\,\text{dB}\angle-5.7°$
    c. $\frac{C(s)}{R(s)} = \frac{\omega^2}{s^2+2\zeta\omega s+\omega^2}$, evaluated at $s = j\omega$ and $0.1 \times j\omega$: $-3\,\text{dB}\angle-90°$ and $-0\,\text{dB}\angle-8.1°$
4. $t_{0.5\%} = 5.3/(f_{BW} \times 2\pi)$
5. a. 1.5 cycles in 20 msec: $F_{RING} = 75\,\text{Hz}$
    b. 75 Hz
    c. In marginally stable systems, the frequency of ringing is the same as the frequency of peaking (this is, in fact, the case!).
6. $\frac{C(s)}{R(s)} = \frac{G_1(s)(1+G_2(s)H_2(s)) + G_2(s)G_3(s)}{1 + G_2(s)H_2(s) + G_3(s)H_3(s) + G_2(s)H_2(s)G_3(s)H_3(s)}$
7. All are LTI except
    f. Violates homogeneity where $k \times (x)^2 \neq (kx)^2$
    g. Violates homogeneity where $k \times (1/x) \neq (1/kx)$
    h. Violates homogeneity where $ke^x \neq e^{kx}$

Chapter 3

1. a. 15 dB, or a factor of 5.6.
    b. 3.36 (5.6×0.6)
    c. 12 dB, or a factor of 4 applied to K_P: $K_P = 2.4$. Step response gives severe ringing; peaking is 12 dB.
    d. 3 dB of GM is insufficient.
2. a. 15 dB, or a factor of 5.6.
    b. 2800

   c. $G = 2000$. Step response gives severe ringing; peaking is 12 dB.

   d. Loss of gain margin from high loop gains or high plant gain gives similar results.

3. a. $K_P = 1$ and $K_I = 50$

   b. $K_P = 0.3$ and $K_I = 25$

   c. $K_P = 1.7$ and $K_I = 100$

   d. Reducing phase lag in the loop supports higher gains.

4. a. PM $= 58°$ and GM $= 11$ dB

   b. PM $= 59°$ and GM $= 13$ dB

   c. PM $= 59°$ and GM $= 11$ dB

   d. Similar limits on square wave performance yield similar margins of stability.

5. a. 178 Hz

   b. 49 Hz

   c. 284 Hz

   d. Higher gains yield higher bandwidths.

Chapter 4

1. a. $60°$

   b. $51°$

   c. $34°$

   d. Slower sampling reduces phase margin.

2. a. $68°$

   b. $34°$

   c. $68° - 34° = 34°$

   d. 1/2 sample time, or 500 μsec, which is $10.3°$ at 57 Hz.

   e. $10.3°$

   f. 65% of $0.001s$, which is $13.3°$

   g. $34°$ in both cases

3. a. $T_{CALC} = 5\%$, Current Loop Hz $= 800$ Hz, $T_{SAMPLE} = 0.0005s$, $K_{VI} = 0$. Raise K_{VP} to 1.3, producing 218 Hz BW.

   b. $60°$

   c. Gain crossover $= 85$ Hz. Phase lag from the sample/hold $= 0.00025s = 8°$, from velocity estimation $= 8°$, and from calculation delay $= 1°$. So about $17°$ of $27°$, or 63%, of nonintegrator lag is from sampling in the "aggressive assumption." Comment: The "aggressive" assumption is that there is little phase lag from any source except sampling.

Chapter 5

1. a. $0.01z/(z - 1)$

   b. $C_N = C_{N-1} + 0.01 \times R_N$

   c. $0.005(z + 1)/(z - 1)$
   d. $C_N = C_{N-1} + 0.005 \times (R_N + R_{N-1})$
2. a. $1\angle 18°$
   b. $1\angle 36°$
   c. $1\angle 90°$
3. a. $0.118z/(z - 0.882)$
   b. $C_N = 0.882 \times C_{N-1} + 0.118 \times R_N$
   c. $0.118, 0.222, 0.314, 0.395$
   d. $0.118/(1 - 0.882) = 1$
4. a. $0.9397 \times (z^2 - 1.902z + 1)/(z^2 - 1.79z + 0.882)$
   b. $-1.7\,\text{dB} \angle -34°, -6.6\,\text{dB} \angle -62°, -18\,\text{dB} \angle -82°, 0, -18\,\text{dB} \angle +82°$
5. a. 10 Hz
   b. $n \times 20\,\text{Hz}$, where n is any integer
6. a. 819.2 counts/volt
   b. 3276.8 counts/volt
   c. 0.00061 volts/count

Chapter 6

1-6.

	P (1)	PI (2)	PI+(3)	PID (4)	PID+(5)	PD (6)
Overshoot	0%	15%	15%	15%	10%	0%
Bandwidth	43 Hz	49 Hz	40 Hz	68 Hz	50 Hz	65 Hz
Phase lag at the bandwidth	99°	118°	115°	116°	103°	102°
Peaking	0 dB	1.6 dB	1.4 dB	1.7 dB	1.1 dB	0 dB
PM	68°	55°	41°	56°	45°	69°
GM	14 dB	13.1 dB	12.2 dB	14.7 dB	14.2 dB	15 dB
K_P	0.3	0.3	0.3	0.42	0.42	0.42
K_I	—	35	76	50	94	—
K_D	—	—	—	0.00094	0.00094	0.00094
K_{FR}	—	—	0.65	—	0.65	—

7. Allow more overshoot when setting K_P and allow more overshoot when setting K_I.

Chapter 7

1. Load current
2. Increasing the amount of capacitance, inertia, and thermal mass
3. Integral gain, obtaining a measurement of the disturbance

4. a. $K_{VP} = 0.42$ and $K_{VI} = 35$
   b. ± 20 RPM
   c. $K_{VP} = 1.2$ and $K_{VI} = 100$
   d. ± 8 RPM
   e. Little effect in the high zone (above 100 Hz). In the middle range, as shown by the peaks of the plots, disturbance response with $K_{VP} = 0.42$ is 27.8 dB, compared to $K_{VP} = 1.2$ is 18.8 dB; raising K_{VP} $1.2/.42 = 9$ dB improves middle-frequency disturbance response by 9 dB. In the low-frequency area, difference of low gains to high gains is 18 dB, or a factor of 8.2, about the ratio of $K_{VP} \times K_{VI}$ in the low- and high-gain sets.
   f. Faster sampling supports higher loop gains, which provide better disturbance response.
5. a. ± 5 RPM
   b. Disturbance decoupling provides much more benefit (5-RPM excursions compared to 15 RPM when speeding the sample rate four times).

Chapter 8

1. a. 1 (perfect command result)
   b. Delays in the power converter and feedback filtering usually cause excessive overshoot.
2. a. Ls
   b. Ls (the controller function normally does not influence the feed-forward function).
3. No effect on disturbance response, GM, or PM.
4. a. $K_P = 0.707$ and $K_I = 60$
   b. 119 Hz
   c. $K_P = 1.2$ and $K_I = 100$, BW=198 Hz
   d. Faster sampling supports higher gains, which provide faster command response.
5. a. $K_F = 50\%$
   b. 230 Hz.
   c. Feed-forward improved the command response more than raising the sample rate allowed. Adding feed-forward is usually much easier to do since significant increases in the sample rate normally require faster processor hardware.

Chapter 9

1. 1 kHz
2. a. $9.86 \times 10^6/(s^2 + 6.28 \times 10^3 s + 9.86 \times 10^6)$
   b. $(s^2 + 9.86 \times 10^6)/(s^2 + 6.28 \times 10^3 s + 9.86 \times 10^6)$
   c. $1.98 \times 10^9/(s^3 + 2.51 \times 10^3 s^2 + 3.16 \times 10^6 s + 1.98 \times 10^9)$

3. Set $\omega_D = \omega_N$ and $\zeta_N = 0$
4. b and c

Chapter 10

1. a. $K_P = 0.6$, $K_I = 10.0$
   b. Approximately $0.25s$
   c. $K_P = 1.7$, $K_I = 30.0$. Settles in approximately 0.1 sec.
   d. Luenberger observer removes phase lag. This allows higher control law gains, which provide faster settling time.
2. a. $K_{DO} = 0.05$, $K_{PO} = 17$, $K_{IO} = 2000$
   b. 108 Hz
   c. No significant difference. Tuning values do not depend on observer band-width, at least when the observer models are very accurate.
3. a. Yes, nearly identical
   b. No. Phase lag can cause instability in the loop, even when the amount of lag is so small that it is difficult to see comparing signals by eye.
4. a. 50
   b. That the procedure finds the correct K_{Est} if the estimated sensor dynamics are just reasonably close to representing the actual sensor

Chapter 11

1. $dx_1/dt = 2\pi200(x_2 - x_1)$

   $dx_2/dt = 50(x_3)$

   $dx_3/dt = 2\pi100(K_Ix_4 + K_P(R(t) - x_1) - x_3)$

   $dx_4/dxt = K_P(R(t) - x_1)$

   $C(t) = x_2$

This is based on the following block diagram:

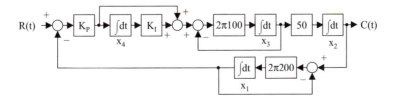

2. $dx_1/dt = 2\pi 200(x_2 - x_1)$

$dx_2/dt = 50(x_3)$

$dx_3/dt = 2\pi 100(K_I x_4 + K_P(R(t) - x_1)) + 2\pi 60(K_D K_P(R(t) - x_1) - x_5) - x_3)$

$dx_4/dt = K_P(R(t) - x_1)$

$dx_5/dt = 2\pi 60(K_D K_P(R(t) - x_1) - x_5)$

$C(t) = x_2$

This is based on the following block diagram:

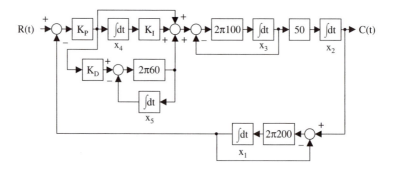

Note that the K_D term and the two blocks that connect to the summing junction to its immediate right form the transfer function $K_D 2\pi 60 s/(s + 2\pi 60)$, which is a derivative scaled by K_D and filtered with a single-pole low-pass filter with a break frequency of 60 Hz.

3. Control equations (run once every 0.001 sec):

$\text{Err}_N = R_N - x_{1N}$

$\text{Int}_N = \text{Int}_{N-1} + \text{Err}_N K_P K_I/1000$

$\text{Out}_N = \text{Int}_N + \text{Err}_N K_P$

Continuous equations:

$dx_1/dt = 2\pi 200(x_2 - x_1)$

$dx_2/dt = 50(x_3)$

$dx_3/dt = 2\pi 100(\text{Out}_N - x_3)$

This is based on the following block diagram:

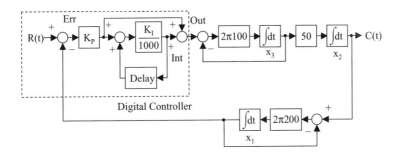

Chapter 12

1. About 160/s, or 10 dB × 50/s.
2. $\omega = 800$ rad/sec. (*Hint: Find the range of t where the inequality $|\omega(t) \times dJ(t)/dt| < 0.25(dJ(t) \times \omega(t))/dt)$ is true.*)
3. Gain correction:

   1       $i < 2$ A

   $2/i$ $2A < i < 4A$

   0.5 $4A < i$

4. $[J_M + J_L(1 + (2 \sin (\Theta) \times R)^2)]/(J_M + J_L)$
5. Gain scheduling (both reversal error correction and deadband compensation work by changing the gain as a deterministic function of a system parameter).

Chapter 14

1. $360/(5000 \times 4) = 0.018°$, or $1.08 \, \text{min}^{-1}$
2. Lower-cost, maximum rotational speed is not limited by resolution, and the bandwidth of the conversion process can be changed at any time.
3. Higher resolution and higher accuracy
4. $(1/8000 \text{ revolutions})/(1/5000 \text{ sec}) \times 60 \text{ sec}/ \text{min} = 37.5$ RPM
5. a. 0.3%
   b. Ripple is 3 RPM, or 0.3% of 1000-RPM, the default speed of Experiment 14B. (*Hint: Turn servo gains down to $K_{VP} = 0.72$ and $K_{VI} = 10$; this reduces servo bandwidth to 7 Hz, well below the 66-Hz ripple frequency (4 × 1000 RPM/60), so the motor speed will be smooth and almost all velocity error will be seen in the feedback signal.*)

6. a. $K_{VP} = 1$, bandwidth $= 8.3\,Hz$, torque ripple $= 0.12\,Nm$ (peak)

    $K_{VP} = 2$, bandwidth $= 17\,Hz$, torque ripple $= 0.25\,Nm$ (peak)

    $K_{VP} = 5$, bandwidth $= 46\,Hz$, torque ripple $= 0.5\,Nm$ (peak)

    $K_{VP} = 10$, bandwidth $= 115\,Hz$, torque ripple $= 0.5\,Nm$ (peak)

    $K_{VP} = 20$, bandwidth $= 356\,Hz$, torque ripple $= 0.5\,Nm$ (peak)

  b. $T_{E-PEAK}/P_{E-PEAK} = Js^2 = 877.3$; $P_{E-PEAK} = 3\,min^{-1} = 8.74 \times 10^{-4}\,Rad$:

    $T_{E-PEAK} = 0.51\,Nm$

  c. Equation 14.13 is accurate when servo bandwidth is above ripple frequency.

7. a. Less sensitive to heat, less sensitive to shock and vibration, less sensitive to contamination, and inexpensive

  b. Broad range of accuracy and resolution available, signals easy to process

Chapter 15

1. The power block drive. The single-axis controller.
2. One reason is that increased losses in the motor, such as friction and windage, reduce output torque.
3. 90° (See Equation 15.7)
4. One reason is saturation (increase of permeability) in the steel.
5. Increasing efficiency. Because the transistor is either off (zero current) or on (low voltage) almost all of the time, the losses (current × voltage) remain low.
6. $0.3 \times 330 = 99\,VDC$
7. Advantages of brush: simple control, single current sensor, fewer power transistors, can be very smooth at low speed, low cost (especially at low power)

    Advantages of brushless: no brush wear, no brush debris, less electric noise, smaller (space for commutator not required), lower frictional losses, easier to cool (windings are stationary), speed not limited by commutator, lower inertia rotor
8. Sine wave
9. To use current to reduce the magnetic field strength in order to reduce back-EMF and allow higher speed
10. The current loop does not pass the commutation frequency; thus current loop responsiveness does not limit torque at high speeds.
11. Hall sensors (also, specialized encoders called comcoders provide commutation tracks very much like Hall sensors). Inadequate resolution.

Chapter 16

1. Tuned resonance. There are two indications, both of which are shown by Bode plots. First, the resonance and antiresonance peaks in the "Motor" plot (V_M/T_E) are sharp. Second, the frequency of instability shown in the closed-loop plot (533 Hz) matches the mechanical resonance shown in the motor plot.

2. Inertial reduction. There are two indications in the Bode plots. First, the resonance and antiresonance peaks in the motor plot are broad. Second, the frequency of instability shown in the closed-loop plot (467 Hz) is much larger than the frequency of the mechanical resonance shown in the motor plot (170 Hz)

3.

Case	J_m (kg-m^2)	$J_L{:}J_m$	K_{VP}	BW (Hz)
1	0.0018	10:1	2	15
2	0.0045	4:1	4.2	26
3	0.009	2:1	8.5	35
4	0.018	1:1	17	44

Conclusion: lower load-inertia-to-motor inertia ratios support higher band-width.

4. Increasing the motor inertia increases the current and affects system cost by increasing the size of the drive (to provide more current) and motor (to dissipate more losses).

5. a. $K_{VP} = 4.2$, bandwidth $= 26$ Hz
   b. *Den Freq* $= 50$ Hz, $K_{VP} = 4.2$, bandwidth $= 31$ Hz
   c. *Den Freq* $= 50$ Hz, *Num Freq* $= 200$ Hz, $K_{VP} = 6$, bandwidth $= 37$ Hz
   d. In many cases, reduced-inertia instability can improved by low-pass filters and even more by lag filters.

6. The notch filter does not substantially improve performance of systems affected by reduced-inertia instability.

7. a. $K_{VP} < 0.7$, bandwidth < 5 Hz (the DSA is not configured to give good readings with such low gains)
   b. *Den Freq* $= 30$ Hz, $K_{VP} = 2.5$, bandwidth $= 26$ Hz
   c. *Den Freq* $= 10$ Hz, *Num Freq* $= 40$ Hz, $K_{VP} = 1.2$, bandwidth $= 10$ Hz
   d. In many cases, tuned resonance can be improved by low-pass filters and not as well by lag filters.

8. *Notch Freq* $= 392$ Hz, $K_{VP} = 8.5$, and bandwidth $= 76$ Hz

Chapter 17

1. a. $K_{VP} = 1.4$, $K_{VI} = 40$, and $K_{PP} = 350$
   b. 210 Hz
   c. 106 Hz

2. a. $K_{VF} = 0.7$ and $K_{FA} = 0.001$
   b. 432 Hz
3. a. $K_{VF} = 0.5$ and $K_{PP} = 250$
   b. 129 Hz
4. a. Slight rounding at corners
   b. Slight rounding at corners and high-frequency noise
   c. Slight rounding at corners
   d. Effect of inertia variation on command response with and without feed-forward are about the same, but acceleration feed-forward generates high-frequency noise.
5. a. $K_{VP-PI/P} = K_{VP-P/PI}$

   $K_{PP-PI/P} = K_{PP-P/PI} + K_{VI-P/PI}$

   $K_{PI-PI/P} = K_{VI-P/PI} \times K_{PP-P/PI} / K_{PP-PI/P}$

   b. $K_{PD-PID} = K_{VP-P/PI}$

   $K_{PP-PID} = K_{VP-P/PI} \times (K_{PP-P/PI} + K_{VI-P/PI})$

   $K_{PI-PID} = K_{VP-P/PI} \times K_{PP-P/PI} \times K_{VI-P/PI}$

   c. $K_{AF-PI/P} = K_{AF-P/PI}$

   $K_{VF-PI/P} = K_{VF-P/PI}$

   d. $K_{AF-PID} = K_{AF-P/PI}$

   $K_{VF-PID} = K_{VP-P/PI} \times (K_{VF-P/PI} - 1)$

   e. Compared to the s^1 term in PI/P and PID, the s^1 term in the numerator of P/PI is smaller by the amount of $K_{VP-P/PI} \times K_{VI-P/PI} \times (1 - K_{VF-P/PI})$. For large values of K_{VF}, the difference is not noticeable.

Chapter 18

1. First phase crossover disappears when $K_{IO} = 0$; without the additional 90° phase lag contributed by K_{IO}, there is no crossover at (or near) 200 Hz.
2. a. Both signals are accurate.
   b. V_O provides the best response. V_S provides similar response except for strong peaking at 400 Hz caused by the too high control law gains for the phase lag of the sensed signal.
3. The noise sensitivities V_O and V_S are similar.
4. a. 48°
   b. 24°
   c. 51°
   d. 40°

5. a.

T_{SAMPLE}	Maximum K_{AFB}
0.00025	1.2
0.0002	2.5
0.00015	6.0

b.

K_{DD}	Disturbance response at 10 Hz (dB)	Improvement offered by acceleration feedback (dB)
0	−15	(reference)
1.2	−22	7 (2.2×)
2.5	−26	11 (3.5×)
6	−32	17 (7×)

c. Ideal result is $1 + K_{AFB}$ improvement:

Case 1: $1 + 1.2 = 2.2$

Case 2: $1 + 2.5 = 3.5$

Case 3: $1 + 6 = 7$

The results in part b are similar to the ideal results.

References

1. Antoniou, A., *Digital Filters Analysis, Design, and Applications*, 2nd ed., McGraw-Hill, 1993.

2. Armstrong-Helouvry, B., *Control of Machines with Friction*, Kluwer Academic Publishers, 1991.

3. Bassani, R., Piccigallo, B., *Hydrostatic Lubrication*, Elsevier, 1992.

4. Beineke, S., Schutte, F., Wertz, H., Grotstollen, H., "Comparison of Parameter Identification Schemes for Self-Commissioning Drive Control of Nonlinear Two-Mass Systems," *Conf. Rec. IEEE IAS Annual Mtg.*, 1997, New Orleans, pp. 493–500.

5. Bilewski, M., Gordano, L., Fratta, A., Vagati, A., Villata, F., "Control of high performance interior permanent magnet synchronous drives," IEEE Trans. Ind. Appl., Vol. 29, No. 2, March–April 1993, pp. 328–337.

6. Biernson, G., *Principles of Feedback Controls*, Vol. 1–2, John Wiley & Sons, 1988.

7. Boulter, B., "The Effect of Speed Loop Bandwidths and Line-Speed on System Natural Frequencies in Multi-Span Strip Processing Systems," *Conf. Rec. IEEE IAS Annual Mtg.*, 1997, New Orleans, pp. 2157–2164.

8. Brown, R. H., Schneider, S. C., Mulligan, M. G., "Analysis of Algorithms for Velocity Estimation from Discrete Position Versus Time Data," *IEEE Trans. Ind. Elec.*, vol. 39, no. 1, Feb. 1992, pp. 11–19.

9. Bueche, F. J., *Introduction of Physics for Scientists and Engineers*, McGraw-Hill, 1980.

10. Burke, J., "Extraction of High Resolution Position Information from Sinusoidal Encoders," Proc. PCIM-Europe 1999, Nuremberg, pp. 217–222.

11. Carstens, J. R., *Automatic Control Systems and Components*, Prentice-Hall, 1996.

12. Corley, M. J., Lorenz, R. D., "Rotor Position and Velocity Estimation for a Salient-Pole Permanent Magnet Synchronous Machine at Standstill and High Speeds," *IEEE Trans. Ind. Appl.*, vol. 34, no. 4, Jul./Aug. 1998, pp. 784–789.

13. Data Devices Corp., *Manual for 19220 RDC* (19220sds.pdf). Available at www.ddc-web.com (use product search for "19220").

14. D'Azzo, J. and Houpis, C. *Linear Control System Analysis and Design 3rd Edition*, McGraw-Hill, 1988.

15. Del Toro, V., *Electromechanical Devices for Energy Conversion and Control Systems*, Prentice-Hall, 1968.

16. Deliyannis, T. Sun, Y., Fidler, J. K., *Continuous-Time Active Filter Design*, CRC Press, 1999.

17. Deur, J., "A comparative study of servosystems with acceleration feedback," *Proc of IEEE IAS (Rome), 2000*.

18. Demerdash, N. A. O., Alhamadi, M. A., "Three Dimensional Finite Element Analysis of Permanent Magnet Brushless DC Motor Drives - Status of the State of the Art," *IEEE Trans. Ind. Elec.*, vol. 43, no. 2, Apr. 1996, pp. 268–275.

19. Dhaouadi, R., Kubo, K., Tobise, M., "Analysis and Compensation of Speed Drive Systems with Torsional Loads," *IEEE Trans. Ind. Appl.*, vol. 30, no. 3, May/Jun. 1994, pp. 760–766.

20. Dorf, R.C., Bishop, R.H., *Modern Control Systems (9th Ed.)*, Addison-Wesley Publishing Company, 2000.

21. Ellis, G., "Driven by Technology Advancements, Brushless DC Motors Are Displacing Brush Types," PCIM Magazine, March, 1996.

22. Ellis, G., "PDFF: An Evaluation of a Velocity Loop Control Method," *Conf. Rec. PCIM*, 1999, pp. 49–54.

23. Ellis, G., Lorenz, R. D., "Comparison of Motion Control Loops for Industrial Applications," *Conf. Rec. IEEE IAS Annual Mtg.*, 1999, Phoenix, pp. 2599–2605.

24. Ellis, G., Lorenz, R.D., "Resonant Load Control Methods for Industrial Servo Drives," *Proc. of IEEE IAS (Rome), 2000*.

25. Ellis, G., "Cures for Mechanical Resonance in Industrial Servo Systems," PCIM 2001 Proceedings, Nuremberg, pp. 187–192.

26. Ellis, G., Krah, J.O., "Observer-Based Resolver Conversion in Industrial Servo Systems," PCIM 2001 Proceedings, Nuremberg, pp. 311–316.

27. Ellis, G., "Observer-based Enhancement of Tension Control in Web Handling," PCIM 2002 Proceedings, Nuremberg, pp. 147–154.

28. Ellis, G., *Observers in Control Systems*, Academic Press, 2002.

29. Ellis, G., "Comparison of Position-Control Algorithms for Industrial Applications," PCIM 2003 Proceedings, Nuremberg, pp. 71–78.

30. Fassnacht, J. , "Benefits and Limits of Using an Acceleration Senosr in Actively Damping High Frequency Mechanical Oscillations.," *Conf. Rec. of the IEEE IAS, 2001*, pp 2337–2344.

31. Fitzgerald, A. E., Kingsley, C., Umans, S. D., Kingsley, C., Jr., Umans, S. D., *Electric Machinery*, 6th ed., McGraw-Hill, 2002.

32. Franklin, G. F., Powell, D. J., Emami-Naeini, A., *Feedback Control of Dynamic Systems*, 4th ed., Addison-Wesley Publishing Company, 2002.

33. Franklin, G. F., Powell, D. J., Workman, M. L., Powell, D., *Digital Control of Dynamic Systems*, 3rd ed., Addison-Wesley Publishing Company, 1997.

34. Ghausi, M. S., Laker, K. R., *Modern Filter Design: Active RC and Switched Capacitor*, Prentice-Hall, 1981.

35. Ghosh, B.K., Xi, N., Tarn, T.J., *Controls in Robotics and Automation*, Academic Press, 1999.

36. Gopal, M., *Modern Control System Theory*, 2nd ed., John Wiley and Sons, 1993.

37. Gunther, R. C., *Lubrication*, Chilton Book Company, 1971.

38. Hagl, R., Heidenhain, J., "Encoders Keep Linear Motors on Track," *PT Design Magazine*, Mar. 1998, pp. 43–46.

39. Hameyer, K., Belmans, R. J. M., "Permanent Magnet Excited Brushed DC Motors," *IEEE Trans. Ind. Elec.*, vol. 43, no. 2, Apr. 1996, pp. 247–255.

40. Hanselman, D. C., *Brushless Permanent-Magnet Motor Design*, 2nd ed., Writer's Collective, 2003.

41. Hava, A. M., Sul, S., Kerkman, R. J., Lipo, T. A., "Dynamic Overmodulation Characteristics of Triangle Intersection PWM Methods," *IEEE Trans. Ind. Appl.*, vol. 35, no. 4, July/Aug. 1999, pp. 896–907.

42. Haylock, J. A., Mecrow, B. C., Jack, A. G., Atkinson, D. J., "Enhanced Current Control of High-Speed PM Machine Drives through the Use of Flux Controllers," *IEEE Trans. Ind. Appl.*, vol. 35, no. 5, Sep./Oct. 1999, pp. 1030–1038.

43. Hong, K., Nam., K., "A Load Torque Compensation Scheme under the Speed Measurement Delay," *IEEE Trans. Ind. Elec.*, vol. 45, no. 2, Apr. 1998, pp. 283–290.

44. Hori, Y., Sawada, H., Chun, Y., "Slow Resonance Ratio Control for Vibration Suppression and Disturbance Rejection in Torsional System," *IEEE Trans. Ind. Elec.*, vol. 46, no. 1, Feb. 1999, pp. 162–168.

45. Jackson, L. B., *Digital Filters and Signal Processing*, 2nd ed., Kluwer Academic Publishers, 1999.

46. Johnson, C. T., Lorenz, R. D., "Experimental Identification of Friction and its Compensation in Precise, Position Controlled Mechanisms", *IEEE Trans. Ind. Appl.*, Nov./Dec. 1992, pp. 1392–1398.

47. Kamen, E. W., *Industrial Controls and Manufacturing*, 1st ed., Academic Press, 1999.

48. Kang, J.K., Sul, S.K., "Vertical-Vibration Control of Elevator Using Estimated Car Acceleration Feedback Compensation," *IEEE Trans. on Ind. Elec.*, Vol. 47, No. 1, Feb. 2000, pp. 91–99.

49. Kenjo, T., *Electric Motors and Their Controls, an Introduction.*, Oxford University Press, 1994.

50. Kerkman, R. J., Leggate, D., Seibel, B. J., Rowan, T. M., "Operation of PWM Voltage Source-Inverters in the Overmodulation Region," *IEEE Trans. Ind. Elec.*, vol. 43, no. 1, Feb. 1996, pp. 132–141.

51. Konishi, Y., Nakaoka, M., "Current-Fed Three Phase and Voltage-Fed Three-Phase Active Converters with Optimum PWM Pattern Scheme and Their Performance Evaluations," *IEEE Trans. Ind. Elec.*, vol. 46, no. 2, Apr. 1999, pp. 279–287.

52. Krah, J. O., "Software Resolver-to-Digital Converter for High Performance Servo Drives," *Proc. PCIM Europe*, 1999, Nuremberg, pp. 301–308.

53. Krishnan, R., "Control and Operation of PM Synchronous Motor Drives in the Field-Weakening Region," *Proc. IECON 93*, 1993, Maui, pp. 745f.

54. Lee, H. S., Tomizuka, M., "Robust Motion Controller Design for High Accuracy Positioning Systems," *IEEE Trans. Ind. Elec.*, vol. 43, no. 1, Feb. 1996, pp. 48–55.

55. Lee, Y.M., Kang, J.K., and Sul, S.K. ,"Acceleration feedback control strategy for improving riding quality of elevator system," *Proc. of IEEE IAS* (Phoenix), 1999, pp. 1375–1379.

56. Leonhard, W., *Control of Electric Drives*, 3rd ed., Springer Verlag, 2001.

57. Lorenz, R. D., "ME-746. Dynamics of Controlled Systems: A Physical Systems-Based Methodology for Non-Linear, Multivariable, Control System Design," Video Class, University of Wisconsin-Madison, 1998.

58. Lorenz, R. D., "Modern Control of Drives," COBEP'97, Belo Horizonte, MG, Brazil, Dec. 2–5, 1997, pp. 45–54.

59. Lorenz, R. D., "New Drive Control Algorithms (State Control, Observers, Self-Sensing, Fuzzy Logic, and Neural Nets)," *Proc. PCIM Conf.*, Sep. 3–6, 1996, Las Vegas

60. Lorenz, R. D., "Synthesis of State Variable Controllers for Industrial Servo Drives," *Proc. Conf. on Applied Motion Control*, June 10–12, 1986, pp. 247–251.

61. Lorenz, R. D., Lucas, M. O., Lawson, D. B., "Synthesis of a State Variable Motion Controller for High Performance Field Oriented Induction Machine Drives," *Conf. Rec. IEE IAS Annual Mtg.*, 1986, pp. 80–85.

62. Lorenz, R. D., Novotny, D. W., "A Control Systems Perspective of Field Oriented Control for AC Servo Drives," *Proc. Controls Engineering Conf.*, June 6–11, 1988, Chicago, pp. XVIII-1–XVIII-11.

63. Lorenz, R. D., Van Pattern, K., "High Resolution Velocity Estimation for All Digital, AC Servo Drives," *IEEE Trans. Ind. Appl.*, vol. 27, no. 4, Jul./Aug. 1991, pp. 701–705.

64. Mason, S. J., "Feedback Theory: Some Properties of Signal Flow Graphs," *Proc. IRE*, vol. 41, July 1953, pp. 1144–1156.

65. Mason, S. J., "Feedback Theory: Further Properties of Signal Flow Graphics," *Proc. IRE*, vol. 44, July 1956, pp. 920–926.

66. Matsui, N., "Sensorless PM Brushless DC Motor Drives," *IEEE Trans. Ind. Elec.*, vol. 43, no. 2, Apr. 1996, pp. 300–308.

67. McClellan, S., "Electromagnetic Compatibility for SERVOSTAR® S and CD." Available at www.motionvillage.com/training/handbook/cabling/shielding.html.

68. Miller, R. W., *Lubricants and Their Applications*, McGraw-Hill Book Company, 1993.

69. Miron, D. B., *Design of Feedback Control Systems*, International Thomson Publishing, 1997.

70. Moatemri, M. H., Schmidt, P. B., Lorenz, R. D., "Implementation of a DSP-Based, Acceleration Feedback Robot Controller: Practical Issues and Design Limits," *Conf. Rec. IEEE IAS Annual Mtg.*, 1991, pp. 1425–1430.

71. Nasar, S. A., Boldea, I., Unnewehr, L. E., *Permanent Magnet, Reluctance, and Self-Synchronous Motors*, CRC Press, 1993.

72. Natarajan, S., *Theory and Design of Linear Active Networks*, Macmillan Publishing Company, 1987.

73. Nekoogar, F., and Moriarty, G., *Digital Control using Digital Signal Processing*, Prentice-Hall Information and System Sciences Series, 1999.

74. Ohm, D. Y., "A PDFF Controller for Tracking and Regulation in Motion Control," *Proceedings of 18th PCIM Conference, Intelligent Motion*, Philadelphia, pp. 26–36, Oct. 21–26, 1990.

75. Ohm, D. Y., "Analysis of PID and PDF Compensators for Motion Control Systems," *IEEE IAS Annual Meeting*, pp. 1923–1929, Denver, Oct. 2–7, 1994.

76. Orlosky, S., "Improve Encoder Performance," *PT Design Magazine*, Sept. 1996, pp. 43–46.

77. Palm, W., *Modeling, Analysis and Control of Dynamic Systems*, 2nd ed., John Wiley and Sons, 1999.

78. Park, J.W., Koo, D.H., Kim, J.M., Kang, G.H., Park, J.B., "High Performance Speed Control of Permanent Magnet Synchronous Motor with Eccentric Load," *Conf. Rec. IEEE IAS Annual Mtg., 2001*, pp 815–820.

79. Parks, T. W., Burrus, C. S., *Digital Filter Design*, John Wiley & Sons, 1987.

80. Phillips, C. L., Harbor, R. D., *Feedback Control Systems*, 4th ed., Prentice-Hall, 1999.

81. Phelan, R. M., *Automatic Control Systems*, Cornell University Press, 1977.

82. Press, W. H., Teukolsky, S. A., Vettering, W. T., Flannery, B. P., *Numerical Recipes in C*, 2nd ed., Cambridge University Press, 2002.

83. Rahman, M. A., Radwan, T. S., Osheiba, A. M., Lashine, A. E., "Analysis of Current Controllers for Voltage-Source Inverter," *IEEE Trans. Ind. Elec.*, vol. 44, no. 4, Aug. 1997, pp. 477–485.

84. Rahman, M. F., Zhong, L., Lim, K. W., "A Direct Torque-Controlled Interior Permanent Magnet Synchronous Motor Drive Incorporating Field Weakening," *IEEE Trans. Ind. Appl.*, vol. 34, no. 6, Nov./Dec. 1998, pp. 1246–1253.

85. Rubin, O., *The Design of Automatic Control Systems*, Artech House, 1986.

86. Schmidt, C., Heinzl, J., Brandenburg, G., "Control Approaches for High-Precision Machine Tools with Air Bearings," *IEEE Trans on Ind. Elec.*, vol. 46, no. 5, Oct. 1999, pp. 979–989.

87. Schmidt, P. B., Lorenz, R. D., "Design Principles and Implementation of Acceleration Feedback to Improve Performance of DC Drives," *IEEE Trans. Ind. Appl.*, May/Jun. 1992, pp. 594–599.

88. Schmidt, P., Rehm, T., "Notch Filter Tuning for Resonant Frequency Reduction in Dual-Inertia Systems," *Conf. Rec. IEEE IAS Annual Mtg.*, 1999, Phoenix, pp. 1730–1734.

89. Sheingold, Daniel H, *Analog-Digital Conversion Handbook*, 3rd Edition, by Analog Devices, Ed. Prentice Hall

90. Shin, H., "New Antiwindup PI Controller for Variable-Speed Motor Drives," *IEEE Trans. Ind. Elec.*, vol. 45, no. 3, Jun. 1998, pp. 445–450.

91. Suh, Y. S., and Chun, T.W., "Speed Control of a PMSM motor based on the new disturbance observer," *Conf Rec. of IEE IAS Annual Mtg., 2001*, pp. 1319–1326.

92. Trzynadlowski, A. M., Bech, M. M., Blaabjerg, F., Pedersen, J. K., "An Integral Space Vector PWM Technique for DSP-Controlled Voltage-Source Inverters," *IEEE Trans. Ind. Appl.*, vol. 35, no. 5, Sept./Oct. 1999, pp. 1091–1097.

93. Trzynadlowski, A. M., Kirlin, R. L., Legowski, S. F., "Space Vector PWM Technique with Minimum Switching Losses and a Variable Pulse Rate," *IEEE. Trans. Ind. Elec.*, vol. 44, no. 2, Apr. 1997, pp. 173–181.

94. Vukosavic, S. N., Stojic, M. R., "Suppression of Torsional Oscillations in a High-Performance Speed Servo Drive," *IEEE Trans. Ind. Elec.*, vol. 45, no. 1, Feb. 1998, pp. 108–117.

95. Welch, R. H., "Mechanical Resonance in a Closed-Loop Servo System: Problems and Solutions," Tutorial available from Welch Enterprises, Oakdale, MN.

96. Wilson, F. W., *Pneumatic Controls for Industrial Applications*, Books on Demand, 1999.

97. Yang, S.M., and Ke, S.J., "Performance Evaluation of a Velocity Observer for Accurate Velocity Estimation of Servo Motor Drives," *Conf Rec. of IEEE IAS Annual Mtg., 1998*, p1697

98. Yoshitsugu, J., Inoue, K., and Nakaoka, M., "Fuzzy Auto-Tuning Scheme based on α-Parameter Ultimate Sensitivity Method for AC Speed Servo System," *Conf Rec. of IEEE IAS Annual Mtg., 1998*, p1625

99. Younkin, G. W., McGlasson, W. D., Lorenz, R. D., "Considerations for low Inertia AC Drivers in Machine Tool Axis Servo Applications," *IEEE Trans. Ind. Appl.*, vol. 27, no. 2, Mar./Apr. 1991, pp. 262–268.

100. Younkin, G. W., *Industrial Servo Control Systems: Fundamentals and Applications*, 2nd ed., Marcel Dekker, 2002.

Index